# The Sound of Silence

Burkhard Vogel

# The Sound of Silence

## Lowest-Noise RIAA Phono-Amps: Designer's Guide

Springer

Dipl.-Ing. Burkhard Vogel
BUVOCON-Burkhard Vogel Consult GmbH
70180 Stuttgart, Germany

ISBN 978-3-642-09557-3 e-ISBN 978-3-540-76884-5

DOI 10.1007/978-3-540-76884-5


Softcover reprint of the hardcover 1st edition 2008



*Coverdesign*: WMX Design GmbH, Heidelberg

Printed on acid-free paper

9 8 7 6 5 4 3 2 1

springer.com

To Beate

"Noise analysis can be a daunting task at first and is an unfamiliar territory for many design engineers."

*Glen Brisebois*
*Design note 355*
*Linear Technology*

"Noise – it's the amplifier's tinnitus! No chance to fight it!"

*A valve enthusiastic ENT doc in*
*a discussion with the author 2006*

"A kingdom for a piece of wire with gain!"

*Frustrated design engineer*

# Preface

It is still a challenge to develop a low-noise amplifier – despite the fact that nowadays (2007) nearly every solution of an electronic question of the consumer world can be solved by digital means. There is a wide field of tasks left that can only be satisfyingly attacked with the help of old-fashioned analogue technology: sensors that are coupled to the existing and living world around us are always confronted with analogue signals. Those – in most cases – tiny signals have to be amplified and treated with unbelievably high electronic care. Therefore, frustration on noisy devices should always be turned around into motivation for the search of nearly noiseless solutions!

As a producer of such tiny analogue signals the vinyl record (33 1/3 LP and 45 Single/Maxi) is a typical representative of our yesterday – 20th century – life. Despite the nearly 100% digitization of the consumer world it is still alive – with growing sales revenues around the world. One should expect that all secrets of the amplifier chain that transfers the signals out of the record's grooves to our ears are well known. Yes and no! Much is written about distortion, overload matters, noise, phase angles, frequency response, etc[1]. Most technical aspects of amplifiers and sensors were well described.

But simple questions like e.g.: "my moving-magnet cartridge – how much noise does it produce?" or "what's the signal-to-noise-ratio (SN) of my phono-amp after A-weighting?" are still not that easy to answer today.

World-wide, mathematics is the only language that can be understood by nearly everybody, assumed that there exists a certain talent for it, and, not to forget, the right software for calculations. In this book calculations were all carried out with MathCad[2]. An easy to get for free simulation software would help as well, e.g. MicroSim v8.0[3] but, not to increase the necessity for the use of various softwares, this simulation software is not essential to understand and follow the mathematical courses.

---

[1] Inter alia: "Self on Audio", Douglas Self 2000, Newnes, UK, ISBN 0-7506 4765 5

[2] MathSoft Inc., USA

[3] MicroSim Corp., USA (see also footnote 3 on page 17)

Therefore, for mathematics-refusal-free and ambitious amateurs and/or students who want to design their own amplifier for specific cartridges this book will find answers to such simple questions and many others concerning RIAA phono-amps! It's also a collection of articles which were published in a more condensed form in the British magazine ELECTRONICS WORLD (EW, formerly called "Electronics World and Wireless World (EW+WW)" or "Wireless World (WW)").

As a consequence, the content of this book will lead to affordable amplifier design approaches which will end up in lowest-noise solutions not far away from the edge of physical boundaries set by room temperature and given cartridges – thus, fully compatible with very expensive so called "high-end" or "state-of-the-art" offers on today markets – and, from a noise point of view in most cases outperforming them!

With easy to follow mathematical treatment it will be demonstrated as well that theory is not far away from reality. Measured SNs will be found within 1 dB off the calculated ones and deviations from the exact amplifier transfer won't cross the ±0.1 dB tolerance lines. Additionally, measurement set-ups and results will be presented and comparisons with measurement results of test magazine will soon become easier to perform.

Last remark: the presented electronic circuits do not contain extra made or extremely expensive components. They all can be found at component dealers worldwide.

Very last remark: I guess that creativity does not mean to reinvent the wheel again nor to find out absolutely new things. In many cases it's nothing else but simply rearranging well known parts. Therefore, when I started developing the many circuit schemes presented in this book Okham's Razor[4] and one main goal ranked very high: to combine and to re-arrange well known different circuit designs to promising new solutions.

[4] "If you have to choose from some number of competing theories, choose the simplest one because it's most likely to be true",

Sharon Kay, www.royalinstitutephilosophy.org/think/

# Contents

# Part I
# Theory

# Chapter 1
# Amps, Pre-Amps, Pre-Pre-Amps

## Purpose of the Book

The purpose of this book is an economic one: to enable the reader to calculate certain noise related aspects of a phono-amp before building it up, thus, saving a lot of energy and time as well as avoiding needless expenses. That's why I won't debate whether the black box between the output of a turn-table and the input of an amplifier (amp) with the volume control is a pre-amp or a pre-pre-amp or what ever it will be. This book follows the purpose and that will be the only story to tell.

As result of the rotation of a vinyl record (VR) on the platter of a turn-table via the in-groove movement of a stylus of a cartridge the grooves-stylus-coil chain generates a voltage $e_1$ at the output of the coils of that specific cartridge. It is fixed in a head shell at the end of a tonearm of the turn-table and its output leads are connected to the input of an appropriate amplifying chain.

The voltage $e_1$ has to be amplified to a level $e_2$ that can make a loudspeaker sound loud enough to enjoy listening to the content of the VR. Rather often, the quotient of $e_2/e_1$ lies in the range of more than 50,000 provided that a power amplifier with a rated output of e.g. $100\,\mathrm{W}/8\,\mathrm{Ohm}$ is used. Figure 1.1 helps to better understand the situation by giving a general overview of the different amplifying stages of an amplifier chain – from groove to ear. Starting at the top with the basic arrangement of low-level phono amplification we'll see underneath the dotted line that there exists a broad range of handling possibilities of the tiny groove signals. Taking into account all the possible electronic devices that might be useful for amplification purposes (from valves via transformers up to most modern ICs) we'll end up in a rather great number of phono-amp types.

In Fig. 1.1 all indicated values are given as an average of the market products, supposed that the MC cartridge is a low-output one (e.g. Denon DL-103) and the MM cartridge is not a very loud one (e.g. Shure V15V MR). Today (2007), the market offers many different MC cartridges with higher output levels (up to $2\,\mathrm{mV_{rms}}/1\,\mathrm{kHz}$) as well as rather high-output MM cartridges (up to


DOI: 10.1007/978-3-540-76884-5, ©Springer 2008

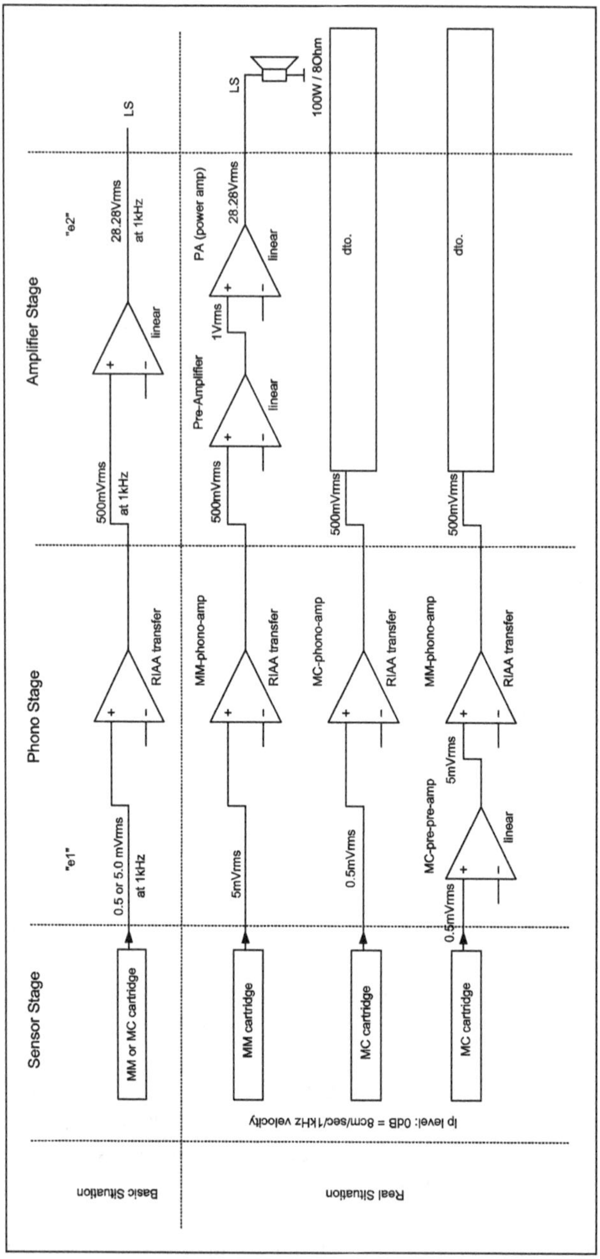


**Fig. 1.1** From groove to ear

$10\,mV_{rms}/1\,kHz$). The reasons why I've chosen as examples the above mentioned types are the following ones: both sound good, both really challenge the phono-amplifiers when talking about noise and both are not too expensive to purchase. Other cartridges with higher output levels need lower overall gains, thus, improving signal-to-noise-ratios (SN) by the same amount. In other words: a cartridge with a rated output voltage of $1\,mV_{rms}/1\,kHz$ offers a 6 dB better SN than the one with an output voltage of only $0.5\,mV_{rms}/1\,kHz$, provided that they have nearly the same impedance and SN is calculated or measured with the same phono stage.

When studying cartridge technical data one often find that output levels are rated at a peak velocity of 5 cm/s at 1 kHz. In fact, that is not the 0 dB level of a VR. Its definition is: 0 dB equals a peak velocity of 8 cm/s at 1 kHz[1]. Applying the rule of three will lead to the required cartridge output and phono-amp input level values. For the Shure V15V MR that means:

- output level at 5 cm/s/1 kHz: $3.5\,mV_{rms}$
- output level at 8 cm/s/1 kHz: $(3.5\,mV_{rms} \times 8\,cm/s)/(5\,cm/s) = 5.6\,mV_{rms}$

Because of the so called – any modern phono-amp equalizing – RIAA transfer[2] [R(f) in Fig. 1.2] one should not forget, that, with reference to 1 kHz, the overall gain of the amplifier chain is app. 10 times higher at 20 Hz and app. 10 times lower at 20 kHz. Details on that will be given in the next chapter.

Another thing should be mentioned as well: besides the nasty noise problems and the very high gains that are needed to produce a high enough output signal to drive loudspeakers via linear pre- and power-amps the design engineers have to pay a lot attention on suppressing any hum interferences by enormous amounts of shielding and decoupling efforts around the whole phono-amp circuitry – including its connection to the turntable/cartridge.

Therefore, the enormous task to design a low-noise AND hum-free phono-amp sould never be underestimated!

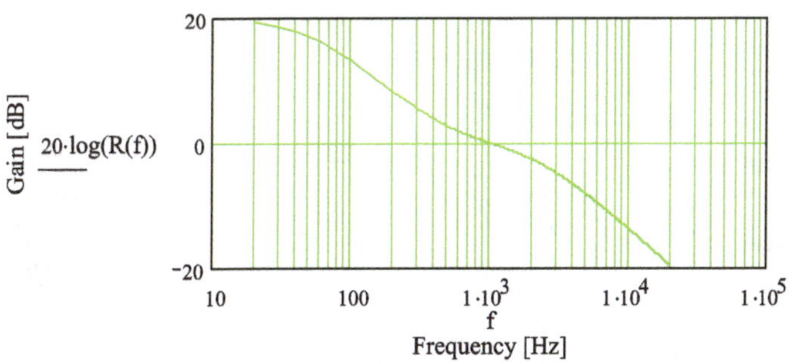


**Fig. 1.2** RIAA transfer relative to 0 dB

[1] Reference, trackability and frequency test record no. 10 99 112, Deutsche Grammophon Gesellschaft

[2] RIAA: Record Industry Association of America

Concerning phono-amp noise problems the following definitions for three different types of phono-amps will lead us through all calculation and design exercises that will be the content of the following chapters:

### Type 1 Phono-Amp

- One amp for MM cartridge purposes only (Fig. 1.3).
- Gain setting components can be valves or solid state components like BJTs, FETs or op-amps (Bipolar Junction Transistors, Field Effect Transistors or Operational Amplifiers).
- With a specified input sensitivity of $5\,\text{mV}_{\text{rms}}/1\,\text{kHz}/0\,\text{dB}$ gain is set to 100 (+40 dB) at 1 kHz.

### Type 2 Phono-Amp

- One amp for MC cartridge purpose only (Fig. 1.4).

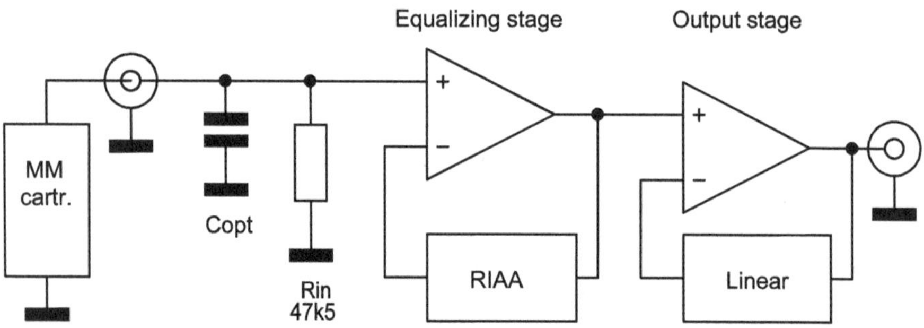


**Fig. 1.3** Type 1 phono-amp – basic circuitry

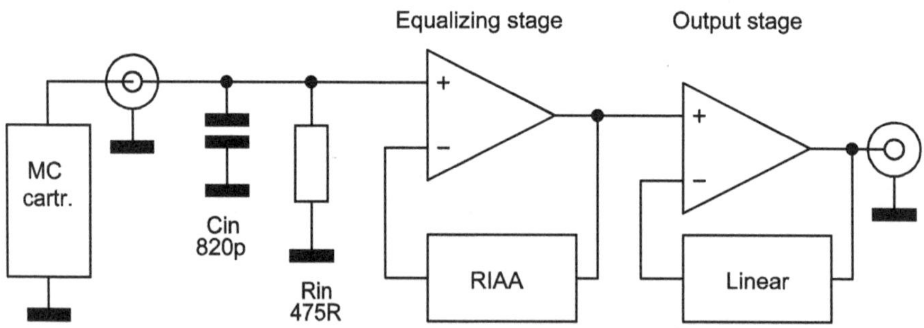


**Fig. 1.4** Type 2 phono-amp – basic circuitry

- Gain setting components can be valves or solid state devices like BJTs, FETs or op-amps, but the ones of the very 1st stage should be BJTs only because of their low-noise uniqueness and superiority over FETs, OPAs and valves[3].
- With a specified input sensitivity of $5\,\mathrm{mV_{rms}}/1\,\mathrm{kHz}/0\,\mathrm{dB}$ gain is set to 1000 (+60 dB) at 1 kHz.

### Type 3 Phono-Amp

- One amp for MM and MC purposes (Fig. 1.5).
- A step-up transformer (alternative A with a real balanced input) or a linear gain pre-pre-amp (alternative B with an un-balanced input) can be switched to the input of a RIAA equalized MM phono-amp stage, thus, allowing to select between MM or MC cartridges.
- Gain of the MM phono-amp stage is set to 100, the gain of the transformer or pre-pre-amp is set to 10 (but very much depending on the output level of the MC cartridge in use).
- Gain setting components of the MM phono stage are the same like the ones of Type 1.
- Gain setting components of the 1st stage of the pre-pre-amp are the same like the ones of Type 2.

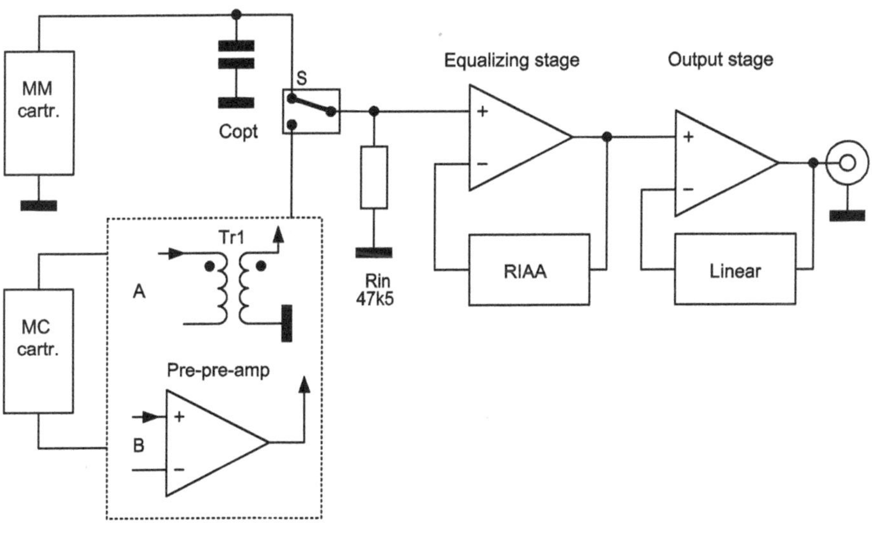


**Fig. 1.5** Type 3 phono-amp – basic circuitry

[3] "Ultra-Low-Noise Preamplifier for Moving-Coil Phono cartridges"
E. H. Nordholt, R. M. van Vierzen, JAES 1980-April, Volume 28, no. 4

To define amps or pre-amps it should be pointed out that there exists no DIN or EU nor any US standard. Of course, one also can find complete audio amplifiers on the market, comprising one of the three types followed by a linear gain pre-amp and power-amp inside one case.

But, to fulfil the purpose of this book, it is sufficiently useful that I only discuss the treatment of noise effects of phono-amps of types 1, 2 and 3 (alternative A only – alternative B would be nothing else but the non-equalized input section of a type 2 phono-amp).

In addition, to get calculation adequate measurement results each type has been built-up in a separate 1HU-19″ Al plug-in box.

# Chapter 2
# RIAA Transfer/Anti-RIAA Transfer

The signal on VRs is coded according to the rules set by the Record Industry Association of America (RIAA). This code is determined by three time constants: $T_1$, $T_2$, $T_3$. Cutting a VR means that the three time constants encode the signal in a specific way. The reason for this is to handle overloading and noise issues the optimal way.

### Cutting Process with Anti-RIAA

Playing a VR on a turntable means that an amplifier with the three time constants has to decode the signal the opposite way. The transfer as result of the decoding process is often called RIAA weighting and the respective transfer function RIAA($f$). Therefore, the result of the process to encode the VR cutting is performed by the anti-RIAA transfer function ARIAA($f$).

To demonstrate the whole process without many words the following charts will help to understand. Figure 2.1 shows the basic circuitry[1] to cut a VR and Fig. 2.2 gives the resulting transfer plot, referenced to 0 dB at 1 kHz.

### Decoding with RIAA Transfer

The RIAA transfer performing amplifier in Fig. 2.3 shows the (active) decoding situation[2] between cartridge and amplifier output. The respective transfer plot is given in and Fig. 2.4.

The result of this rather complex process should be a flat frequency response at the output of the amplifying chain and at the input of the loudspeakers. The horizontal line at 0 dB in Fig. 2.5 shows the result of the sum of the two transfer functions output($f$):

---

[1] See Chap. 12 for more details on the circuitry of an anti-RIAA transfer performing amplifier

[2] See Chap. 8 for more details on active and passive equalization

B. Vogel, *The Sound of Silence*

DOI: 10.1007/978-3-540-76884-5, ©Springer 2008

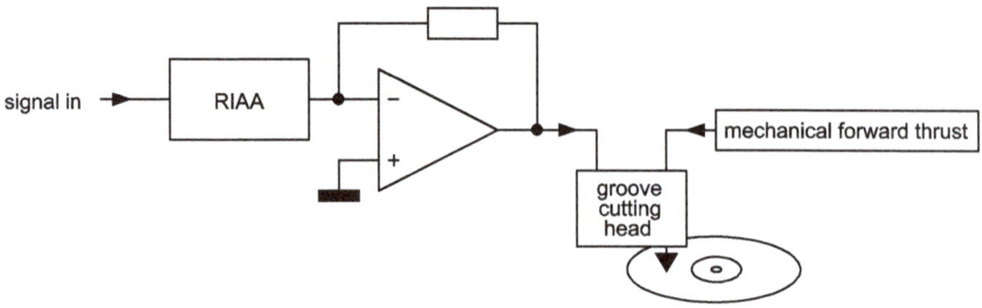


**Fig. 2.1** VR cutting process with Anti-RIAA transfer function ARIAA$(f)$

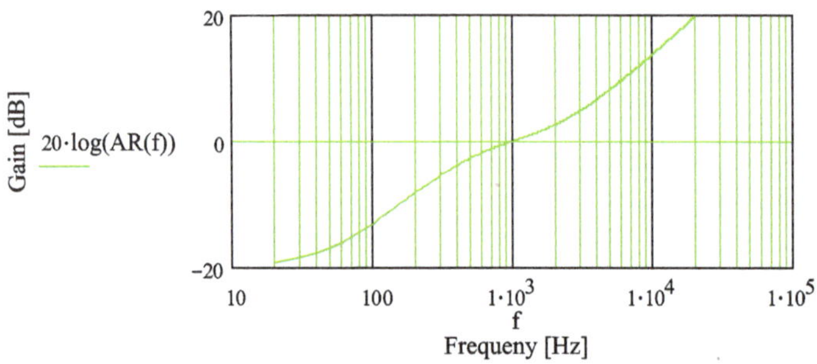


**Fig. 2.2** Anti-RIAA transfer function AR$(f)$ (= ARIAA$(f)$ referenced to 0 dB/1 kHz) used to encode the signal on the VR

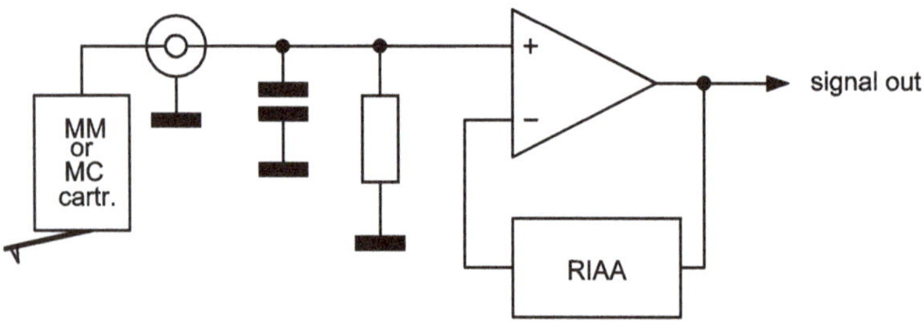


**Fig. 2.3** Cartridge-amplifier chain with decoding elements to perform the RIAA transfer function RIAA$(f)$

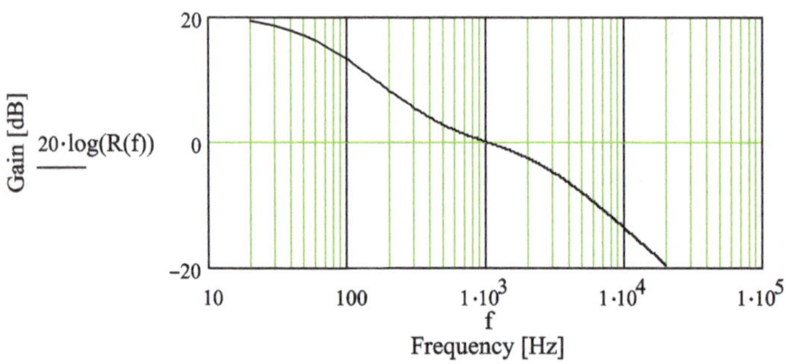


**Fig. 2.4** Decoding transfer function R$(f)$ (= RIAA$(f)$ referenced to 0 dB/1 kHz)

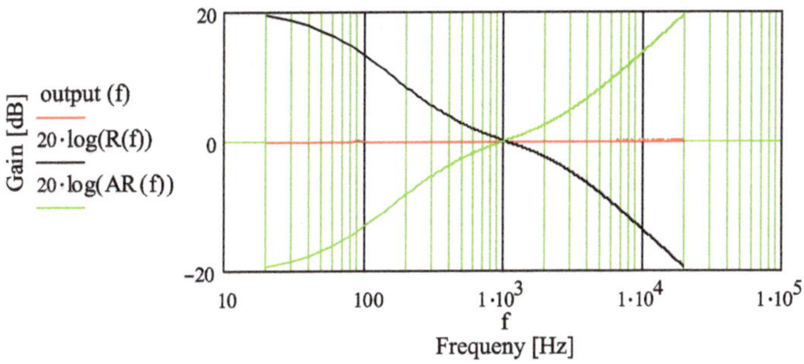


**Fig. 2.5** Plot of all three transfers: $\mathrm{AR}(f) + \mathrm{R}(f) = \mathrm{output}(f)$

Mathematically, the whole process looks a bit difficult, but I guess, relatively easy to understand.

## RIAA Transfer – Ideal Situation

Let's start with the three time constants and how they produce the transfer function of the RIAA weighting. By definition[3,4] they are:

- $T_1 = 3180 \times 10^{-6}$ s with corresponding frequency of 50.05 Hz $= 1/(T_1 \times 2 \times \pi)$
- $T_2 = 75 \times 10^{-6}$ s with corresponding frequency of 2122.1 Hz $= 1/(T_2 \times 2 \times \pi)$
- $T_3 = 318 \times 10^{-6}$ s with corresponding frequency of 500.5 Hz $= 1/(T_3 \times 2 \times \pi)$

Hence, the complex transfer function $H_{\mathrm{E}}(p)$ for the encoding mode $(E)$ looks like:

$$H_{\mathrm{E}}(p) = \frac{(1 + pT_1)(1 + pT_2)}{(1 + pT_3)} \tag{2.1}$$

Therefore, the equation for the decoding mode (D) must look like the inverse of it:

$$H_{\mathrm{D}}(p) = \frac{(1 + pT_3)}{(1 + pT_1)(1 + pT_2)} \tag{2.2}$$

I call the magnitude of $H_{\mathrm{E}}(p)$ Anti-RIAA transfer = ARIAA$(f)$ and the magnitude of $H_{\mathrm{D}}(p)$ RIAA transfer = RIAA$(f)$. Thus, RIAA$(f)$ becomes:

$$\mathrm{RIAA}(f) = \frac{\sqrt{1 + (2\pi f T_3)^2}}{\sqrt{1 + (2\pi f T_1)^2}\sqrt{1 + (2\pi f T_2)^2}} \tag{2.3}$$

---

[3] Standard set by RIAA and DIN 45535/6

[4] In some regions of the world a fourth time constant also plays a role: $T_4 = 7950 \times 10^{-6}$ s (corresponding frequency = 20.02 Hz). This was set by the IEC (Publication 98-1964 + Amendment n. 4, Sep. 1976). It was never standardized and its effects are not described in detail in this book. Nevertheless, an electronic solution will be given in the "RIAA Phono-Amp Engine" chapter

This is nothing else but a sequence of two 6 dB/octave lp filters with time constants $T_1$ and $T_2$ followed by an differentiator with the time constant $T_3$.

Now we can calculate any gain at any frequency. But the result will not be very elegant because it will not produce the picture we are used to live with: gain (in dB with reference to 1 kHz) versus frequency. To get this we have to relate the calculation results to the reference point as well as to show the frequency axis in logarithmic scaling.

The reference point in audio is always 0 dB at 1 kHz. Therefore RIAA(1000 Hz) becomes:

$$\mathrm{RIAA}(10^3\,\mathrm{Hz}) = \frac{\sqrt{1+(2\pi 10^3\,\mathrm{Hz}T_3)^2}}{\sqrt{1+(2\pi 10^3\,\mathrm{Hz}T_1)^2}\sqrt{1+(2\pi 10^3\,\mathrm{Hz}T_2)^2}} \tag{2.4}$$

$$\mathrm{RIAA}(10^3\,\mathrm{Hz}) = 0.101 \tag{2.5}$$

To get the final – plot-ready – transfer function R($f$) which is related to the reference point 1 kHz and to a reference gain of 0 dB Eq. 2.3 and the inverse of Eq. 2.4 at 1 kHz have to be multiplied:

$$R(f) = \left(\frac{\sqrt{1+(2\pi f T_1)^2}}{\sqrt{1+(2\pi f T_2)^2}\sqrt{1+(2\pi f T_3)^2}}\right) \times \left(\frac{\sqrt{1+(2\pi 10^3\,\mathrm{Hz}T_1)^2}}{\sqrt{1+(2\pi 10^3\,\mathrm{Hz}T_2)^2}\sqrt{1+(2\pi 10^3\,\mathrm{Hz}T_3)^2}}\right)^{-1} \tag{2.6}$$

Hence, the plot in Fig. 2.4 can be achieved with the following equation:

$$20\log\{R(f)\} = 20\log\{\mathrm{RIAA}(f)\} - 20\log\{\mathrm{RIAA}(10^3\,\mathrm{Hz})\} \tag{2.7}$$

The plot in Fig. 2.2 is the result of the following equation:

$$20\log\{\mathrm{AR}(f)\} = 20\log\{\mathrm{RIAA}(f)^{-1}\} - 20\log\{\mathrm{RIAA}(10^3\,\mathrm{Hz})^{-1}\} \tag{2.8}$$

Consequently, the 0 dB line plot of Fig. 2.5 becomes:

$$\mathrm{output}(f) = 20\log\{R(f)\} + 20\log\{\mathrm{AR}(f)\} \tag{2.9}$$

Table 2.1 is an EXCEL[5] sheet which includes the respective equations to calculate transfer amplitude data for $R(f)$ with reference to 0 dB/1 kHz. The value in box C3 is calculated with Eq. 2.4. Values in boxes C5 … C13 were calculated with Eq. 2.6. To calculate AR($f$) with reference to 0 dB/1 kHz we only have to inverse the signs in column C, lines 5 … 13.

[5] EXCEL is a registered trade mark of Microsoft Inc., USA

**Table 2.1** Selected frequencies and calculated (Eq. 2.6) transfer amplitudes of $R(f)$ with reference to 0 dB/1 kHz

| 1/A | B | C |
|---|---|---|
| 2 | Frequency [Hz] | Transfer amplitude [dB] |
| 3 | 1000 | – 19.911 |
| 4 | | Transfer amplitude [dB rel. 0 dB] |
| 5 | 20 | 19.274 |
| 6 | 50 | 16.946 |
| 7 | 100 | 13.088 |
| 8 | 500 | 2.648 |
| 9 | 1000 | 0.000 |
| 10 | 2122 | – 2.866 |
| 11 | 5000 | – 8.210 |
| 12 | 10,000 | – 13.734 |
| 13 | 20,000 | – 19.620 |

## RIAA Transfer – Real Situation

Figures 2.2, 2.4, 2.5 were created with MathCad (MCD). A detailed approach to get solutions with this software will be shown on worksheets of the following chapters. But, to demonstrate the very helpful features, I will create a chart showing the deviation from the exact RIAA transfer as the relative error in dB versus frequency for a phono-amp with actual time constants (…$a$) not far away from the exact ones.

- $T_{1a} = 3183\,\mu s$;
- $T_{2a} = 74\,\mu s$;
- $T_{3a} = 321\,\mu s$;

### Deviation Between Ideal and Real Situation – Calculated

Application of Eq. 2.6 with the exact time constants $T_1$, $T_2$, $T_3\{=R(f)\}$ minus Eq. 2.6 with the actual time constants $T_{1a}$, $T_{2a}$, $T_{3a}\{=Ra(f)\}$ will lead to the error plot dev$(f)$ shown in Fig. 2.6:

$$\mathrm{dev}(f) = 20\log\{R(f)\} - 20\log\{\mathrm{Ra}(f)\} \quad (2.10)$$

### Deviation Between Ideal and Real Situation – Simulated

Another possibility is the use of a pSpice simulation. To get the plot of Fig. 2.8 we need to draw a schematic like the one in Fig. 2.7 – with the following content:

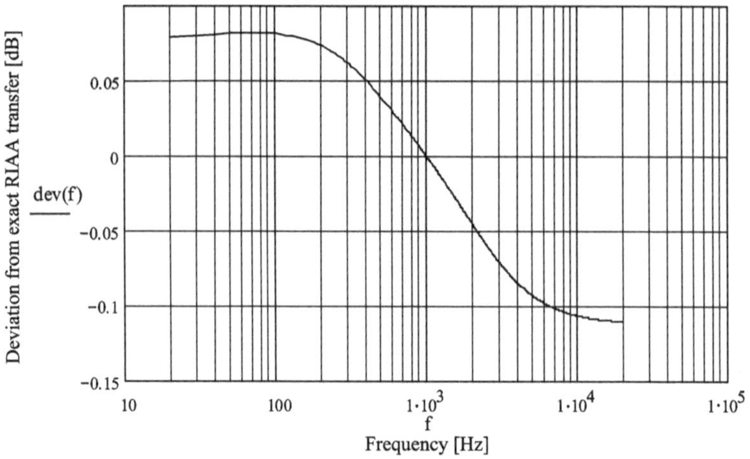


**Fig. 2.6** MathCad calculated deviation dev($f$) [dB] versus frequency: exact RIAA transfer minus actual transfer

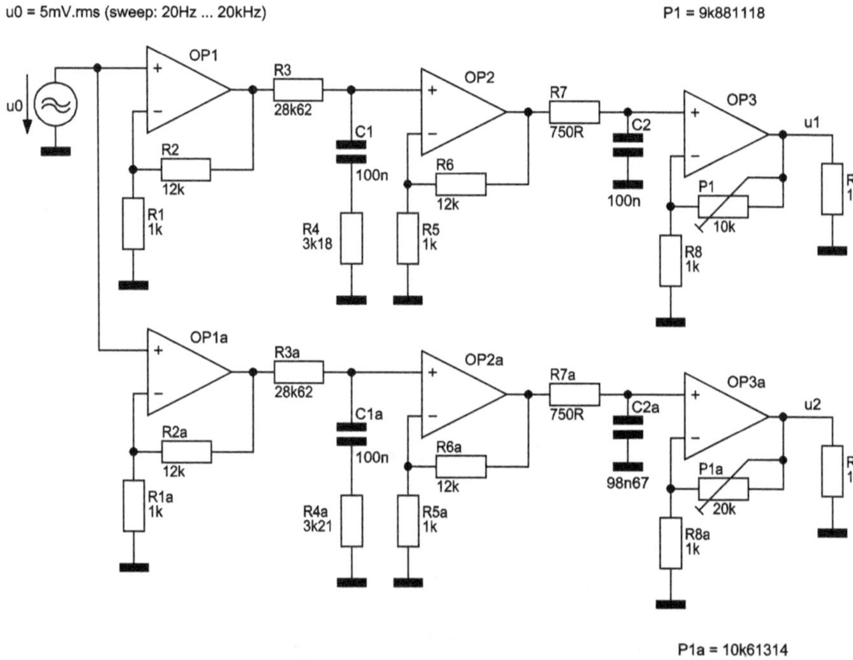


**Fig. 2.7** Creation of Fig. 2.8: pSpice simulation schematic to perform a deviation plot between exact RIAA transfer (output voltage u1) and actual RIAA transfer (output voltage u2)

- a first circuit (top) that performs the ideal RIAA transfer:
  $u_1 = R(20\,\text{Hz}\ldots 20\,\text{kHz})$
- a second circuit (bottom) that performs the RIAA transfer with actual components:
  $u_2 = Ra(20\,\text{Hz}\ldots 20\,\text{kHz})$ with
  - $T_{1a} = (R_{3a} + R_{4a}) \times C_{1a}$
  - $T_{2a} = R_{4a} \times C_{1a}$
  - $T_{3a} = R_{7a} \times C_{2a}$
- the gain of both circuits must be trimmed with $P_1$ and $P_{1a}$ to get equal rms output voltages of 0 dBV for $u_1$ and $u_2$ at 1 kHz

It should not be a surprise that both plots look absolutely equal.

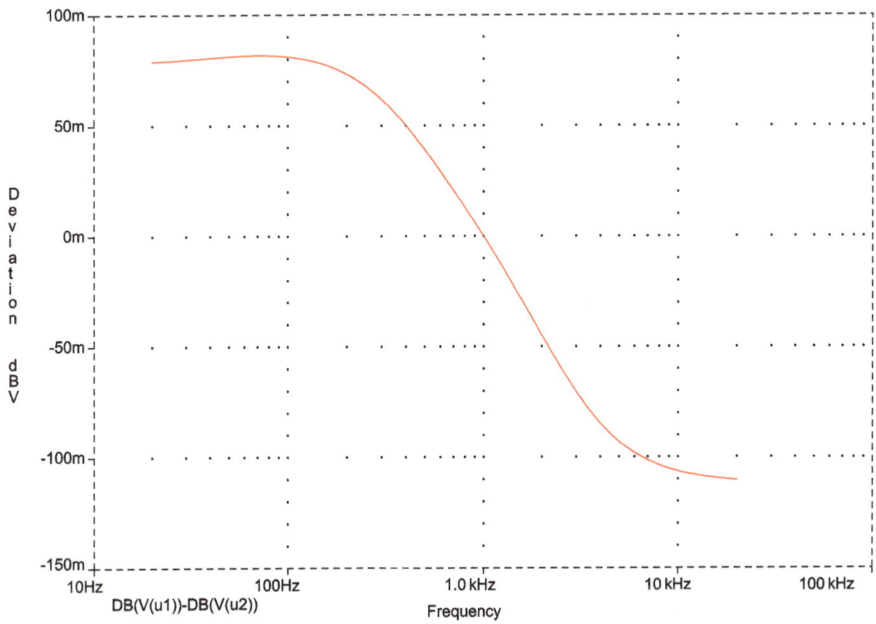


**Fig. 2.8** pSpice (MicroSim v8.0) simulated deviation [dB] versus frequency: exact RIAA transfer minus actual transfer

# Chapter 3
# Noise Basics

## 3.1 Noise in Components and Other General Noise Effects

### Intro

In fact, the reason why I wrote this book is noise, noise from electronic devices as well as noise as the signal-disturbing output of loudspeakers. The music I like should come out of the NOTHING. Clear, undistorted and close to the original.

Distortion matters of amplifiers can be wonderfully tackled and accompanying problems as well be wonderfully solved with the help of Douglas Self's Audio book[1]. Because of the high-gain and low-noise op-amps we find on today's markets it's rather easy to design low-distortion amps of any kind. That's why this book doesn't treat this matter in depth.

But it's still a kind of art to design lowest-noise amplifiers, amps near the boundaries only set by physics and not set by lousy compromises.

Let's examine first some interesting basic noise issues. Issues which were much deeper analysed by Messrs Motchenbacher and Connelly in their fantastic noise hand-book[2] as well as by many other publications[3]. But in these publications it's a hard job to filter out what we need for a fast and handy noise analysis of a design we want to check first in theory before we spent a lot of money for expensive (high-end/high-price) components. That's why this book is a collection of ready to take formulae and circuit design approaches to quickly noise-check any type of amp design in the audio field as well as a collection of design rules for precision RIAA equalized phono-amps.

---

[1] "Self on Audio", Douglas Self, Newnes 2000, ISBN 0 7506 4765 5
Abbreviation for the following footnotes: **D/S**

[2] "Low-Noise Electronic System Design", C. D. Motchenbacher, J. A. Connelly
John Wiley & Sons 1993, ISBN 0 471 57742 1
Abbreviation for the following footnotes: **M/C**

[3] "Electronic Circuits", Handbook for Design and Application, U. Tietze, C. Schenk,
2nd Edition, Springer 2008, ISBN 978-3-540-00429-5 (The accompanying CD-ROM also covers data sheets and different simulation softwares like eg. MicroSim v8.0)
Abbreviation for the following footnotes: **T/S**

B. Vogel, *The Sound of Silence*
DOI: 10.1007/978-3-540-76884-5, ©Springer 2008

When talking about noise I always mean noise mechanisms like thermal noise and low-frequency noise ($1/f$-noise). It sounds like the noise between two FM stations as well as the output of the loudspeakers when turning on the volume knob to max without music signal: these types of noise are totally random signals, random in amplitude and phase. Based on the equivalent heating effect noise voltage $e(t)$ can be expressed in terms of rms:

$$e_{\text{rms}}(t) = \sqrt{\frac{1}{T_{\text{p}}}\int_{0}^{T} e(t)^2 dt} \tag{3.1}$$

$e(t)$ is the time dependent noise voltage, $T_{\text{p}}$ is its period and the formula applied to a sine wave of a peak value $\hat{e}(t)$ results in a rms voltage for $e_{\text{rms}}$ of:

$$e_{\text{rms}} = 0.707\hat{e}(t) \tag{3.2}$$

### Amplifier Noise Model

The most simple noise model of an amplifier is given in Fig. 3.1. It consists of an ideal and noise-free amp with all noise sources transferred to the input, thus, creating only one noise voltage source $e_{\text{NT}}(f)$ and only one noise current source $i_{\text{NT}}(f)$. Of course, any amp is fed by a noisy source. That's why a 3rd noise component has to be added to the circuit. With that it will be relatively easy to calculate Signal-to-Noise-Ratios (*SN*s). The so-called equivalent input noise sources of the amp are totally independent from the amp's gain and input impedance $Z_{\text{in}}(f)$. In the case of BJTs and FETs they only depend on frequency, temperature, collector or drain current, and some physical constants.

### Noise Voltage

Hence, the total input referred noise voltage $e_{\text{N.tot}}(f)$ becomes[4]:

$$|e_{\text{N.tot}}(f)|^2 = |e_{\text{N.amp}}(f)|^2 + |i_{\text{N.amp}}(f)|^2 RS^2 + e_{\text{N.RS}}^2 \tag{3.3}$$

### Noise Current

In data sheets the spectral noise densities $e_{\text{N.xy}}(f)$ and $i_{\text{N,xy}}(f)$ are often given for transistors and op-amps – but – as far as I know – never for valves! A typical example is the well known op-amp OP27:

$$e_{\text{n.op27}}(1\,\text{kHz}) = 3.2\,\text{nV/rtHz} \qquad i_{\text{n.op27}}(1\,\text{kHz}) = 0.4\,\text{pA/rtHz}$$

[4] M/C – Chap. 2

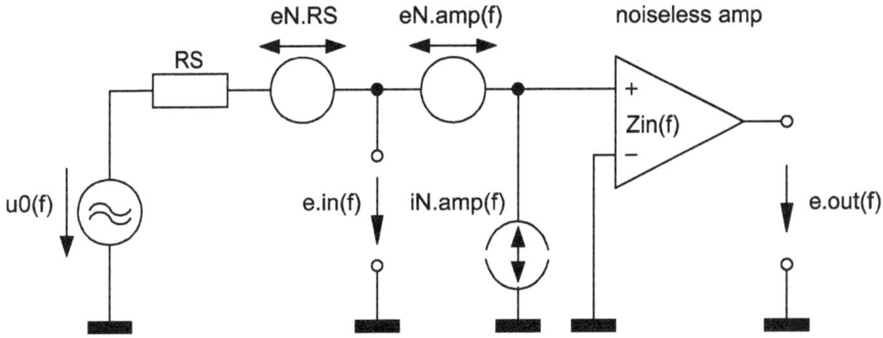


**Fig. 3.1** Amplifier with equivalent noise sources $e_{N.amp}(f)$ and $i_{N.amp}(f)$ and signal source $u_0(f)$ and noise of source resistance $e_{N.RS}$

## $1/f$ Noise

When calculating *SN*s one of the problems that will occur is the fact that most op-amps and transistors have an increase of noise voltage and noise current at the lower end of the frequency range of interest (audio band $= B_{20\,k} = 20\,\text{Hz}\ldots 20\,\text{kHz} = 19{,}980\,\text{Hz}$). It's the so-called $1/f$-noise or flicker noise. It decreases with growing frequency with a slope of 3 db/octave. That slope must sound like pink noise! The cross-point of the two tangents of the constant value plot with the $1/f$-slope plot create the $1/f$-corner-frequency $f_{ce}$ for noise voltage and $f_{ci}$ for noise current. This is not the case at the higher end of $B_{20\,k}$. In Figs. 3.2 and 3.3 this effect is demonstrated for the OP27. With Eqs. (3.4) and (3.5) formulae are given to calculate rms noise voltages and currents for any frequency band $B = f_{high} - f_{low}$.

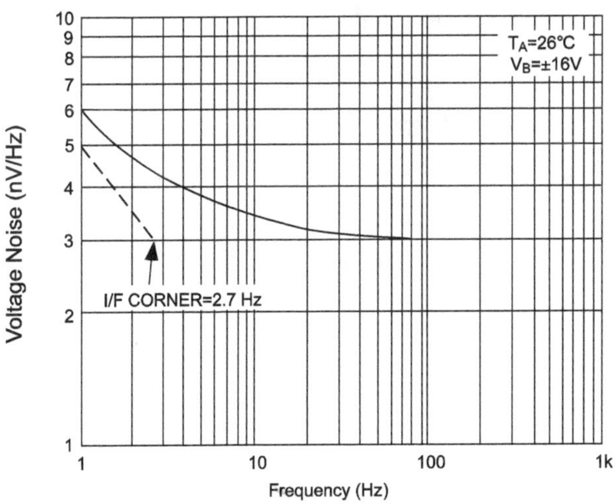


**Fig. 3.2** OP27 spectral voltage noise density[5] with corner frequency $f_{ce}$

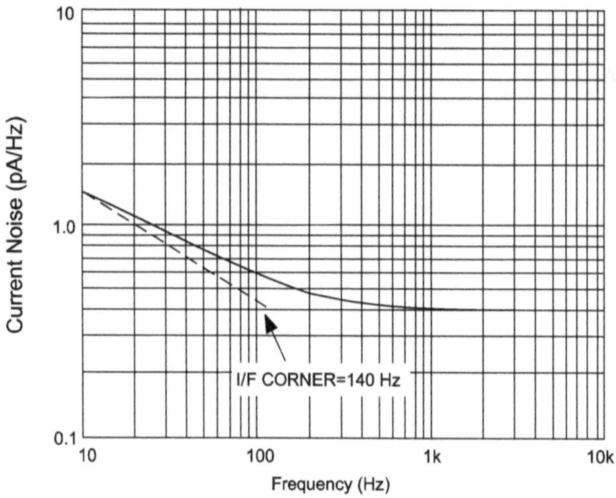


**Fig. 3.3** OP27 spectral current noise density[6] with corner frequency $f_{ci}$

$$e_{N.B} = e_{n.W}\sqrt{f_{ce}\ln\frac{f_{high}}{f_{low}} + (f_{high} - f_{low})} \tag{3.4}$$

$$i_{N.B} = i_{n.W}\sqrt{f_{ci}\ln\frac{f_{high}}{f_{low}} + (f_{high} - f_{low})} \tag{3.5}$$

$e_{n.W} = 3.2\,\mathrm{nV/rtHz}$ at 1 kHz can be taken from the white noise region (suffix W) right of $f_{ce}$ in Fig. 3.2, $i_{n.W} = 0.4\,\mathrm{pA/rtHz}$ at 10 kHz from Fig. 3.3. Both values should be taken from the region where the noise density is most constant.

$f_{ce}$ is outside $B_{20\,k}$, consequently the rms noise voltage inside $B_{20\,k}$ becomes

$$e_{N.B\,20\,k.op27} = e_{n.op27} \times \sqrt{B}_{20\,k} \tag{3.6}$$

$$e_{N.op27} = e_{n.op27} \tag{3.7}$$

$f_{ci}$ is inside $B_{20\,k}$ at 140 Hz. Hence, $i_{N.B.op27}$ becomes:

$$i_{N.B\,20\,k.op27} = \frac{0.4 \times 10^{-12}\mathrm{A}}{\sqrt{\mathrm{Hz}}}\sqrt{140\,\mathrm{Hz} \times \ln\frac{20{,}000}{20} + (20{,}000 - 20)} \tag{3.8}$$

$$i_{N.B\,20\,k.op27} = \frac{0.4 \times 10^{-12}\mathrm{A}}{\sqrt{\mathrm{Hz}}} \times 144.73\sqrt{\mathrm{Hz}} = 57.89\,\mathrm{pA} \tag{3.9}$$

$$\sqrt{B_{20\,k}} = 141.35\sqrt{\mathrm{Hz}} \tag{3.10}$$

[5] Analog Devices OP27 data sheet

[6] dto.

Hence, the new noise current $i_{N.op27}$ referenced to $B_1 = 1\,\text{Hz}$ can be calculated as follows:

$$i_{N.op27} = i_{n.W.op27} \times \frac{144.73}{141.35} = 0.41\,\text{A}/\sqrt{\text{Hz}} \tag{3.11}$$

or:

$$i_{N.op27} = \frac{i_{N.B\,20\,k.op27}}{\sqrt{B_{20\,k}}} = \frac{57.89\,\text{pA}}{141.35\sqrt{\text{Hz}}} = 0.41\,\text{pA}/\sqrt{\text{Hz}} \tag{3.12}$$

## Resistor (Johnson) Noise

But back to the roots of noise. Not only active semiconductor components create noise at their junctions. In consequence of random motion of charge carriers inside their matter any passive component does it as well[7]. Their resistive part is a noise generator of thermal noise and the noise voltage $e_{N.xy}$ and noise current $i_{N.xy}$ – called Johnson noise – can be calculated as follows – according to the rules M. Nyquist has set long time ago:

$$e^2_{N.R} = 4kTRB \tag{3.13}$$

$$i^2_{N.R} = \frac{4kTB}{R} \tag{3.14}$$

- $k$ is Boltzmann's constant: $k = 1.38065 \times 10^{-23}\,\text{V\,A\,s/K}$
- $T$ is the temperature in K (Kelvin)
- $R$ is the resistor in Ω (Ohm)
- $B$ is the frequency range of interest. In most data sheets it's $B_1 = 1\,\text{Hz}$. That's why the unit of the noise voltage $e_N$ becomes V/rtHz and the unit of the noise current $i_N$ becomes A/rtHz.
- rtHz means $\sqrt{\text{Hz}}$ or $\text{Hz}^{0.5}$. It's easier to write rtHz in a text.

$$e_{N.R} = \sqrt{4kTRB} \qquad i_{N.R} = \sqrt{\frac{4kTB}{R}}$$

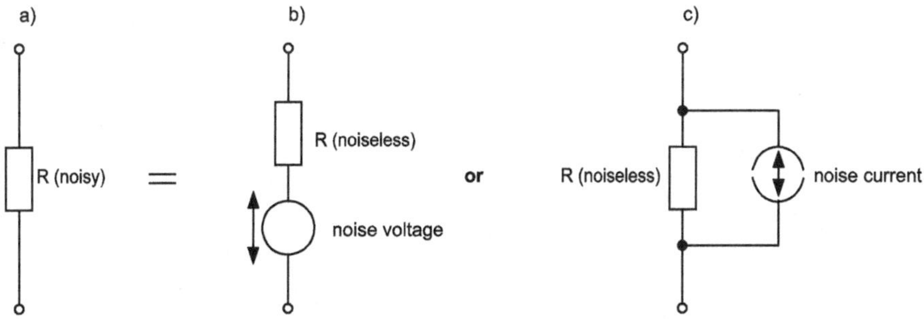


**Fig. 3.4** Equivalent circuits for thermal noise (Johnson) in resistors

[7] M/C – Chap. 12

The noise voltage of a resistor at room temperature of 300 K and in $B_1$ can simply be calculated as follows:

$$e_{N.R} = 4.07 \times \sqrt{\text{resistance in k}\Omega}\,[\text{nV/rtHz}] \tag{3.15}$$

Therefore the rms noise voltage of a 1 kΩ resistor in the frequency band $B_{20\,k}$ looks like:

$$e_{N.rms.1k.B\,20\,k} = 4.07\,\text{nV/rtHz} \times \sqrt{B_{20\,k}} = 575.3\,\text{nV} \tag{3.16}$$

**Noise Voltage Sources Series-Connected**

In Fig. 3.5 the sum $e_{N.tot}$ of two or $n$ noise voltages $e_{N1}$, $e_{N2}$ or $e_{N.n}$ series- or sequence-connected is not $e_{N1} + e_{N2} + \ldots + e_{N.n}$!!! This would only be the case if the voltages were absolutely (100%) correlated. When talking about noise this is a rather seldom fact.

Generally, noise voltages and currents are 100% uncorrelated, unless they were generated from the same source. Their uncorrelated sum of e.g. two noise voltages can be calculated according to the so-called rms sum format:

$$(e_{N.1+N.2})^2 = e_{N.1}^2 + e_{N.2}^2 \tag{3.17}$$

or:

$$e_{N.1.2.seq} = \sqrt{e_{N1}^2 + e_{N2}^2} \tag{3.18}$$

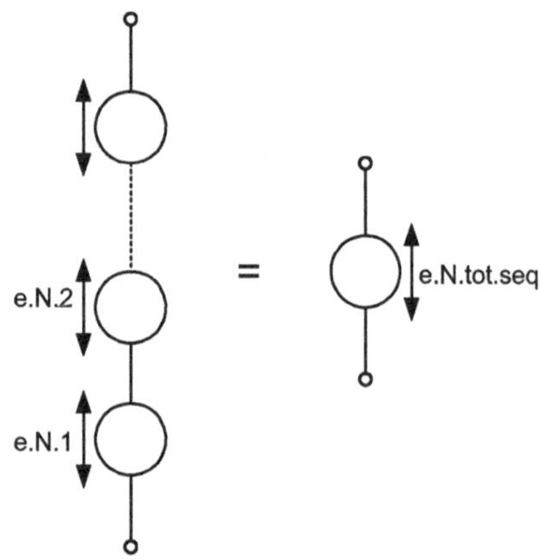


**Fig. 3.5** Noise voltage sources sequence-connected

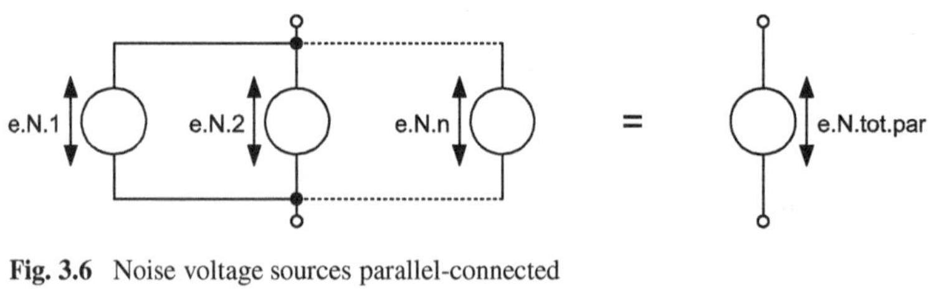


**Fig. 3.6** Noise voltage sources parallel-connected

if both noise sources have identical values $e_{\mathrm{N}}$, than the formula as above becomes:

$$e_{\mathrm{N.1.2.seq}} = \sqrt{2}e_{\mathrm{N}} \tag{3.19}$$

$n$ noise voltage sources sequence-connected result in:

$$e_{\mathrm{N.tot.seq}} = \sqrt{n}e_{\mathrm{N}} \tag{3.20}$$

### Noise Voltage Sources Parallel-Connected

Parallel-connection of noise voltage sources is given in Fig. 3.6.

The sum of $n$ parallel-connected noise voltage sources $e_{\mathrm{N.1}} \ldots e_{\mathrm{N.n}}$ becomes:

$$e^2_{\mathrm{N.tot.par}} = \left(\frac{1}{e^2_{\mathrm{N.1}}} + \frac{1}{e^2_{\mathrm{N.2}}} + \ldots + \frac{1}{e^2_{\mathrm{N.n}}}\right)^{-1} \tag{3.21}$$

if all $n$ noise voltage sources are of identical value $e_{\mathrm{N}}$, than, the equation as above changes to:

$$e_{\mathrm{N.tot.seq}} = \frac{e_{\mathrm{N}}}{\sqrt{n}} \tag{3.22}$$

### Noise Current Sources Parallel-Connected

Mathematically, noise current sources parallel-connected become treated like the sequence-connected noise voltage sources:

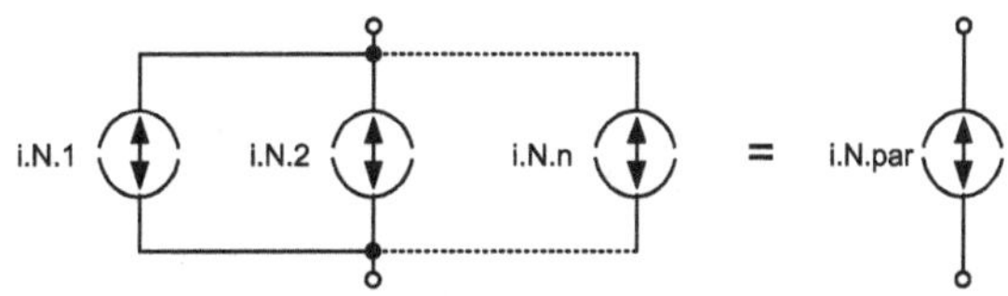


**Fig. 3.7** Noise current sources parallel-connected

hence,

$$i_{\mathrm{N.tot.par}} = \sqrt{i^2_{\mathrm{N.1}} + i^2_{\mathrm{N.2}} + \ldots + i^2_{\mathrm{N.n}}} \tag{3.23}$$

with identical current noise sources $i_{\mathrm{N.1\ldots n}} = i_{\mathrm{N}}$

$$i_{\mathrm{N.tot.par}} = \sqrt{n}i_{\mathrm{N}} \tag{3.24}$$

### Noise Current Sources Series-Connected

Mathematically, noise current sources sequence-connected become treated like the parallel-connected noise voltage sources:

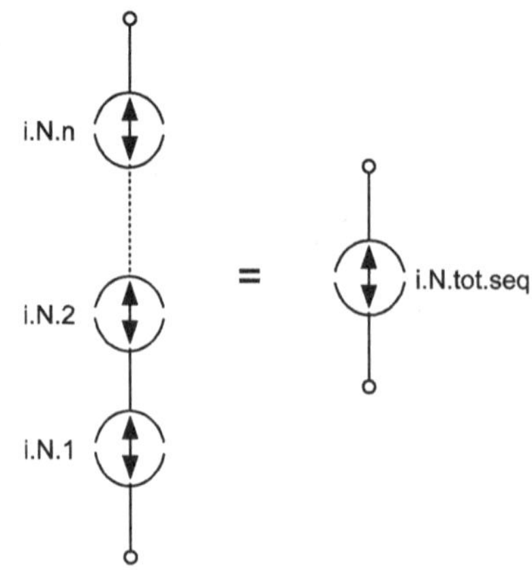


**Fig. 3.8** Noise current sources sequence-connected

hence,

$$i^2_{\mathrm{N.tot.par}} = \left(\frac{1}{i^2_{\mathrm{N.1}}} + \frac{1}{i^2_{\mathrm{N.2}}} + \ldots + \frac{1}{i^2_{\mathrm{N.n}}}\right)^{-1} \tag{3.25}$$

with identical current noise sources $i_{\mathrm{N.1...n}} = i_{\mathrm{N}}$

$$i_{\mathrm{N.tot.seq}} = \frac{i_{\mathrm{N}}}{\sqrt{n}} \tag{3.26}$$

The noise voltage sources of $n$ resistors in a sequence have to be summed up as follows[8]:

$$e^2_{\mathrm{N.tot.seq}} = 4kTB(R_1 + R_2 + \ldots + R_n) \tag{3.27}$$

The noise sources of $n$ resistors in parallel have to be calculated as follows:

$$e^2_{\mathrm{N.tot.par}} = 4kTB\left(\frac{1}{R_1} + \frac{1}{R_2} + \ldots + \frac{1}{R_n}\right)^{-1} \tag{3.28}$$

Impedances like capacitors or inductances are treated the same way: concerning noise their resistive parts follow the above shown rules. But it must be taken into account that the noise voltages and currents of these types of impedances change according to the frequency dependency of these components. In detail, this will be demonstrated later in the MM cartridge chapter.

## Paralleling of Active Devices

What about paralleling active devices like op-amps and transistors? Which effect has it on noise voltage and noise current? To make a long story short I only will show the results of paralleling $n$ devices as of Fig. 3.9 – and not the detailed calculations.

[8] M/C – Chap. 2

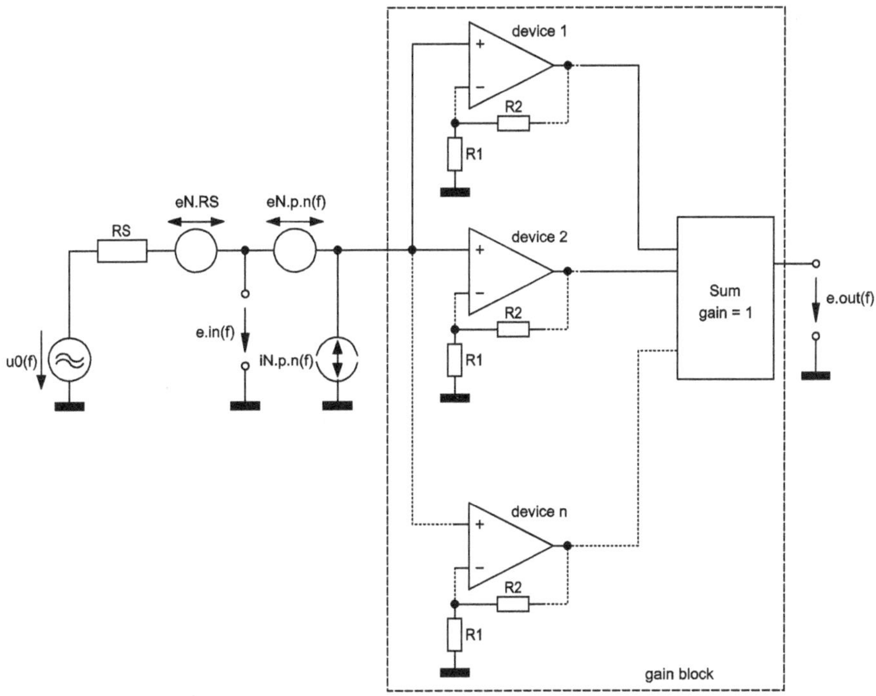


**Fig. 3.9** Paralleling of $n$ active devices

Supposed that all devices were equal, especially their gain and their equivalent noise voltages and noise currents, for the newly created gain block inside the dashed box the following formulae will lead to the right results:

$$e^2_{\mathrm{N.npar}} = \frac{e^2_{\mathrm{N.1device}}}{n} \tag{3.29}$$

or:

$$e_{\mathrm{N.npar}} = \frac{e_{\mathrm{N.1device}}}{\sqrt{n}} \tag{3.30}$$

$$i^2_{\mathrm{N.npar}} = n \times i^2_{\mathrm{N.1device}} \tag{3.31}$$

or:

$$i_{\mathrm{N.npar}} = \sqrt{n} \times i_{\mathrm{N.1device}} \tag{3.32}$$

The dotted lines between device and resistors $R_1$, $R_2$ are a reminder for additional biasing components, being noise sources around any active device. As part of a real circuit noise calculation they have to be taken into any noise calculation as well.

Total gain $G_{\mathrm{par.tot}}(f)$ of the paralleling gain block becomes:

$$|G_{\mathrm{par.tot}}(f)| = \frac{|e_{\mathrm{out}(f)}|}{|e_{\mathrm{in}(f)}|} = |G_{\mathrm{device}_1}(f)| + |G_{\mathrm{device}_2}(f)| + \ldots + |G_{\mathrm{device}_n}(f)| \tag{3.33}$$

With equal gain $G_{\text{device}}(f)$ for all devices Eq. (3.33) becomes:

$$|G_{\text{tot}}(f)| = n|G_{\text{device}}(f)| \tag{3.34}$$

With $R_1 = Z_1(f)$, $R_2 = Z_2(f)$ $G_{\text{device}}(f)$ becomes:

$$\left|G_{\text{device}}(f)\right| = 1 + \left|\frac{Z_2(f)}{Z_1(f)}\right| \tag{3.35}$$

### Sequence of Two Amplifying Stages

What about sequencing amplifiers? Which effect does this have on noise voltages and currents? The answers look relatively simple. The two amps form a new gain block and their noise sources can be transferred to the input of that newly created amp.

Figure 3.10 and the following formulae will lead to the right solutions:

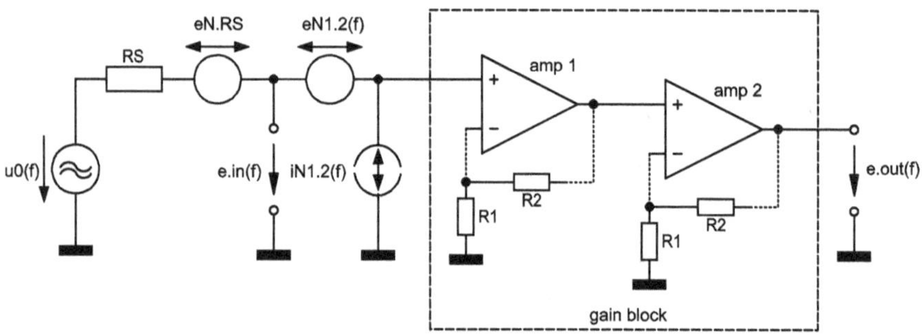


**Fig. 3.10** Sequence of two amplifying devices

The gain of amp 1 should be $G_1(f)$, the gain of amp 2 should be $G_2(f)$. All noise sources around the amps are transferred into the respective equivalent input noise sources of each amp[9]. Than,

$$|e_{\text{N1.2}}(f)|^2 = \left(\frac{|e_{\text{N.amp2}}(f)|}{|G_1(f)|}\right)^2 + |e_{\text{N.amp1}}(f)|^2 + |i_{\text{N1.2}}(f)RS|^2 + e_{\text{N.RS}}^2 \tag{3.36}$$

$$i_{\text{N1.2}}(f) = i_{\text{N.amp1}}(f) \tag{3.37}$$

$$|G_{\text{seq.tot}}(f)| = \frac{|e_{\text{out}(f)}|}{|e_{\text{in}(f)}|} = |G_1(f)| \times |G_2(f)| \tag{3.38}$$

But real life is not that simple: supposed

[9] Detailed analysis for that will be given in the following Sects. 3.2–3.5 on noise in solid-state devices, valves, transformers, etc.

- amp1 has an output impedance > 0 R

and/or

- the connection between output amp1 and input amp2 is a resistor which forms a voltage divider together with the magnitude of the input impedance of amp2

and/or

- the magnitude of the input impedance of amp1 is not very much bigger than RS, thus, at the input of amp1, creating another counting voltage divider for $u_0(f)$

than, the whole exercise to calculate the total noise voltage and noise current of that amplifier chain becomes much more complex. Figure 3.11 and the following equations show the results:

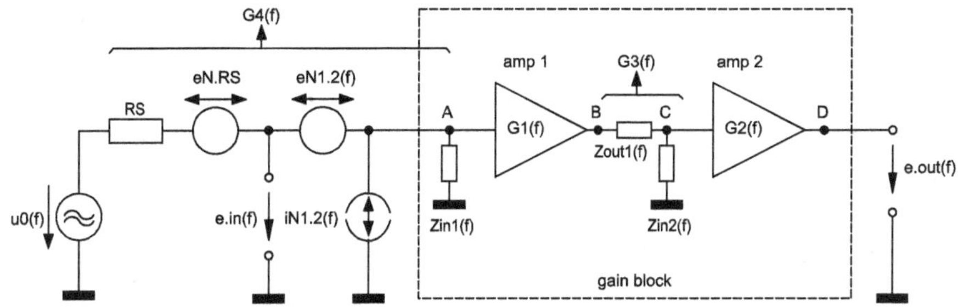


**Fig. 3.11** Real life situation of a sequence of two amplifying stages or of two separate amps

Generally, Eq. (3.36) is still valid. But, because of $G_3(f)$ and $G_4(f)$, it needs certain additions! Hence, concerning $e_{\text{in}(f)}$ the total gain $G_{\text{seq.tot}}[e_{\text{in}(f)}]$ of the gain block becomes:

$$|G_{\text{seq.tot}}[e_{\text{in}(f)}]| = \frac{|e_{\text{out}(f)}|}{|e_{\text{in}(f)}|} = |G_1(f)| \times |G_2(f)| \times |G_3(f)| \tag{3.39}$$

Concerning $u_0(f)G_{\text{seq.tot}}[u_0(f)]$ becomes:

$$|G_{\text{seq.tot}}[u_0(f)]| = \frac{|e_{\text{out}(f)}|}{|u_0(f)|} = |G_1(f)||G_2(f)||G_3(f)||G_4(f)| \tag{3.40}$$

$$|G_3(f)| = \left|\frac{Z_{\text{out1}}(f) + Z_{\text{in2}}(f)}{Z_{\text{in2}}(f)}\right| \tag{3.41}$$

$$|G_4(f)| = \left|\frac{RS + Z_{\text{in1}}(f)}{Z_{\text{in1}}(f)}\right| \tag{3.42}$$

The equivalent input noise current of amp 2 is $i_{\text{N.amp2}}(f)$. It flows through the parallel impedance $Z_{\text{in2}}(f)||Z_{\text{out1}}(f)$, thus, creating a new noise voltage source $e_{\text{N.G}_3}(f)$. By creating another new noise voltage source $e_{\text{N.G}_4}(f)$ the noise current of amp1 flows through the parallel impedance $Z_{in1}(f)||RS$. $e_{N.Z_3}(f)$ is the noise voltage of $RS||Z_{in1}(f)$ and $e_{N.Z_4}(f)$ is the noise voltage of $Z_{out1}(f)||Z_{in2}(f)$. Taking this into account we can rewrite Eq. (3.36) ff as follows:

$$|e_{N1.2out(f)}|^2 = \left(\sqrt{(|e_{N.amp1}(f)|^2 + |e_{N.G_4}(f)|^2 + |e_{N.Z_4}(f)|^2)}|G_1(f)G_2(f)G_3(f)|\right)^2 + \left(\sqrt{(|e_{N.amp2}(f)|^2 + |e_{N.G_3}(f)|^2 + |e_{N.Z_3}(f)|^2)}|G_2(f)|\right)^2 \quad (3.43)$$

$$|e_{N1.2}(f)| = \frac{|e_{N1.2out}(f)|}{|G_1(f)G_2(f)G_3(f)|} \quad (3.44)$$

$$i_{N1.2}(f) = i_{N.amp1}(f) \quad (3.45)$$

### Noise Factor and Noise Figure

Quite often, in noise charts of transistors and op-amps, we'll find certain plots with noise figure versus frequency or source resistance or collector current etc. What does noise figure mean?

Noise figure $NF$ is the logarithmic expression of the noise factor $F$ and $_P$ indicates that it is based on power:

$$NF_P = 10\log F_P \quad (3.46)$$

If we have a noise power creating source (e.g. a resistor or op-amp) followed by a noise power creating amp, than, the noise factor of the source-amp-chain is the factor with which the amp increases the noise power of the source:

$$F_P = \frac{\text{total available output noise power}}{\text{portion of output noise power caused by the source}} \quad (3.47)$$

For our purposes a more practical formula for $F$ is based on noise voltages alone, indicated with $_e$:

$$F_e = \frac{\text{real noise voltage at the output of the amp}}{\text{noise voltage at the output of the noiseless amp}} \quad (3.48)$$

and $NF_e$ becomes:

$$NF_e = 20\log F_e \quad (3.49)$$

With the respective data sheet figures and charts like the ones shown below it is quite easy to select the right input device.

For example: we're in search of a low-noise input transistor for source resistances in the range of 700 R–20 k (which are typical values for a MM cartridge operating in $B_{20\,k}$), than, the plots provide us with the optimal value for the collector current for $NF_e \leq 0.5$ dB: $I_{C.opt} = 100\,\mu A$. It makes no sense to search for a transistor which is

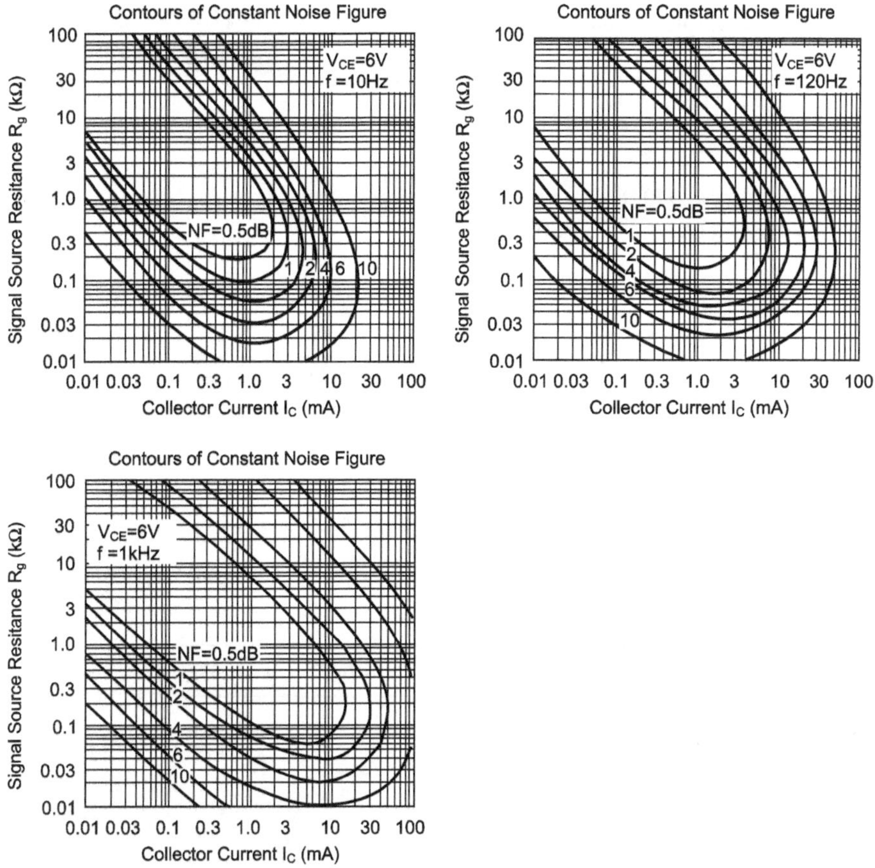


**Fig. 3.12** Noise figures of the low-noise transistor 2SC2546 NF vs. source resistance and collector current[10] at three different frequencies

favourable for a source resistance of only 1 k. In that case a collector current of 1 mA would do as well. But noise-wise this doesn't work good enough because the coil inductance of the cartridge ($200\,\mu\text{H} \leq L \leq 700\,\mu\text{H}$) parallel to the input resistance of the amp (47 k) creates a maximal source resistance of $\approx 40\,\text{k}$ at app. 15 kHz!

Assumed, that the spectral noise density of all components were of constant value in the frequency band of interest, than, expressed in a practical equation for a typical example as of Fig. 3.1 plus input resistance $R_i$, for any amplifier that follows a source with source resistance $RS$ $NF_{\text{e,amp}}$ looks as follows:

$$NF_{\text{e.amp}} = 20\log\left(\frac{\sqrt{e^2_{\text{N.RS}} + e^2_{\text{N.amp}} + (i_{\text{N.amp}}RS)^2}}{\sqrt{e^2_{\text{N.RS}}}}\right) \tag{3.50}$$

[10] Hitachi Data Sheet

$NF_{\text{e.amp}}$ example calculation: with

- $e_{\text{N.amp}} = 3.2\,\text{nV/rtHz}$
- $i_{\text{N.amp}} = 0.41\,\text{pA/rtHz}$
- $R_0 = 1\,\text{k}$
- $R_{\text{i}} = 47\,\text{k}$
- $RS = R_{\text{i}}||R_0 = 979\,\text{R}$
- $e_{\text{N.RS}} = (4\,\text{kT}RSB_1)^{0.5} = 4.03\,\text{nV/rtHz}$
- $B_1 = 1\,\text{Hz}$

$NF_{\text{e.amp}}$ becomes:

$$NF_{\text{e.amp}} = 20\log\left(\frac{\sqrt{4.03\,\text{nV}^2 + 3.2\,\text{nV}^2 + (979R \times 0.41\,\text{pA})^2}}{4.03\,\text{nV}}\right) = 2.15\,\text{dB} \tag{3.51}$$

With changed values for

- $R_0 = 50R$
- $RS = R_{\text{i}}||R_0 = 49R95$
- $e_{N.RS} = (4kTRSB_1)^{0.5} = 0.91\,\text{nV/rtHz}$

$$NF_{\text{e,amp}} = 20\log\left(\frac{\sqrt{0.91\,\text{nV}^2 + 3.2\,\text{nV}^2 + (49R95 \times 0.41\,\text{pA})^2}}{0.91\,\text{nV}}\right) = 11.26\,\text{dB} \tag{3.52}$$

With changed values for

- $R_0 = 47\,\text{k}$
- $RS = R_{\text{i}}||R_0 = 23k5$
- $e_{\text{N.R}_{\text{in}}} = (4kTRSB_1)^{0.5} = 19.7\,\text{nV/rtHz}$

$$NF_{\text{e.amp}} = 20\log\left(\frac{\sqrt{19.7\,\text{nV}^2 + 3.2\,\text{nV}^2 + (23k5 \times 0.41\,\text{pA})^2}}{19.7\,\text{nV}}\right) = 1.02\,\text{dB} \tag{3.53}$$

These results lead to the question if there exists an optimum source resistance $R_{\text{opt}}$ for given equivalent input noise sources $i_{\text{N}}$ and $e_{\text{N}}$? It does:

$$R_{\text{opt}} = \frac{e_{\text{N}}}{i_{\text{N}}} \tag{3.54}$$

With the above shown figures of the OP27 its $R_{\text{opt}}$ becomes 7.8 k. With that, its $NF_{\text{e.op27.opt}}$ becomes 0.64 dB! Another approach is the development of a plot showing the relationship between source resistance and NF of a given device like the OP27:

The below shown graph is made with MCD together with Eq. (3.50). With a specific tool we can pick any value out of the plot, e.g. OP27's minimum $NF_{\text{e.op27}} = 0.64\,\text{dB}$ at $RS = 7.77\,\text{k}$ to $7.84\,\text{k}$.

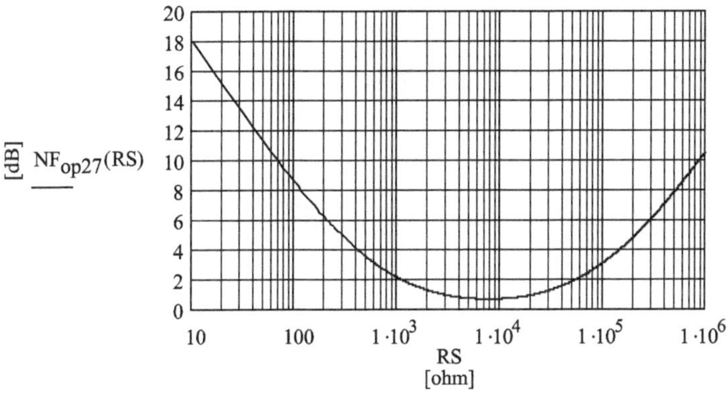


**Fig. 3.13** OP27: $NF_e$ vs. source resistance *RS*

To get adequate informations about the noise performance of any type of active device the easiest way would be plots like the ones of Figs. 3.14–3.16. Unfortunately, it's rather seldom to find such plots in data sheets for elder types of solid state devices.

To sum up: $NF_{amp}$ becomes relatively small if collector current is high and source resistance is low or source resistance is high and collector current is low. But $NF_{amp}$ does not say anything about the absolute noise voltage and noise current situation of an amp nor does it say anything about the Signal-to-Noise-Ratio SN of an amp!

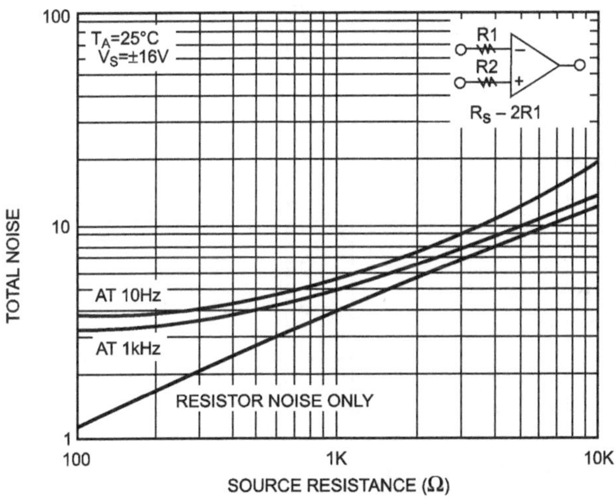


**Fig. 3.14** OP27: total noise vs. source resistance[11]

[11] Analog Devices Data Sheet OP27

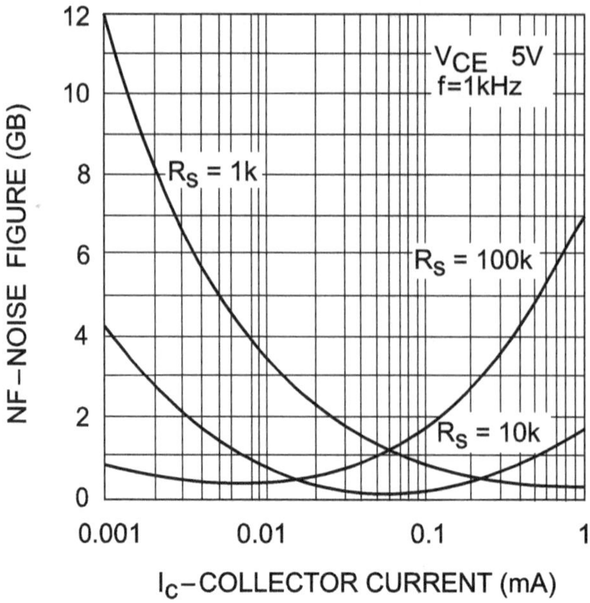


**Fig. 3.15** LM394: $NF_e$ vs. source resistance and collector current[12]

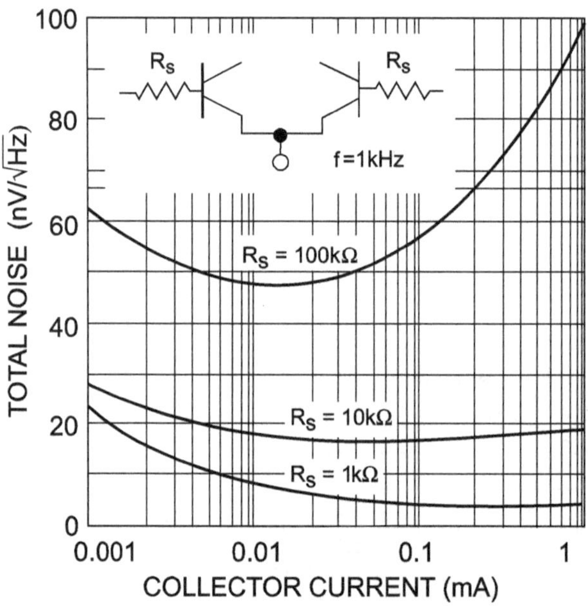


**Fig. 3.16** SSM2210: total noise vs. source resistance[13]

---

[12] National Semiconductor Data Sheet LM394

[13] Analog Devices Data Sheet SSM-2210

## Signal-to-Noise Ratio (SN)

How can we calculate SN – and what is it?

Generally spoken, SN is the quotient of total output power $P_{out.tot}$ vs. total output noise power $P_N$. It can be expressed in dB as follows[14]:

$$SN_p = 10\log\left(\frac{P_{out.tot}}{P_N}\right) \tag{3.55}$$

Power is proportional to the square of the rms value of a voltage. If $e_{N.out}(f)$ is the calculated noise density in a 1Hz band, than, expressed in dB, at the output of an amp, in a certain frequency band $f_{high} - f_{low}$, SN with reference to the nominal output rms voltage $e_{out.rms}$ becomes:

$$SN_{out.dB} = 20\log\left(\frac{\int_{low}^{f_{high}} |e_{N.out}(f)|^2 \, df}{e_{out.rms}^2}\right) \tag{3.56}$$

## Weighted and/or Equalized SN

*SN*s of amps with equalizing effects and/or weighting filters at their output (expressed by a transfer function $X(f)$) have to be calculated as follows:

$$SN_{out.weight.dB} = 20\log\left(\frac{\int_{low}^{f_{high}} |e_{N.out}(f)|^2 |X(f)|^2 \, df}{e_{out.rms}^2}\right) \tag{3.57}$$

If $e_{N.out}(f)$ is expressed as rms voltage in a specifically defined frequency band $> 1$ Hz($e_{N.out.B}(f), B = f_{high} - f_{low}$), than, SN becomes:

$$SN_{out.weight.dB} = 20\log\left(\frac{\sqrt{\frac{1}{f_{high}-f_{low}} \int_{low}^{f_{high}} |e_{N.out.B}(f)|^2 |X(f)|^2 \, df}}{e_{out.rms}}\right) \tag{3.58}$$

$X(f)$ might be the RIAA equalization transfer $R(f)$ or the A-weighting filter transfer $A(f)$ or both $R(f) \times A(f)$ or any other transfer function.

There might be cases when it's better to calculate *SN*s with reference to the nominal input rms voltage of an amp. Than, simply replace $e_{out.rms}$ and $e_{N.out}$ in Eqs. (3.57) and (3.58) by $e_{in.rms}$ and $e_{N.in}$. If we change the denominator with the numerator we'll get a positive SN result.

---

[14] T/S Chap. 4.2

### Excess Noise of Resistors

At the end of this section I must come back to a resistor noise problem which only occurs when putting a resistor between two different voltage potentials. Doing so this means that a DC-current flows through the resistor, thus, producing a special kind of noise. It's the so-called excess noise voltage $e_{\text{N.R.ex}}$. This excess noise voltage exists in excess to the thermal noise voltage of a resistor $e_{\text{N.R}}^2 = 4kTRB$ ($B = f_{\text{high}} - f_{\text{low}}$). It has to be rms-added to this thermal noise voltage.

Thus, with $f_{\text{low}} \leq f \leq f_{\text{high}}$ and for each frequency $f$ in $B$ total noise voltage of such a resistor between two different potentials becomes:

$$e_{\text{N.R.tot}}(f) = \sqrt{\left(\frac{e_{\text{N.R}}}{\sqrt{B}}\right)^2 + e_{\text{N.R.ex}}(f)^2} \tag{3.59}$$

The nature of this excess noise is one which follows the $1/f$-law ($= -10\,\text{dB/decade}$) and its plot voltage vs. frequency looks similar to the plot of an op-amp with a specific $1/f$ corner frequency $f_{\text{ce}}$ or $f_{\text{ci}}$ (see Figs. 3.2, 3.3, 3.17). In Eq. (3.59) the first term underneath the root is the resistor's noise voltage density in a 1 Hz bandwidth. It represents the white noise part. The $1/f$-part is the 2nd term.

To calculate the 2nd term we must know how to handle resistor excess noise. For any resistor there exists a so-called noise index NI which is defined as follows[15]:

- NI is the rms voltage of the noise of a resistor expressed in µV for each volt of DC drop across this resistor in one frequency decade.

  Hence, for

$$NI = 1\,\mu\text{V}/1\,\text{V}/1\,\text{decade} \tag{3.60}$$

  NI expressed in dB becomes:

$$NI_{\text{e}} = 20 \log NI = 0\,\text{dB} \tag{3.61}$$

If not indicated in data sheets the NI or $\text{NI}_{\text{e}}$ values for many types of resistors can be found in one of VISHAY's application notes[16].

If in data sheets $\text{NI}_{\text{dB}}$ is given in dB, than NI in µV/V/decade can be calculated as follows:

$$NI = 10^{\frac{N_{\text{e}}}{20}}\,[\mu\text{V/V/decade}] \tag{3.62}$$

If NI is given in µV/V/decade, than, $\text{NI}_{\text{e}}$ expressed in dB can be calculated as follows:

$$NI_{\text{e}} = 20 \log(NI)\,[\text{dB}] \tag{3.63}$$

---

[15] M/C Chap. 12

[16] VISHAY application note AN0003 – table 1 (attention: in this AN it is not specifically indicated that the shown NI figures are given for 1 decade only – but from their values and definition point of view they are all referenced to 1 decade)

$e_{\text{N.R.ex}}$ at a specific frequency $f$ can be calculated as follows:

$$e_{\text{N.R.ex}}(f) = \sqrt{\left(\frac{10^{\frac{NI_e}{10}} \times 10^{-12}}{\ln 10}\right)\left(\frac{V_{\text{DC}}^2}{f}\right)} \text{V/rtHz]} \tag{3.64}$$

Example to calculate $e_{\text{N.R.ex.f}}$. Given:

- $R = 100\,\text{k}$ metal film resistor
- $NI_e = -30\,\text{dB}$, with Eq. (3.62) $NI$ becomes: $NI = 0.03162\,\mu\text{V/V}_{\text{DC}} = 31.62\,\text{nV/V}_{\text{DC}}$
- $V_{\text{DC}} = 100\,\text{V}$ across $R$
- $f = 1000\,\text{Hz}$
- $f_{\text{low}} = 20\,\text{Hz}$
- $f_{\text{high}} = 20\,\text{kHz}$
- $B_{20\,\text{k}} = 20\,\text{kHz}{-}20\,\text{Hz} = 19{,}980\,\text{Hz}$

thus, $e_{\text{N.R.ex.1kHz}}$ becomes:

$$e_{\text{N.R.ex.1kHz}} = \sqrt{\left(\frac{10^{\frac{-30}{10}} \times 10^{-12}}{\ln 10}\right)\left(\frac{100\,\text{V}^2}{1000\,\text{Hz}}\right)} [\text{V/rtHz}] = 65.9\,\text{nV/rtHz} \tag{3.65}$$

In the 3 decades (3d) frequency range of the audio spectrum $B_{20\,\text{k}}$ the rms voltage of that example resistor $e_{\text{N.R.ex}}(3\text{d})$ becomes:

$$\begin{aligned} e_{\text{N.R.ex.3d}} &= NI \times \sqrt{3(=\text{number of decades})} \times V_{\text{DC}} \\ &= 31.62\,\text{nV/V/d} \times \sqrt{3} \times 100\,\text{V} \\ &= 5.48\,\mu\text{V} \end{aligned} \tag{3.66}$$

With Eq. (3.59) and with $e_{\text{N.100 k}} = 40.7\,\text{nV/rtHz}$ total rms voltage $e_{\text{N.R.B20k}}$ of the resistor $R$ in $B_{20\,\text{k}}$ becomes:

$$\begin{aligned} e_{\text{N.R.B20k}} &= \sqrt{\left(40.7\,\text{nV/rtHz} \times \sqrt{B_{20\,\text{k}}}\right)^2 + 5.48\,\mu\text{V}^2} \\ &= 7.94\,\mu\text{V} \end{aligned} \tag{3.67}$$

With the given conditions the spectral noise voltage density plot of that resistor is shown in the following Fig. 3.17.

Referenced to a bandwidth $B_1$ of 1 Hz $e_{\text{N.R.B1}}$ becomes:

$$e_{\text{N.R.B1}} = \frac{e_{\text{N.R.B20k}}}{\sqrt{B_{20\,\text{k}}}} = 56.2\,\text{nV/rtHz} \tag{3.68}$$

which is app. 38.1% more noise than the 100 k resistor's white noise of 40.7 nV/rtHz!

I do not want to over-complicate things, but, as long as they were designed between different potentials, excess noise also appears with the resistive parts of ca-

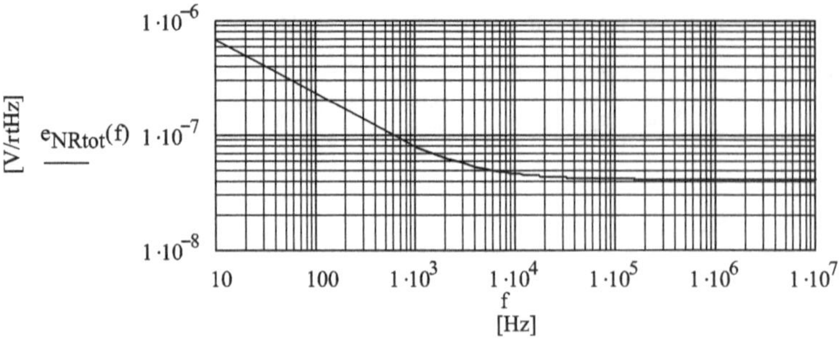


**Fig. 3.17** Resistor spectral noise voltage density ($R = 100$ k, DC-voltage across $R = 100$ V)

pacitors and inductances as well. Principally, it can be calculated with the above given equations as well – but I guess the respective influence on the noise situation of an amp can totally be ignored.

Because of the $1/f$-character of excess noise this kind of noise is "loud" in the lower regions of the audio frequency spectrum <1 kHz. Fortunately, in the rather low-DC-voltage driven solid-state designs it's relevance is marginal. It can be ignored in most cases. This will be explained in the BJT section. But, with rather high DC-voltages in valve designs it makes sense to take the respective influences into account. These effects will be reflected in the valve section of this book.

## 3.2 Noise in Bipolar Junction Transistors (BJTs)

### BJT Noise Model

The circuit of Fig. 3.1 becomes a bit more complex when dealing with a real life amplifier with a transistor as the first amplifying stage. I'm mainly concentrating on BJTs and not so much on FETs or valves. Because their noise is nearly always higher than BJT noise. Exceptions will be discussed later.

I do not dive very deep into all the theories of modelling transistors. This can be done by studying some other publications[17,18]. After many trials I found that the following adapted model for audio frequencies is very useful for practical noise calculations. It is derived from the so-called $\pi$-model and looks – for the audio band – as follows:

### Base Spreading Resistance – Calculation Approach

The real transistor consists of a noise voltage source $e_{n.T}$ and noise current source $i_{n.T}$. Unfortunately for the production of noise inside the whole transistor there exist

[17] Noise in transistor circuits, Baxandall, WW 11&12-1968

[18] T/S and M/C

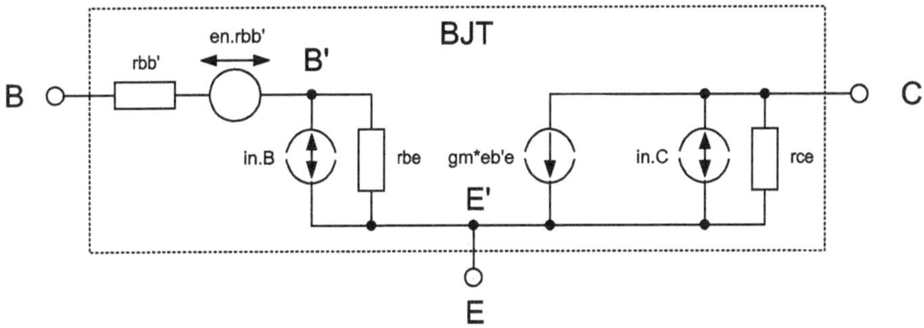


**Fig. 3.18** General BJT noise model for the audio band

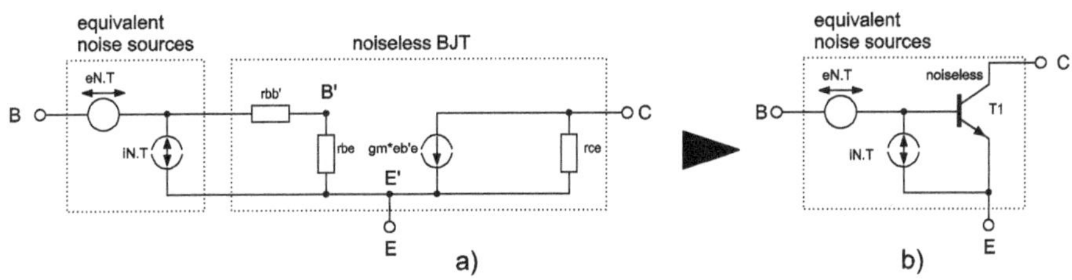


**Fig. 3.19** a) BJT noise model with equivalent noise sources, b) simplified model

resistors between the external and internal terminals of the base ($r_{bb'}$), the collector ($r_{cc'}$) and the emitter ($r_{ee'}$) which, of course also generate noise. Assuming, that in the audio band $B_{20\,k}$ the noise is white, $e_n$ and $i_n$ become:

$$e_{n.T} = kT\sqrt{\frac{2}{qI_C}B} \tag{3.69}$$

$$i_{n.T} = \sqrt{\frac{2qI_C}{h_{FE}}B} \tag{3.70}$$

Fortunately, $r_{cc'}$ doesn't play a noise-making role. In most cases it's the same with $r_{ee'}$, often called emitter bulk resistor. It only plays a role with very low impedances for the input load as well as for low-valued emitter resistors $R_E$. Than, for noise calculations it can be treated like being placed between points $E$ and $E'$. In the above shown noise model its noise making effects were covered by $r_{bb'}$. The base spreading resistor $r_{bb'}$ is the one who plays the most negative role. Unfortunately, the value of $r_{bb'}$ is not given in data sheets. We must find ways to get it. Methods to calculate it will be given further down in this chapter.

Additional equations will mathematically demonstrate how this internal resistors influences the whole noise calculations. But first of all, the equivalent noise current $i_{n.T}$ doesn't change! It is not dependent on $r_{bb'}$.

$$i_{N.T} = i_{n.T} \tag{3.71}$$

$i_{N.T}$ expressed in terms of the mutual conductance $g_{m.T}$ of the BJT looks like:

$$i_{N.T} = \sqrt{\frac{2kT}{h_{FE}} g_{m.T} B} \tag{3.72}$$

$$g_{m.T} = \frac{qI_C}{kT} \tag{3.73}$$

The suffix $_T$ indicates it's relation with a BJT – not to change with the temperature $T$!

Secondly, all internal noise voltages have to be summed up to form the equivalent noise voltage $e_{N.T}$ of the BJT. With

$$e^2_{N.rbb'} = 4kTr_{bb'}B \tag{3.74}$$

$e_{N.T}$ becomes:

$$e^2_{N.T} = e^2_{n.T} + i^2_{N.T} r^2_{bb'} + e^2_{N.rbb'} \tag{3.75}$$

In terms of $g_m$ $e_{N.T}$ looks as follows ($h_{FE}$ is the small signal current gain):

$$e^2_{N.T} = 4kT \left( r_{bb'} + \frac{1}{2g_m} + \frac{r^2_{bb'}}{2h_{FE}} g_m \right) B \tag{3.76}$$

In the $I_C$-dependent form $e_{N,T}$ becomes:

$$e_{N.T}(I_C)^2 = B \left( \frac{2k^2T^2}{qI_C} + \frac{2qI_C r^2_{bb'}}{h_{FE}} + 4kTr_{bb'} \right) \tag{3.77}$$

This equation permits us to calculate $r_{bb'}$. It's a quadratic equation for $r_{bb'}$ which can easily be solved with MCD under the following conditions: we must know $e_{N.T}$ at a certain collector current $I_{K.T}$ of a specific BJT.

Example: with the respective values of the 2SC2546 a calculation will demonstrate how it goes:

$$e_{N.T} = 0.5\,\text{nV}/\sqrt{\text{Hz}} \quad \text{at} \quad I_{K.T} = 10\,\text{mA} \tag{3.78}$$

Hence, with the MCD solving approach "solve, $x \rightarrow r_{bb'}$" $r_{bb'}$ becomes (without units and $B = B_1 = 1$ Hz):

$$(0.5 \times 10^{-9})^2 = B \left( \frac{2kT}{q10^{-2}} + \frac{2q10^{-2} r^2_{bb'}}{600} + 4kTr_{bb'} \right) \text{solve,}$$
$$r_{bb'} \rightarrow (-3115.94, +13.74) \tag{3.79}$$

It's obvious that only the positive result makes sense:

$$r_{bb'.2SC2546} = 13.74\,\Omega\,.$$

This value doesn't change very much when calculation-wise "playing around" with $h_{FE}$ from 500 to 800! In addition: with the value for $r_{bb'}$ we can calculate any equivalent noise voltage and current at any collector current. With changing collector current by application of Eqs. (3.77), (3.70) and (3.71) the following two graphs will demonstrate how it's done. $h_{FE}$ will be kept at a constant 600:

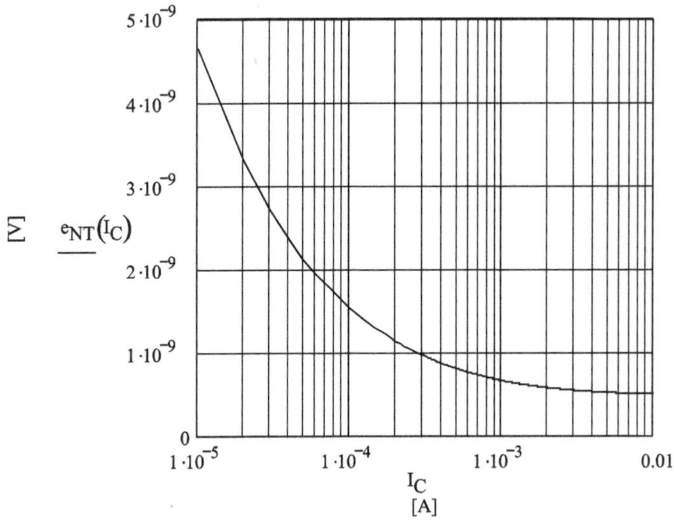


**Fig. 3.20** 2SC2546 noise voltage $e_{N.T}$ vs. collector current $I_C$

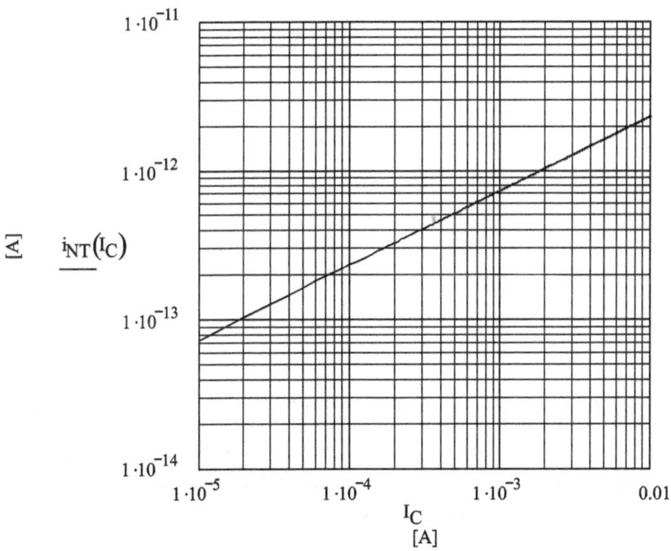


**Fig. 3.21** 2SC2546 noise current $i_{N.T}$ vs. collector current $I_C$

Instead of calculating noise voltage and current for each $I_C$ of interest with Eqs. (3.70) and (3.75) for further development steps and noise calculations we can pick from the above given figures any value for $e_{N.T}$ or $i_{N.T}$ at a specific collector current with the respective MCD tool.

In 2006 I had an e-mail conversation with Mr W. Adam[19] about his question on how to evaluate the value for $r_{bb'}$. I couldn't help him because he wanted to find a direct measurement method and not an indirect one via measurement of total noise and calculation back to get the value he was sought after. In May, 1992 this type of method was described in a letter to Mr Adam by Mr T. McCormick. I will describe it at the end of this chapter.

### Noise Model for BJT Plus Source

Based on the following figure that shows the connection of a source to a BJT the noise calculation expands into Eq. (3.81).

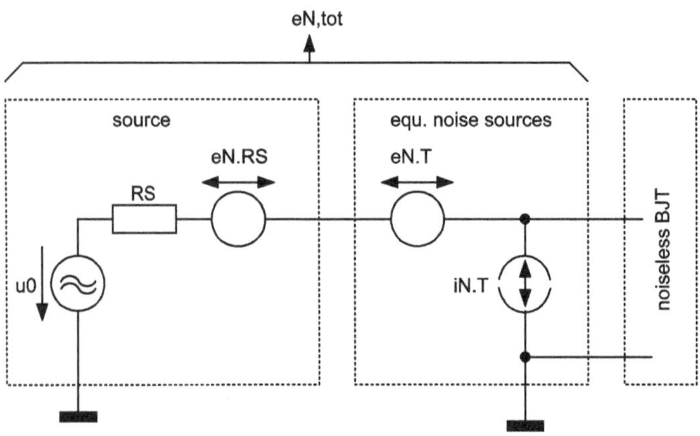


**Fig. 3.22** BJT plus source

With

$$e^2_{N.RS} = 4kTRSB \tag{3.80}$$

$e_{N,tot}$ becomes:

$$e_{N.tot} = \sqrt{e^2_{N.T} + e^2_{N.RS} + i^2_{N.T}RS^2} \tag{3.81}$$

As we know, a transistor gain stage can be configured in three different ways: in the common emitter (CE), the common collector (CC) and the common base configuration (CB). Because of the fact that there is no difference in noise treatment

[19] "Designing low-noise audio amplifiers", Wilfried Adam, E&WW June 1989

in the three configurations we will concentrate on the one which is used in most audio circuitry cases: the CE configuration. As a 1st stage of a low-noise amplifier a separate transistor gain stage is a very strong tool to overcome weaknesses in noise performance of e.g. op-amps. Two typical examples on how to put the BJTs in place are given in the following figures:

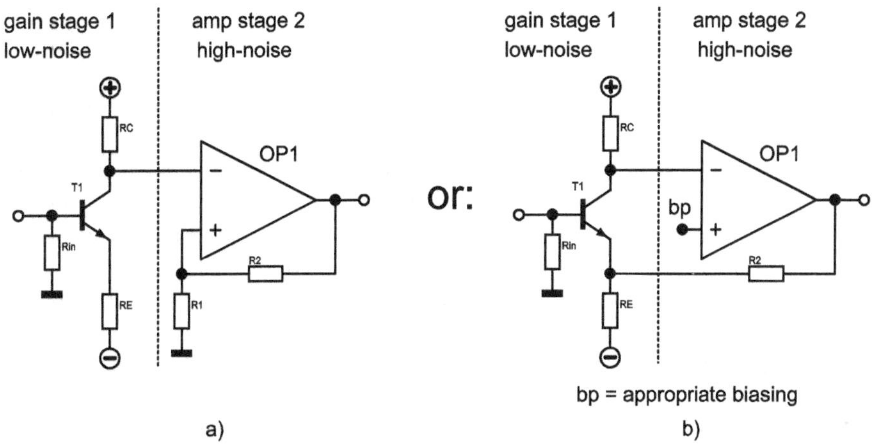


**Fig. 3.23** Low-noise BJT (stage 1) to improve noise performance of the following amp (stage 2): **a** as a stand alone stage, **b** inside the overall negative feedback loop of that op-amp

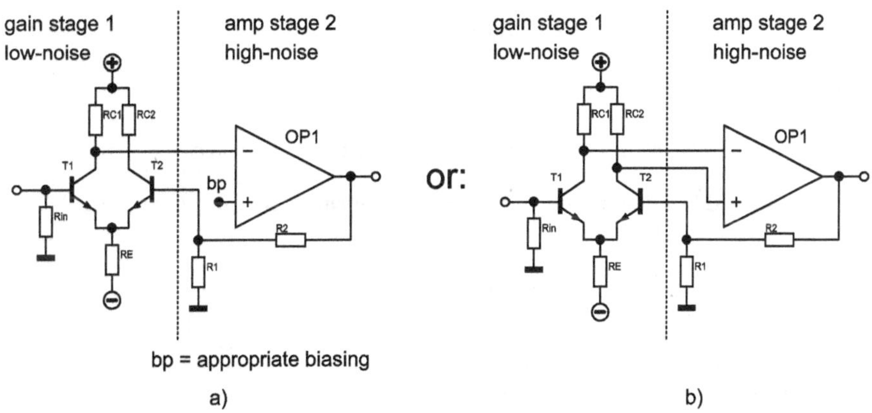


**Fig. 3.24** Long-tailed pair of 2 low-noise BJTs (stage 1) to improve noise performance of the following amp (stage 2), situated inside the overall negative feedback loop

### Noise Contribution of a 2nd Stage (Contribution Allowed)

To "over-write" the noise of the 2nd stage the 1st stage must have a high gain. The 1st stage noise voltage multiplied with the gain of the 1st stage plus noise voltage of the 2nd stage should end up within certain limits set by the designer, e.g. noise contribution of 2nd stage $\leq$0.05 dB.

If the 1st stage has no gain, it's obvious that the stage after the 1st stage contributes significantly to the noise situation of the whole amp. It must be designed for low-noise as well. This is not the case as long as the 1st stage has a minimum gain that allows a maximal specific noise contribution of a 2nd stage. Table 3.1 shows how much gain we need in the 1st stage to nearly ignore the additional noise created by a 2nd or following stage.

Assumptions for the calculation of Table 3.1 were the following ones: 1st stage total input equivalent noise voltage is 5 nV/rtHz, 2nd stage total input equivalent noise voltage is 5 nV/rtHz. Frequency response of noise voltages and currents are equal and flat inside $B_{20\,\mathrm{k}}$. These assumptions are not so easy to achieve if you think for example of a sequence of a low-noise transistor like the LM394 or MAT02/03 ($e_{\mathrm{N.T.tot}}(f) = 1\,\mathrm{nV/rtHz}$) and a following op-amp like one of the OP27 family ($e_{\mathrm{N.op.tot}}(f) = 3.2\,\mathrm{nV/rtHz}$). The transistor gain must be higher than 26.4 dB: according to Table 3.2 it's +36.5 dB to touch the 0.01 dB contribution level.

The formulae in the two tables work with the following settings:

$$e_{\mathrm{N.1st.tot}} = \text{equivalent input noise voltag of 1st stage [nV/rtHz]}$$

$$e_{\mathrm{N.2nd.tot}} = \text{equivalent input noise voltage of the 2nd stage [nV/rtHz]}$$

$$G = \text{required gain of 1st stage [times \& dB]}$$

$$ca = \text{contribution allowed [dB]}$$

Therefore, the basic equation becomes:

total noise of the two stages minus noise of the 1st stage

should be equal to the allowed noise contribution of the 2nd stage.

**Table 3.1** Noise contribution of the 2nd stage of any amp

| 1/A | B | C | D | E | F |
|---|---|---|---|---|---|
| 2<br>3 | Total input noise of 1st stage [nV/rtHz] | Total input noise of 2nd stage [nV/rtHz] | Amount of contribution allowed [dB] | Required minimum gain [times] | Required minimum gain [dB] |
| 4 | 5 | 5 | 1.00 | 1.97 | 5.87 |
| 5 | 5 | 5 | 0.10 | 6.55 | 16.33 |
| 6 | 5 | 5 | 0.01 | 20.83 | 26.37 |

**Table 3.2** Gain requirements for a transistor 1st stage with noise contribution of an OP27 op-amp as the 2nd stage

| 1/A | B | C | D | E | F |
|---|---|---|---|---|---|
| 2<br><br><br>3 | Total input noise of 1st stage [nV/rtHz] | Total input noise of 2nd stage [nV/rtHz] | Amount of contribution allowed [dB] | Required minimum gain [times] | Required minimum gain [dB] |
| 4 | 1 | 3.2 | 1.00 | 6.29 | 15.97 |
| 5 | 1 | 3.2 | 0.10 | 20.97 | 26.43 |
| 6 | 1 | 3.2 | 0.01 | 66.65 | 36.48 |

After some transformations the 1st stage gain $G_e$ in [dB] becomes:

$$\left(e_{\text{N.1st.tot}} \times 10^{\left(\frac{ca}{20}\right)} \times G\right)^2 = (e_{\text{N.1st.tot}} \times G)^2 + e^2_{\text{N.2nd.tot}} \tag{3.82}$$

$$G = \sqrt{\frac{e^2_{\text{N.2nd.tot}}}{\left(e_{\text{N.1st.tot}} \times 10^{\left(\frac{ca}{20}\right)}\right)^2 - e^2_{\text{N.1st.tot}}}} \tag{3.83}$$

$$G_e = 20 \times \log(G) \tag{3.84}$$

After a long search to find the right input device as well as to develop the right input circuitry in the audio frequency range for MC-cartridge purposes I can sum up the following minimum requirements for a BJT 1st stage in a BJT-amp (op-amp)-chain as of Figs. 3.23 and/or 3.24:

1. lowest noise possible – based on the formulae shown above $r_{bb'}$ should be very, very small (<10 R), to get low noise current $h_{FE}$ should be as high as possible (>500) and collector current as low as possible, to get low noise voltage the collector current should be as high as possible
2. gain ≥75 (37.5 dB)
3. adaptable to source resistances between 4 R and 50 R in case of MC cartridges – consequently the collector current should be made adjustable to guarantee lowest NF
4. low distortion
5. temperature stable.

## Selection of Low-Noise BJTs and Ranking via Noise Factor Calculation

A selection of several low-noise BJTs is given in the following three tables. The selection is not complete but I'm sure it will be hard to find many better ones. I tried to get as much data as possible, but there are still question marks. NF calculations were made according to Fig. 3.1 and the following equation (see also Eq. (3.50)):

$$NF_{\text{e.T}}(I_C) = 20\log\left(\frac{\sqrt{e_{\text{N.RS}}^2 + e_{\text{N.T}}(I_C)^2 + \{i_{\text{N.T}}(I_C)RS\}^2}}{e_{\text{N.RS}}}\right) \tag{3.85}$$

For MC purposes – in comparison with the BFW16A – the boxes marked bold show my favourite transistors for source resistances in the range of 20 – 50 R. To get a high SN for source resistances lower than 20 R I prefer the transformer solution (see the transformers chapter further down in Part I). Although, the BFW16A looks as if it could solve the low source resistance task I don't like several facts with it: collector current very high, not so easy to get with relatively high $h_{FE}$ (out of 50 I only found 5 with $h_{FE} > 35$, the rest was in the range $5 < h_{FE} < 20$, and to find a pair is a real tough task).

To get $NF$s for MM purposes that are low enough the collector current must be decreased up to approximately 100 μA, or even less. As measurement source resistances I've chosen 1 k and 12 k. For many MM cartridges the resistive part of their coil starts in the region of ≥700 R and – depending on the coil inductance – ends somewhere ≥100 k. Together with the 47 k input resistance of a MM phono-amp the two source resistances perfectly mirror the input situation of that kind of amp connected to a MM cartridge. As will be shown in the MM cartridge chapter any capacitor parallel to the source resistance, additionally, will increase the noise voltage level.

The results are shown in Table 3.5. It is obvious that high current gain and low $r_{bb'}$ automatically lead to low NF. I also calculated NF for 4 BJTs in parallel. The results look not very advantageous. In addition, if the input configuration consists of a long tailed pair and it's configured like Fig. 3.24 than, the phono-amp really becomes a transistor grave. For all other 1-BJT cases: if we put them into a long-tailed pair configuration we have to multiply $e_{\text{N.T}}$ with $\sqrt{2}$, $i_{\text{N.T}}$ keeps the same value and – as shown in Figs. 3.23a, 3.24a, b – a high valued $R_1$ increases NF drastically as well as a high valued $R_E$ in Fig. 3.23a, b!

As long as they were not calculated the data for the transistors came from the following sources:

- BFW16A: "Ultra-low-noise Preamplifier for MC phono cartridges", Nordholt & van Vierzen, 1980
- 2N4403: "Low-Noise electronic system design", Motchenbacher & Connelly, 1993
- all other: respective data sheets from Hitachi, Analog Devices, Texas Instruments, National Semiconductors, THAT Corp.

To be mentioned as well:

- MAT02 noise current looks strange. Its data sheet values are much higher than the calculated ones. All its other values look equal to the ones from SSM 2210

**Table 3.3** Selection of low-noise BJTs with all relevant $e_N$ and $i_N$ data to select the right device for MC purposes

| 1/A | B | C | D | E | F | G | H | I | J | K | L | M | N | O | P | Q | R |
|---|---|---|---|---|---|---|---|---|---|---|---|---|---|---|---|---|---|
| 2 | Device | | $r_{bb'}$ | $r_{ee'}$ or $r_{BE}$ | $h_{FE}$ | | $e_N$ | | $f_{ce}$ | $i_N$ | | | $e_N$ | | $i_N$ | | $f_{ci}$ |
| 3 | all data at 300 K | | | | | at $I_C$ | | at $I_C$ | | | | at $I_C$ | at $I_C = 1$ mA & white noise region | | | | |
| 4 | | | Ω | Ω | | mA | nV/rtHz | mA | Hz | pA/rtHz | | mA | nV/rtHz | | pA/rtHz | | Hz |
| 5 | | | | | | | data sheet: lowest available noise voltage | | | calculated (∗) or data sheet | | | data sheet or calculated (∗) or measured (∗∗) at $I_C = 4$ mA (∗∗∗) | | | | |
| 7 | BFW16A | N | 4.0 | ? | 35 | 25 | 0.28 | 25 | 250 | 15.12 | ∗ | 25 | 0.53 | ∗ | 3.02 | ∗ | ? |
| 8 | 2SC2545,6.7E | N | 14.0 | ? | 600 | 1 | 0.50 | 10 | ? | 2.31 | ∗ | 10 | 0.67 | ∗ | 0.73 | ∗ | ? |
| 9 | 4 × 2SC2546E | N | 4.5 | 0.0 | 600 | 4 | 0.25 | 40 | ? | 4.62 | ∗ | 40 | 0.33 | ∗∗∗ | 1.46 | ∗∗∗ | ? |
| 10 | 2SC2545,6F | N | 14.0 | ? | 900 | 1 | 0.50 | 10 | ? | 1.89 | ∗ | 10 | 0.68 | ∗ | 0.60 | ∗ | ? |
| 11 | 2SA1083,4,5E | P | 14.0 | ? | 600 | 1 | 0.50 | 10 | ? | 2.31 | ∗ | 10 | 0.67 | ∗ | 0.73 | ∗ | ? |
| 12 | 1/2 MAT03 | P | 16.1 | 0.3 | 165 | 1 | 0.55 | 7 | <10 | 3.69 | | 7 | 0.70 | | 1.39 | ∗ | ? |
| 13 | 1/2 SSM2220 | P | 16.1 | 0.3 | 165 | 1 | 0.55 | 7 | <10 | 3.69 | | 7 | 0.70 | | 1.39 | ∗ | ? |
| 14 | 1/4 THAT 320 | P | 25.0 | 2.0 | 75 | 1 | 0.75 | 1 | ? | 2.07 | ∗ | 1 | 0.75 | | 2.07 | ∗ | ? |
| 15 | 4 × 1/4 THAT 320 | P | 6.2 | 0.5 | 75 | 1 | 0.38 | 1 | ? | 4.14 | ∗ | 1 | 0.38 | ∗∗∗ | 4.14 | ∗ | ? |
| 16 | 1/2 MAT02 | N | 29.7 | 0.3 | 700 | 1 | 0.75 | 3 | <10 | 5.00 | | 3 | 0.85 | | 3.00 | ∗ | <10 |
| 17 | 4 × 1/2 MAT02 | N | 7.5 | 0.0 | 700 | 4 | 0.38 | 12 | <10 | 10.00 | | 12 | 0.43 | ∗∗∗ | 6.00 | ∗∗∗ | <10 |
| 18 | 1/2 SSM2210 | N | 29.7 | 0.3 | 700 | 1 | 0.75 | 3 | <10 | 1.17 | | 3 | 0.85 | | 1.00 | | <10 |
| 19 | 4 × 1/2 SSM2210 | N | 7.5 | 0.0 | 700 | 4 | 0.38 | 12 | <10 | 2.34 | | 12 | 0.43 | ∗∗∗ | 2.00 | ∗∗∗ | <10 |
| 20 | 1/4 THAT 300 | N | 30.0 | 2.0 | 100 | 1 | 0.80 | 1 | ? | 1.79 | ∗ | 1 | 0.80 | ∗∗∗ | 1.79 | ∗ | ? |
| 21 | 4 × 1/4 THAT300 | N | 7.5 | 0.5 | 100 | 1 | 0.40 | 1 | ? | 3.58 | ∗ | 1 | 0.40 | | 3.58 | ∗ | ? |
| 22 | 1/2 LM394 | N | 40.0 | ? | 680 | 1 | 0.94 | 1 | <20 | 0.69 | ∗ | 1 | 0.94 | | 0.69 | ∗ | 100 |
| 23 | 2N4403 | P | 40.0 | ? | 200 | 1 | 0.94 | 1 | 50 | 1.26 | | 1 | 0.94 | ∗∗ | 1.26 | | 2000 |

For Tables 3.3 … 3.5:
* = calculated
** = measured
*** = specified at 4 mA

**Table 3.4** Selection of low-noise BJTs with all relevant NF data to select the right device for MC purposes (best, 2nd and 3rd: *bold*)

| 1/S | T | U | V | W | X | Y | Z | AA | AB | AC | AD |
|---|---|---|---|---|---|---|---|---|---|---|---|
| 2 | Device | | | *NF* | *NF* | *NF* | *NF* | *NF* | *NF* | *NF* | *NF* |
| 3 | all data at 300 K | | | $I_C$ at lowest available noise voltage | | | | $I_C = 1$ mA or 4 mA (***) | | | |
| 4 | | | | dB/RS (Ω) | | | | dB/RS (Ω) | | | |
| 5 | | | *RS* | 5 | 10 | 20 | 40 | 5 | 10 | 20 | 40 |
| 6 | | | $e_{N.RS}$ | $2.88 \times 10^{-10}$ | $4.07 \times 10^{-10}$ | $5.76 \times 10^{-10}$ | $8.14 \times 10^{-10}$ | $2.88 \times 10^{-10}$ | $4.07 \times 10^{-10}$ | $5.76 \times 10^{-10}$ | $8.14 \times 10^{-10}$ |
| 7 | BFW16A | N | | **3.0** | **2.1** | 1.8 | 2.2 | 6.4 | 4.3 | 2.7 | 1.6 |
| 8 | 2SC2545,6,7E | N | | 6.1 | 4.0 | 2.5 | 1.5 | 8.1 | 5.7 | 3.7 | 2.2 |
| 9 | **4 × 2SC2546E** | N | *** | **2.5** | **1.5** | **0.9** | **0.6** | **3.7** | **2.2** | **1.3** | **0.7** |
| 10 | 2SC2545,6F | N | | 6.1 | 4.0 | 2.5 | 1.4 | 8.2 | 5.8 | 3.8 | 2.3 |
| 11 | 2SA1083,4,5E | P | | 6.1 | 4.0 | 2.5 | 1.5 | 8.1 | 5.7 | 3.7 | 2.2 |
| 12 | 1/2 MAT03 | P | | 6.7 | 4.6 | 2.9 | 1.7 | 8.4 | 6.0 | 3.9 | 2.4 |
| 13 | 1/2 SSM2220 | P | | 6.7 | 4.6 | 2.9 | 1.7 | 8.4 | 6.0 | 3.9 | 2.4 |
| 14 | 1/4 THAT 320 | P | | 8.9 | 6.4 | 4.3 | 2.7 | 8.9 | 6.4 | 4.3 | 2.7 |
| 15 | 4 × 1/4 THAT 320 | P | *** | 4.4 | **2.7** | **1.6** | **1.0** | 4.4 | **2.7** | **1.6** | **1.0** |
| 16 | 1/2 MAT02 | N | | 8.9 | 6.4 | 4.4 | 2.8 | 9.9 | 7.3 | 5.0 | 3.2 |
| 17 | 4 × 1/2 MAT02 | N | *** | **4.4** | 2.9 | 1.9 | 1.6 | **5.1** | 3.3 | 2.0 | 1.4 |
| 18 | 1/2 SSM2210 | N | | 8.9 | 6.4 | 4.3 | 2.7 | 9.9 | 7.3 | 5.0 | 3.2 |
| 19 | 4 × 1/2 SSM2210 | N | *** | **4.3** | **2.7** | **1.6** | **0.9** | **5.0** | 3.2 | 1.9 | **1.1** |
| 20 | 1/4 THAT 300 | N | | 9.4 | 6.9 | 4.7 | 3.0 | 9.4 | 6.9 | 4.7 | 3.0 |
| 21 | 4 × 1/4 THAT 300 | N | *** | 4.7 | 3.0 | **1.8** | **1.0** | 4.7 | **3.0** | **1.8** | **1.0** |
| 22 | 1/2 LM394 | N | | 10.7 | 8.0 | 5.6 | 3.7 | 10.6 | 8.0 | 5.6 | 3.7 |
| 23 | 2N4403 | P | | 10.7 | 8.0 | 5.6 | 3.7 | 8.1 | 8.0 | 5.6 | 3.7 |

**Table 3.5** Selection of low-noise BJTs with all relevant $e_N$, $i_N$ and NF data to select the right device for MM purposes (best, 2nd and 3rd: *bold*)

| 1/AE | AF | AG | AH | AI | AJ | AK | AL | AM | AN | AO |
|---|---|---|---|---|---|---|---|---|---|---|
| 2 | Device | | $h_{FE}$ | $e_N$ | | $i_N$ | | | *NF* | *NF* |
| 3 | all data at 300 K | | $I_C = 0,1$ mA or $0,4$ mA (∗∗∗) & white noise region | | | | | | | |
| 4 | | | | nV/rtHz | | pA/rtHz | | dB/RS (Ω) | | |
| 5 | | | data sheet or calculated (∗) or measured (∗∗) | | | | | RS | 1000 | 12,000 |
| 6 | | | | | | | | $e_{N,RS}$ | $4.07 \times 10^{-09}$ | $1.41 \times 10^{-08}$ |
| 7 | BFW16A | N | 20 | 1.50 | ∗ | 1.27 | ∗ | | 0.9 | 3.4 |
| 8 | 2SC2545,6,7E | N | 500 | 1.54 | ∗ | 0.25 | ∗ | | **0.6** | **0.2** |
| 9 | **4 × 2SC2546E** | N | 500 | 0.77 | ∗+∗∗∗ | 0.56 | ∗+∗∗∗ | ∗∗∗ | **0.2** | 0.9 |
| 10 | 2SC2545,6F | N | 800 | 1.54 | ∗ | 0.20 | ∗ | | **0.6** | **0.2** |
| 11 | 2SA1083,4,5E | P | 500 | 1.54 | ∗ | 0.25 | ∗ | | **0.6** | **0.2** |
| 12 | 1/2 MAT03 | P | 150 | 1.55 | | 0.46 | | | **0.6** | 0.7 |
| 13 | 1/2 SSM2220 | P | 160 | 1.55 | | 0.46 | | | **0.6** | 0.7 |
| 14 | 1/4 THAT 320 | P | 75 | 1.61 | ∗ | 0.65 | ∗ | | 0.7 | 1.2 |
| 15 | 4 × 1/4 THAT 320 | | 75 | 1.80 | ∗ | 1.30 | ∗ | ∗∗∗ | 1.1 | 3.5 |
| 16 | 1/2 MAT02 | N | 700 | 2.00 | | 1.00 | | | 1.1 | 2.4 |
| 17 | 4 × 1/2 MAT02 | N | 700 | 1.00 | ∗+∗∗∗ | 2.00 | ∗+∗∗∗ | ∗∗∗ | 1.1 | 5.9 |
| 18 | 1/2 SSM2210 | N | 700 | 1.63 | | 0.21 | | | 0.7 | **0.2** |
| 19 | 4 × 1/2 SSM2210 | N | 700 | 0.81 | ∗+∗∗∗ | 0.43 | ∗+∗∗∗ | ∗∗∗ | **0.2** | **0.6** |
| 20 | 1/4 THAT 300 | N | 100 | 1.64 | ∗ | 0.57 | ∗ | | 0.7 | 1.0 |
| 21 | 4 × 1/4 THAT 300 | | 100 | 0.82 | ∗ | 1.14 | ∗ | ∗∗∗ | **0.5** | 2.9 |
| 22 | 1/2 LM394 | N | 580 | 1.68 | | 0.24 | | | 0.7 | **0.2** |
| 23 | 2N4403 | P | 140 | 1.68 | ∗ | 0.48 | ∗ | | **0.6** | 0.7 |

- In the calculations $1/f$-effects were not considered
- Low NF of a transistor tells a lot about its ability not to increase the noise of a source too much. But this is not an absolute noise voltage or noise current value and it doesn't say anything about signal-to-noise ratios. This has to be calculated in detail for each case.

## CE Gain Stage

For an excellent 1st stage circuitry that fulfils the above (page 43) set requirements 2–5 a search led to two different circuit set-ups. The first one is recommended by T/S[20]. It's a voltage feedback (via OP1) configured CE stage fed by a current generator ($T_{2,3}$) and looks as follows:

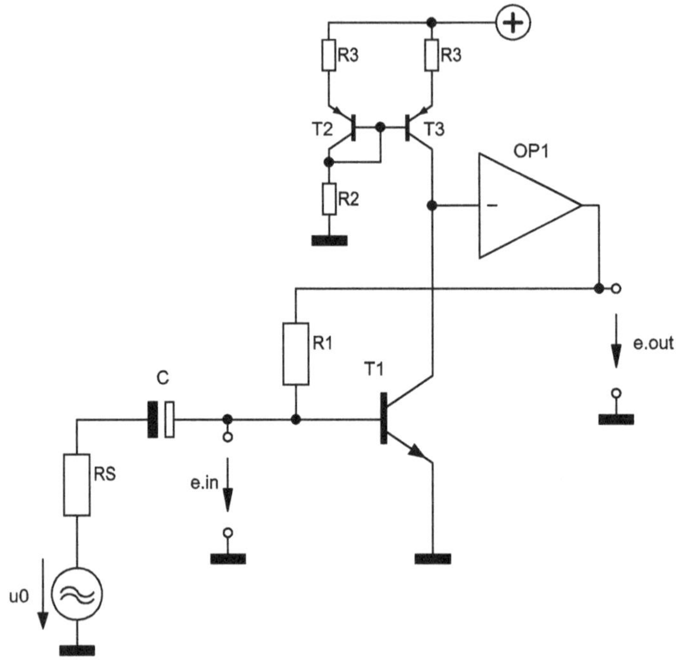


**Fig. 3.25** T/S approach for a low-noise CE BJT stage

The second version could be found in M. Douglas Self's study about his MC phono-pre-pre-amp[21]. It's also a CE configured transistor with voltage and current feedback. The advantage of this circuit: with an adjustable $R_C$ this circuit can easily

---

[20] T/S, Chap. 4

[21] "Design of MC head amplifiers", D/S, EW&WW 12-1987

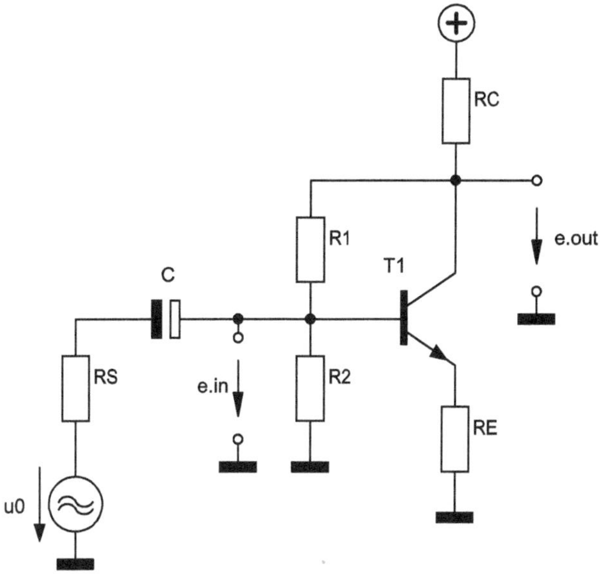


**Fig. 3.26** Transistor 1st stage in CE configuration with voltage and current feedback à la Douglas Self

be adapted to different sources by changing the equivalent input noise sources with changing collector current.

Voltage feedback goes back from the output at the collector of $T_1$ to the base via $R_1$, current feedback is performed by the emitter resistor $R_E$. $R_1$ and $R_2$ bias the base-emitter voltage to create a certain collector current. For both versions: the corner frequency of the hp formed by $C$ and the input impedance of the stage ($\approx R_1$ or $\approx R_1||R_2$) should be 100 times lower than the lowest frequency of interest. Than, the frequency and phase response of the stage will be flat within $B_{20\,k}$.

Now, which one to chose? Okham's Razor gives a quick answer – complexity of T/S design is much higher than that of the D/S design because of the following facts:

- T/S: 3 transistors + 1 op-amp + 4 resistors + much work to design an operating circuit

versus

- D/S: 1 transistor + 4 resistors + working circuit

I think the D/S design should be investigated further!

With simple measures the gain of such a stage can be found as follows:

- pSpice simulation
- gain calculation for the current feedback mode
- gain calculation for the voltage feedback mode

Hoping that the differences are not to big let's check the different gain approaches with

- $R_1 = 190\,\mathrm{K}$
- $R_2 = 40\,\mathrm{k}$
- $R_C = 10\,\mathrm{k}$
- $R_E = 13\,\mathrm{R3}$
- $R_S = 200\,\mathrm{R}$
- $T_1 = \mathrm{BC547B}$ ($h_{FE} = 312$)
- $C$: value high enough that it could be taken as a bridge in the audio frequency range
- $V_{cc} = +15\,\mathrm{V}$

thus, the gain $G_{cf}$ for the current feedback mode becomes:

$$G_{cf} = 20\log(-g_{m.red} \times R_C) \tag{3.86}$$

Because of the existence of $R_E$ the mutual conductance of $T_1(g_m)$ has to be reduced to $g_{m.red}$ by application of the following equation:

$$g_{m.red} = \frac{g_m}{1 + g_m \times R_E} \tag{3.87}$$

Without phase shift of the output signal $G_{cf}$ becomes:

$$G_{cf} = 48.3\,\mathrm{dB} \tag{3.88}$$

Gain $G_{vf}$ for the voltage feedback mode will be:

$$G_{vf} = \frac{-R_1}{RS + \frac{RS+R_1}{g_{m.red} \times R_C}} \tag{3.89}$$

thus, without phase shift of the output signal $G_{vf}$ becomes:

$$G_{vf} = 46.2\,\mathrm{dB} \tag{3.90}$$

and the simulation result $G_s$ figures as:

$$G_s = 47.0\,\mathrm{dB} \tag{3.91}$$

Point 3. of the requirements states, that $I_C$ should be made variable. This can be done by changing the value of $R_C$. Now we take $R_C = 2\,\mathrm{k}$, thus making $I_C = 4.1\,\mathrm{mA}$. With these new values the results for all three gain evaluation methods look as follows:

$$\begin{aligned} G_{cf} &= 40.2\,\mathrm{dB} \\ G_{vf} &= 39.3\,\mathrm{dB} \\ G_S &= 39.6\,\mathrm{dB} \end{aligned} \tag{3.92}$$

Hence, point 2. of the requirements is met too. Because we've chosen a CE configuration points 4. and 5. were also met, especially when choosing an amp configuration as of Fig. 3.23b!

## CE Gain Stage Noise Model

By taking 4 transistors in parallel plus adaptation of the resistors accordingly (division by 4) the gains won't change! Now, let's check what this means for noise calculations for the whole 1st gain stage. No matter if we would take 1 or 4 transistors: the equivalent noise model of Fig. 3.26 looks as follows:

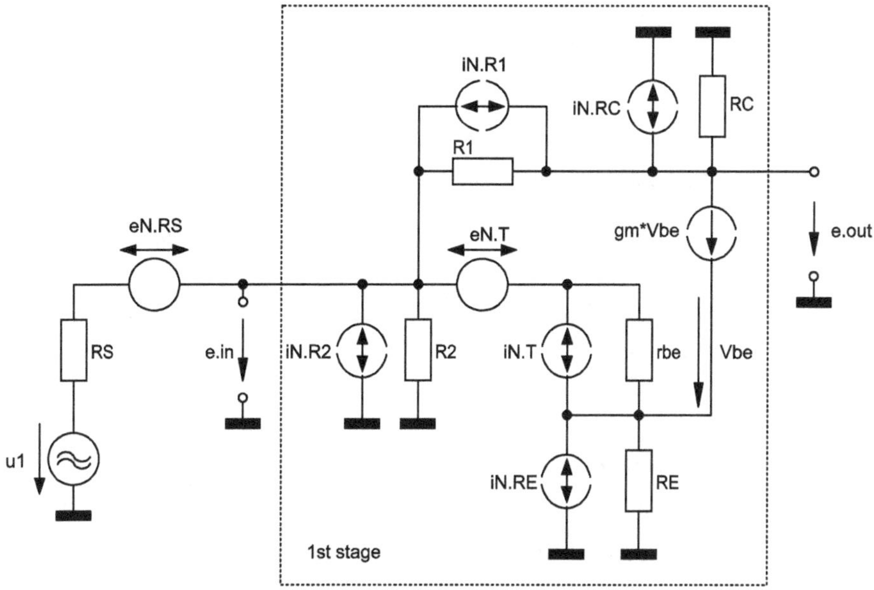


**Fig. 3.27** Noise model of a BJT in a CE configured gain stage

Summary for this model:

$R_C$ doesn't play a significant noise role, unless it becomes very small values, thus making gain small and collector current high, consequently, excess noise becomes dominant. This is not the case here. A calculation example of noise effects of $R_C$ with gain $\ll 100$ will be given further down these pages at the end of this chapter.

Without potential noise effects from the collector resistor $R_C$ and with $B = 1\,\text{Hz}$ the equivalent input noise voltage $e_{N.1st}$ becomes[22]:

$$e_{N.1st} = \sqrt{\frac{2k^2T^2}{qI_C}B + 4kT(r_{bb'} + R_E)B} \tag{3.93}$$

[22] T/S Chap. 4.2

and the equivalent input noise current $i_{N.1st}$ becomes:

$$i_{N.1st} = \sqrt{\frac{2qI_C}{h_{FE}}B + \frac{4kT}{\left(\frac{R_1 R_2}{R_1+R_2}\right)}B} \tag{3.94}$$

$r_{bb'}$ has to be taken from Table 3.3 or it should be calculated according to Eq. (3.79). A n-transistor set-up brings down the value for $r_{bb'(n)}$ to $1/n$-th of the value of $r_{bb'(1)}$ of 1 transistor, hence, the noise voltage goes down by multiplication with $1/\sqrt{n}$ and noise current increases by the factor $\sqrt{n}$ (see also Sect. 3.1 of this chapter and Eqs. (3.17 ... 3.24)).

### CE Gain Stage Example Calculation Incl. Excess Noise

A calculation example with a source resistance of 50 R looks as follows:

$$e_{N.50} = \sqrt{4kT50\Omega B} = 0.91\,\mathrm{nV/rtHz} \tag{3.95}$$

$T_1 = 4 \times 1/2\,\mathrm{LM394}$

| | | |
|---|---|---|
| $h_{FE} = 680$ | $I_C = 6.7\,\mathrm{mA}$ (1.67 mA for each device) | |
| $r_{bb'} = 40\,\mathrm{R}/4 = 10\,\mathrm{R}$ | $B = 1\,\mathrm{Hz}$ | |
| $T = 300\,\mathrm{K}$ | $k = 1.38 \times 10^{-23}\,\mathrm{VAsK^{-1}}$ | $q = 1.6 \times 10^{-19}\,\mathrm{As}$ |
| $R_1 = 47\mathrm{k}5$ | $R_2 = 10\mathrm{k}$ | |
| $R_E = 3\mathrm{R}3$ | $R_C = 1\mathrm{k}44$ | |

Application of Eqs. (3.93) and (3.94) will lead to:

$$e_{N.1st} = 0.5023\,\mathrm{nV/rtHz} \tag{3.96}$$
$$i_{N.1st} = 2.269\,\mathrm{pA/rtHz} \tag{3.97}$$

Remember, from Table 3.3 the equivalent noise sources for $4 \times 1/2$ LM394 look like:

$$e_{N.4} = 0.94/\sqrt{4} = 0.47\,\mathrm{nV/rtHz}$$
$$i_{N.4} = 0.69 \times \sqrt{4} = 1.38\,\mathrm{pA/rtHz} \tag{3.98}$$

Of course, these figures must be lower than those of the gain stage as a whole. The differences come from the resistor noise!

Hence, inclusion of the source resistance into the calculation course and with application of Eqs. (3.80) and (3.81) $e_{N.tot}$ becomes:

$$e_{N.tot} = 1.046\,\mathrm{nV/rtHz} \tag{3.99}$$

With reference to an input voltage of 0.5 $mV_{rms}$ in a frequency band of $B_{20\,k}$ $SN_{ne}$ (ne = not equalized) can be calculated with Eq. (3.56):

$$SN_{ne} = -70.586\,\text{dB} \tag{3.100}$$

and with Eq. (3.50) $NF_e$ becomes:

$$NF_e = 1.207\,\text{dB} \tag{3.101}$$

At the minimal gain of the 1st stage $G_{min} = 39.3\,\text{dB}$ (see Eq. (3.92)) the additional noise of a 2nd stage is set by the contribution allowed $ca = +0.05\,\text{dB}$. After rearrangement of Eq. (3.82) the maximal allowed 2nd stage noise voltage $e_{N.2nd.tot}$ will become:

$$e_{N.2nd.tot} = \sqrt{\left(e_{N.1st.tot} \times 10^{\frac{ca}{20}} \times 10^{-\frac{G_{min}}{20}}\right)^2 - \left(e_{N.1st.tot} \times 10^{-\frac{G_{min}}{20}}\right)^2} \tag{3.102}$$

$$e_{N.2nd.tot} = 8.211\,\text{nV/rtHz} \tag{3.103}$$

Now, one question is left: What are the additional noise effects of collector resistor $R_C$ and emitter resistor $R_E$?

If $R_C$ and/or $R_E$ would be set to values that make the stage gain $\ll$100, than, excess noise of $R_C$ and $R_E$ becomes significant and has to be included into the noise calculation course as well. Equation (3.93) changes to:

$$e_{N.1st.ex} = \sqrt{\frac{2k^2T^2}{qI_C}B + 4kT(r_{bb'} + R_E)B + \left(\frac{e_{N.RCex}}{G}\right)^2 + e_{N.REex^2}} \tag{3.104}$$

With Eqs. (3.66), (3.67), (3.68) and the following example values

- $V_{cc} = 15\,\text{V}$
- $V_{DC.C} = 10\,\text{V}$ (= DC-voltage across $R_C$)
- $V_{DC.E} = 0.6\,\text{V}$ (= DC-voltage across $R_E$)
- $NI_e = -30\,\text{dB}$
- $NI = 31.62\,\text{nV}/\text{V}_{DC}$
- $G = 14$ (gain)
- $R_C = 250\,\text{R}$
- $R_E = 15\,\text{R}$
- $I_C = 40\,\text{mA}$
- $T = 4\times$ 2SC2546 parallel ($r_{bb'} = 6\text{R}87$)

the input referred excess noise $e_{N.RCex.in}$ of $R_C$ becomes:

$$e_{N.RCex.3d} = NI \times \sqrt{3} \times V_{DC.C} = 547.67\,\text{nV} \tag{3.105}$$

$$e_{N.RCex.B20k} = \sqrt{4kTR_CB_{20\,k} + e^2_{N.RCex.3d}} = 618.7\,\text{nV} \tag{3.106}$$

$$e_{N.RCex} = \frac{e_{N.RCex.B20k}}{\sqrt{B_{20\,k}}} = 4.38\,\text{nV/rtHz} \tag{3.107}$$

$$e_{N.RCex.in} = \frac{e_{N.RCex}}{G} = 0.31\,\text{nV/rtHz} \tag{3.108}$$

The calculation of the excess noise $e_{N.REex}$ for $R_E$ goes as follows:

$$e_{N.REex.3d} = NI \times \sqrt{3} \times V_{DC.E} = 32.86\,\text{nV} \tag{3.109}$$

$$e_{N.RE.ex} = \frac{e_{N.REex.3d}}{\sqrt{B_{20\,k}}} = 0.23\,\text{nV/rtHz} \tag{3.110}$$

Thus, as of Eq. (3.104) the input referred noise voltage $e_{N.1st.ex}$ of the 4-2SC2546-example becomes:

$$e_{N.1st.ex} = 0.71\,\text{nV/rtHz} \tag{3.111}$$

Comparison with the input referred noise voltage $e_{N.1st}$ without excess noise effect (= Eq. (3.104) without terms 4 and 5)

$$e_{N.1st} = 0.61\,\text{nV/rtHz} \tag{3.112}$$

leads to the realization that in the above given case excess noise worsens the input noise voltage by 1.5 dB, hence, all *SN*s as well by the same amount.

Conclusions:

- If possible, we have to avoid rather high DC-voltages across $R_E$; e.g. 200 mV$_{DC}$ would increase the noise voltage of $R_E = 15\,\text{R}$ by only 1.2%, thus, excess noise for $R_E$ can be ignored.
- The gain of the amp-stage under development should be chosen as high as the contribution allowed of the following stage can be fulfilled (including any excess noise of $R_C$). Than, with the right amount of feedback the required gain for the whole amp can easily be set.
- An excess noise calculation for the base resistors (e.g. for $R_1$ or for $R_2$ or for both in Fig. 3.26) makes no sense, because this noise voltage will be drastically cut down via voltage divider effect by a rather low source resistance RS:

$$e_{N.R1orR2ex.eff} = e_{N.R1orR2ex}\left(\frac{RS}{RS + R_1 \text{ or } R_2}\right) \tag{3.113}$$

### Base Spreading Resistance $r_{bb'}$ – Measurement Approach[23]

The $r_{bb'}$ measurement set-up is given in Fig. 3.28. The device under test is operated at about 1 mA. The total gain $G_1$ of the circuit is about 50 dB, with $S_2$ closed. With $S_2$ open total gain $G_2$ will be

$$G_2 = G_1 + 20 \times \log\left(\frac{R_6 + R_7 + R_8}{R_5}\right)$$
$$= G_1 + 2.997\,\text{dB} \tag{3.114}$$

[23] Tom McCormick 1992 in a letter to Wilfried Adam – as a follow-up of Mr Adams 1989 EW&WW article on "Designing low-noise audio amplifiers"

All capacitors should have a value large enough to ensure a flat frequency response in $B_{20\,\mathrm{k}}$. For NPN devices power supply should be $+15\,\mathrm{V}$ and all capacitors reversed.

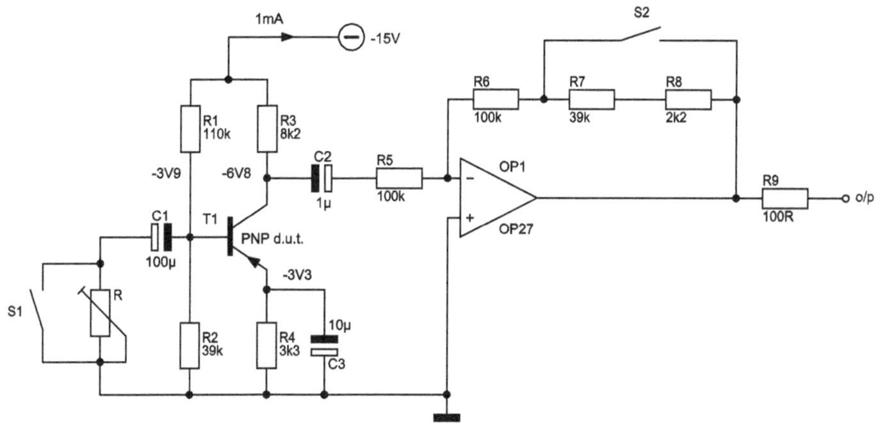


**Fig. 3.28** McCormick $r_{\mathrm{bb}'}$ measurement set-up

The measurement approach goes like follows:

Step 1: measurement of noise in $B_{20\,\mathrm{k}}$ at the output of $\mathrm{OP}_1$ with $S_1$ closed and $S_2$ opened, resulting in a total gain of $G_2$ and an output noise voltage of $e_{\mathrm{N.out.1}}$

Step 2: decrease of circuit gain to $G_1$ by opening of $S_1$ and closing of $S_2$

Step 3: adjustment of $R$ as long as the test circuit produces the Step 1 output noise voltage $e_{\mathrm{N.out.1}}$

Step 4: measurement of $R$: this will be the value of $r_{\mathrm{bb}'}$ of the d.u.t.

The principal method behind this measurement result comes from Eqs. (3.17–3.19) that describe the noise voltage sum of two resistors – in our case of equal valued $R$ (after adjustment) and $r_{\mathrm{bb}'}$.

## 3.3 Noise in Field Effect Transistors (FETs)

### Intro

When watching at FETs I only mean JFETs (Junction Field Effect Transistors) because – when talking about low-noise – they are the only workable ones in the audio field. Whereas they are very good for MM cartridge purposes they are a very bad choice for MC amplification. Like the BJTs, with their three leads to the outer world (Gate, Source, Drain), they can be placed in a circuitry in so-called common source CS, common drain CD and common gate CG configurations.

The device with one of the lowest noise voltages I could find on the market is Toshiba's[24] N-channel 2SK170 (or its P-channel complementary 2SJ74). The noise voltage at 1 kHz and $I_D = 10\,\text{mA}$ is specified as follows:

$$e_{N.2sk170} = 0.85\,\text{nV/rtHz} \tag{3.115}$$

which also includes any $1/f$-effect of the lower end of the audio frequency range. In addition, as a simple rule for JFETs, as of Fig. 3.29, the higher the drain current $I_D$ the lower the noise voltage.

### Noise Voltage Relevant Data Sheet Plots

Fortunately, JFETs have very low noise currents $\leq 50\,\text{fA/rtHz}$, thus, making it not absolutely necessary to calculate total noise voltages including $i_N$ – effects. In addition, there is nearly no $1/f$-noise for noise currents. This current is white in the audio band.

But there exists one major disadvantage: with low source resistances the noise factor of a JFET is very much higher when comparing it with BJTs, thus, in principle, making JFETs less workable for MC amplification than BJTs. The reason for that lies in the fact that JFETs generate massive $1/f$ noise voltage and, in most cases its $f_{ce}$ is quite high. For the 2SK170 this figure is not indicated in the data sheet but

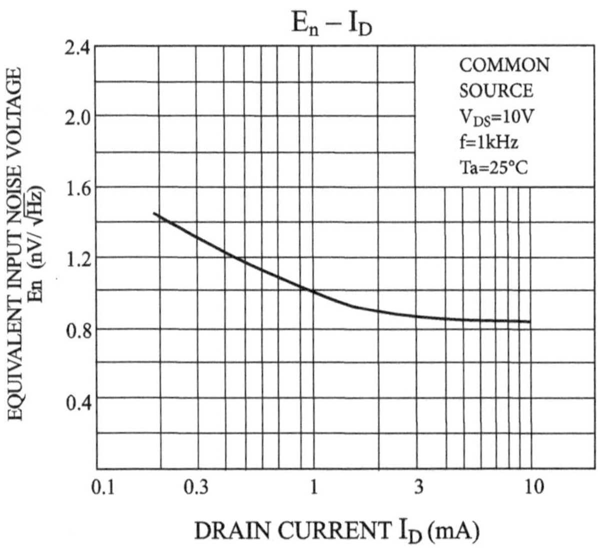


**Fig. 3.29** Equivalent input noise voltage of 2SK170

[24] Toshiba data sheet 2SK170

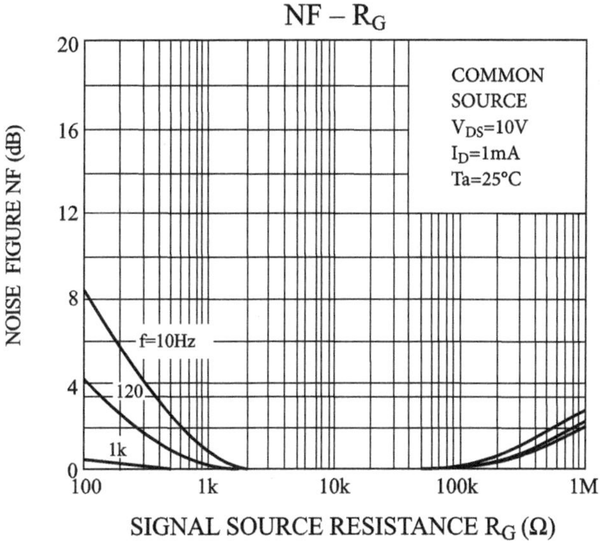


**Fig. 3.30** *NF* for 2SK170 versus source resistance

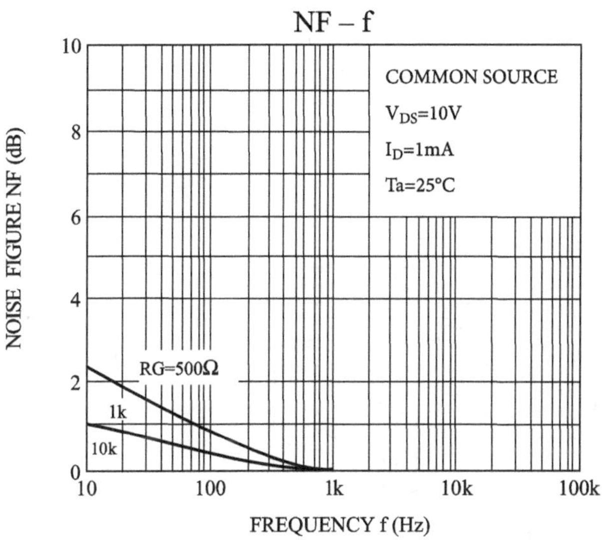


**Fig. 3.31** *NF* for 2SK170 versus frequency

as we can see from the next figures it must be located somewhere in the region between 1 kHz and 10 kHz. This is also in line with the measurement results for many other FETs in M/C's noise book[25].

[25] M/C Appendix B

## JFET Noise Model

In the audio band the small signal noise model – as equivalent circuit for a JFET – looks as follows:

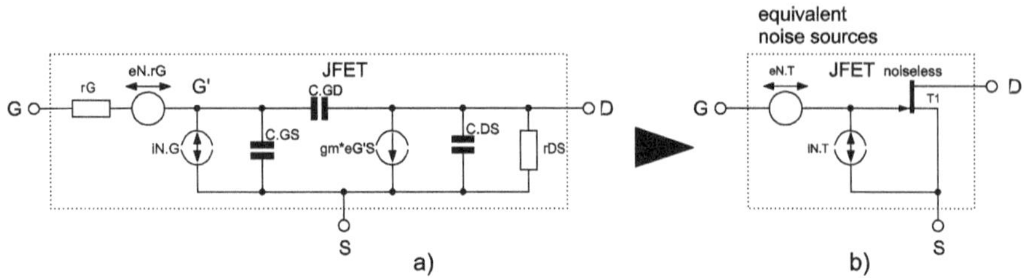


**Fig. 3.32** JFET: **a** Small-signal noise model, **b** simplified version

The following formulae will allow to further approach to the noise secrets of a JFET[26]:

The input noise resistance $r_G$ can be calculated with:

$$r_G \approx \frac{2}{3g_{m.F}} \tag{3.116}$$

$g_{m.F}$ $(= y_{21,s} = y_{fs} = h_{21,s})$ in common source configuration is given in data sheets as forward transfer admittance at a certain drain current:

**Forward Transfer Admittance**

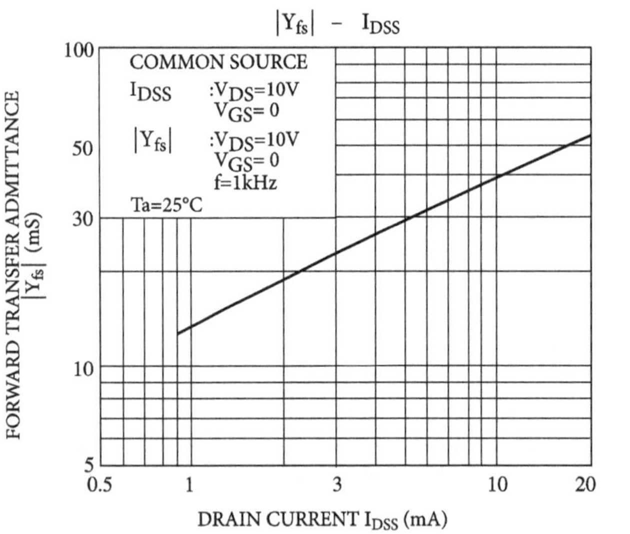


**Fig. 3.33** Forward transfer admittance of a low-noise JFET 2SK170[27]

[26] M/C Chap. 6

[27] Toshiba data sheet 2SK170

### Noise Current Calculation

The equivalent gate noise current $i_{N.G}$ is white in the audio band and it becomes:

$$i_{N.G} = \sqrt{2qI_{GSS}B} \tag{3.117}$$

Normally, the gate cut-off current $I_{GSS}$ is given in data sheets as well. A typical value is 1 nA maximal.

### Noise Voltage Calculation

If not specified in the data sheet the equivalent input noise voltage $e_{N.rG}$ in the white noise region becomes:

$$e_{N.rG} = \sqrt{4kTr_GB} \tag{3.118}$$

or

$$e_{N.rG}^2 \approx \frac{8}{3}\frac{kT}{g_{m.F}}B \tag{3.119}$$

Supposed, the corner frequency $f_{ce}$ is given, the $1/f$-effect should be calculated with Eq. (3.4).

### JFETs and MC Input Stages

With a typical MC cartridge source resistance of $R_0 = 20\,\text{R}(e_{N.20R} = 0.576\,\text{nV}/\text{rtHz})$ the following example will demonstrate how the above shown formulae can be used to calculate noise effects of a 2SK170 low-noise JFET, how much they differ from data sheet values and how they can be compared with BJT results (for the circuit set-up see also Fig. 3.22 and Eqs. (3.80) & (3.81)).

### $1/f$ Noise

With $f_{ce} = 1\,\text{kHz}$, without $1/f$-effect at 10 mA and $B = 1\,\text{Hz}$ application of the rearranged Eq. (3.4) brings $e_{N.2sk170}$ down to:

$$e_{n.2sk170.fce} = 0.73\,\text{nV}/\text{rtHz} \tag{3.120}$$

With $g_{m.F} = 39\,\text{mS}$ and $B = 1\,\text{Hz}$ plus application of Eq. (3.119) $e_{n.2sk170.fce}$ results in:

$$e_{n.2sk170} = 0.53\,\text{nV}/\text{rtHz} \tag{3.121}$$

These two values – Eqs. (3.120) and (3.121) – have a difference of 0.2 nV/rtHz. Something must be wrong in the calculation or in the assumptions. Assumed that the formulae were correct the only variable we have is the noise voltage corner frequency $f_{ce}$. I'm sure, if we shift it to somewhere >1000 Hz we will touch the data sheet result of the FET's noise voltage $e_N$. With $B = 19{,}980$ Hz and after the re-arrangement of Eq. (3.4) $f_{ce}$ becomes:

$$f_{ce} = \frac{\frac{(e_{N.2sk170} \times \sqrt{B})^2}{e^2_{n.2sk170.(3.121)}} - B}{\ln\left(\frac{20{,}000\,\text{Hz}}{20\,\text{Hz}}\right)} = 4486\,\text{Hz} \tag{3.122}$$

This value looks not bad, because, at the beginning of this section I've guessed to find this non-data-sheet-specified frequency between 1 kHz and 10 kHz.

### MC Case Noise Figures

Now, the calculations for the competing BJT looks as follows:

At a collector current $I_C = 10$ mA according to the data sheet one 2SC2546 offers an equivalent input noise voltage density of:

$$e_{N.2sc2546} = 0.5\,\text{nV/rtHz} \tag{3.123}$$

and, with $h_{FE} = 700$, a calculated (Eq. 3.70) equivalent input noise current density of:

$$i_{N.2sc2546} = 2.14\,\text{pA/rtHz} \tag{3.124}$$

thus, with $R_0 = 20$ R $NF_{e.2sc2546.20}$ becomes:

$$NF_{e.2sc2546.20} = 20\log\left(\frac{\sqrt{e^2_{N.2sc2546} + e^2_{N.R0} + (i_{N.2sc2546}R_0)^2}}{e_{N.R0}}\right) \tag{3.125}$$

$$NF_{e.2sc2546.20} = 2.455\,\text{dB} \tag{3.126}$$

and $NF_{e.2sk170.20}$ becomes:

$$NF_{e.2sk170.20} = 20\log\left(\frac{\sqrt{e^2_{N.2sk170} + e^2_{N.R0}}}{e_{N.R0}}\right) \tag{3.127}$$

$$NF_{e.2sk170.20} = 5.025\,\text{dB} \tag{3.128}$$

The difference between the two $NFs$ is 5.025 dB − 2.455 dB = 2.57 dB. At the end, when calculating *SN*s, in comparison with the BJT, by the same amount this difference worsens any result with a FET as 1st stage of an amplifier for low source resistances smaller than approximately 50 – 100 R, e.g.:

- $SN_{ne.BJT} = 71.00$ dB
- $SN_{ne.FET} = 71.00\,\text{dB} - 2.57\,\text{dB} = 68.43\,\text{dB}$!

Similar calculations for the 2SK389 result in even higher $NF_e$ values. Because its $e_N$ at 10 mA is 1.12 nV/rtHz this JFET creates a $NF_e$ of 6.8 dB for a source resistance of 20 R.

These JFET/BJT $NF_e$-differences will continue to worsen the JFET *SN* results after the RIAA equalization or after the A-filter-weighting (or with both together).

### JFETs and MM Input Stage

A totally different picture comes up in the MM cartridge case. There, we find source resistances between 700 R and 40 k, thus, making FETs like these N-channel types 2SK170 and 2SK389 (dual) or the P-channel devices 2SJ94/2SJ109 (dual and complementary to 2SK389) an excellent choice as 1st stage input transistors. To get low NFs with the above mentioned high values of source resistances the BJTs are forced to operate at rather low collector currents (app. 100 μA), thus, increasing drastically their equivalent input noise voltage, but, at the same time, also decreasing drastically their equivalent input noise current. An example calculation will show the tiny NF-difference between FET (2SK170) and BJT (2SC2546) input with a source resistance of 1 k.

$$e_{N.1k} = 4.07\,\text{nV/rtHz} \tag{3.129}$$

$$e_{N.2sc2546} = 1.54\,\text{nV/rtHz} \tag{3.130}$$

$$i_{N.2sc2546} = 0.23\,\text{pA/rtHz} \tag{3.131}$$

### MM Case Noise Figures

Thus, $NF_{e.2sc2546.1k}$ becomes:

$$NF_{e.2sc2546.1k} = 20\log\left(\frac{\sqrt{e_{N.1k}^2 + e_{N.2sc2546}^2 + (i_{N.2sc2546} \times 1\,\text{k})^2}}{e_{N.1k}}\right) \tag{3.132}$$

$$NF_{e.2sc2546.1k} = 0.593\,\text{dB} \tag{3.133}$$

Application of Eqs. (3.115), (3.127), (3.129) for the FET brings its $NF_{e.2sk170.1k}$ down to:

$$NF_{e.2sk1701k} = 0.185\,\text{dB} \tag{3.134}$$

making a difference of 0.4 dB. Advantage FET!

With a source resistance of 10 k ($e_{N.10k}$ =12.87 nV/rtHz) the $NF_e$-difference looks as follows:

$$NF_{e.2sc2546.10k} = 0.196 \tag{3.135}$$

$$NF_{e.2sk170.10k} = 0.019\,\text{dB} \tag{3.136}$$

Of course, another advantage FET! But the difference, indeed, is rather tiny now, only 0.177 dB.

Unfortunately the 2SK170 is hard to find on the markets and it's rather expensive. The other two types are easier to get. With their equivalent input noise voltages and currents they are very near the values for the best BJTs, but paid for it with rather high drain currents, thus making it much more difficult to develop a low-noise and hum-free power supply unit. But that's another task, one for a chapter further down the pages.

Now, from a noise point of view we can sum up:

As input devices

- FETs are a choice for MM cartridge phono-amps. The best ones beat the best BJTs with a tiny advantage.
- FETs are absolutely not good enough for MC cartridge phono-amps. The best BJTs beat the best FETs by far!

### Optimal Input Resistance

These findings can also be backed up with the calculation of the optimal input resistance for the MM phono-amp configuration:

$$R_{\mathrm{in.opt.2sc2546}} = \frac{1.54\,\mathrm{nV/rtHz}}{0.23\,\mathrm{pA/rtHz}} = 6\,\mathrm{k}7 \tag{3.137}$$

$$R_{\mathrm{in.opt.2sk170}} = \frac{0.85\,\mathrm{nV/rtHz}}{18\,\mathrm{fA/rtHz}} = 47\,\mathrm{k} \tag{3.138}$$

and calculated for the MC phono-amp configuration:

$$R_{\mathrm{in.opt.2sc2546}} = \frac{0.5\,\mathrm{nV/rtHz}}{2.14\,\mathrm{pA/rtHz}} = 234\,\mathrm{R} \tag{3.139}$$

$$R_{\mathrm{in.opt.2sk170}} = 47\,\mathrm{k} \tag{3.140}$$

It is unknown to me if the 2SK170 JFET or other low-noise audio FETs have an increase in equivalent noise current in the upper frequency range of the audio band. This would be a +6 dB increase/octave. We cannot expect this and I couldn't find any indication in any data sheet. But if this would be the case, than, the results of Eqs. ((3.138) & (3.140)) would come down, according to a corner frequency $f_{\mathrm{hi}} < 20\,\mathrm{kHz}$.

For readers who prefer watching images, graphs or figures: the next figure shows why JFETs are as well suitable for MM purposes as BJTs with high $h_{\mathrm{FE}}$ – but never for MC amplification:

Because of the $1/f$-effects of the equivalent noise voltage of the above shown JFET (not included in the calculation) $NF_{\mathrm{e}}$ for low source resistances becomes even worse.

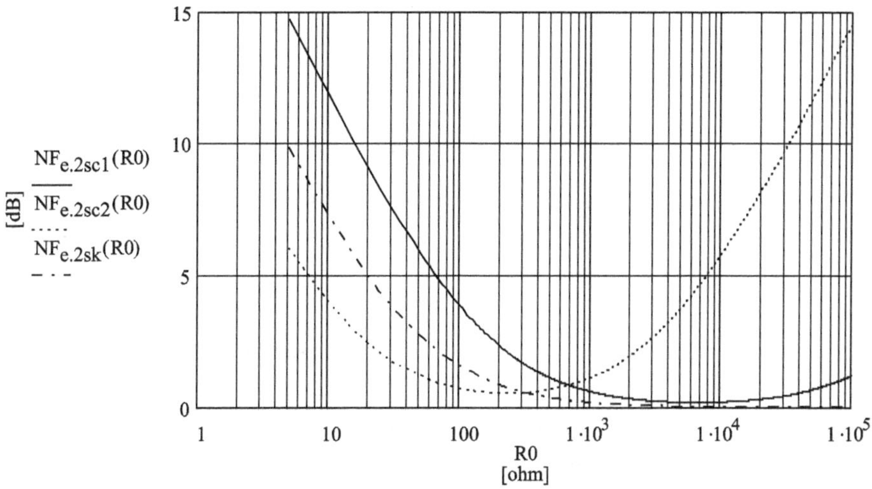


**Fig. 3.34** $NF_e$ of 2SC2546 at $I_C = 100\,\mu A$ ($= NF_{e.2sc1}$ = – the top-plot at 10 R)
$NF_e$ of 2SC2546 at $I_C = 10\,\mu A$ ($= NF_{e.2sc2}$ = – the bottom-plot at 10 R)
$NF_e$ of 2SK170 at $I_D = 10\,\mu A$ ($= NF_{e.2sk}$ = – the mid-plot at 10 R)

That's why it makes no sense to calculate NFs for a low source resistance driven 1st stage with JFETs in CS-configuration any further – like it could be successfully done with BJTs.

## CS Gain Stage Calculations

Now, let's have a noise check on a typical JFET driven gain stage for audio purposes:

Like with BJTs and valves we have to deal with two different situations:

- the source resistor $R_2$ is bypassed by $C_3$ in place (b) or keeps un-bypassed (u) without $C_3$. All $Cs$ in the below figure have values that should not touch the flat frequency and phase response in $B_{20\,k}$.

Thus, the gains $G_{F.b}$ and $G_{F.u}$ of this configuration can be expressed as follows: bypassed version:

$$G_{F.b} = -g_{m.F} R_3 \tag{3.141}$$

with the reduced mutual conductance of the JFET:

$$g_{m.F.red} = \frac{g_{m.F}}{1 + g_{m.F} R_2} \tag{3.142}$$

the un-bypassed version becomes:

$$G_{F.u} = -g_{m.F.red} R_3 \tag{3.143}$$

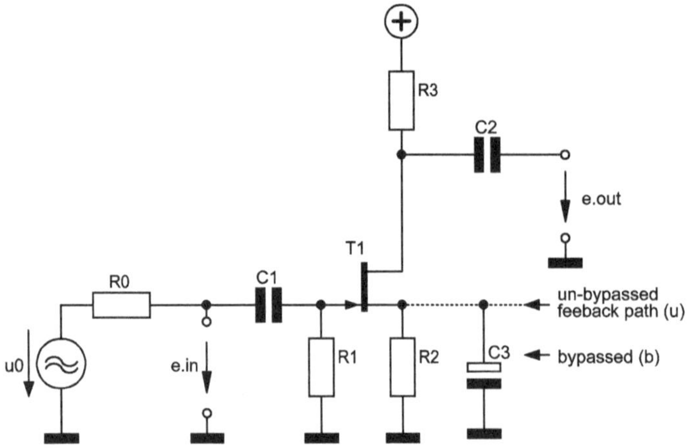


**Fig. 3.35** JFET gain stage in CS configuration

I leave it to the reader to calculate example gain stages. The previous (BJT) and the following chapter (valves) give enough advices on how to do it – and we shouldn't forget the excess noise effect of $R_3$! With proper adjustment of all related components, voltages and currents BJTs in Figs. 3.23 … 3.24 can easily be replaced by JFETs.

But these proper adjustments include activities that might only work in-between rather narrow boundaries. One of these boundaries is the so-called Miller-capacitance the other one is the question about the contribution allowed of a 2nd amp stage.

### Miller Capacitance

What is Miller-capacitance?

A look at Figs. 3.32 & 3.36 and we'll find a capacitance between drain and gate called Gate-Drain-Capacitance $C_{GD}$. We can find it in data sheets. If not, we take the so-called Reverse-Transfer-Capacitance $C_{rss}$. This capacitance is the JFET's Miller capacitance $C_{M.F}$.

Transfered to the input it has to be multiplied with the magnitude of the gain of the JFET plus 1[28]:

$$C_{M.F.in} = C_{GD} \times (1 + |G_F|)$$

or

$$= C_{rss} \times (1 + |G_F|) \tag{3.144}$$

Its negative effect on the frequency response of the gain stage will be demonstrated with the next figure:

[28] T/S Chap. 4

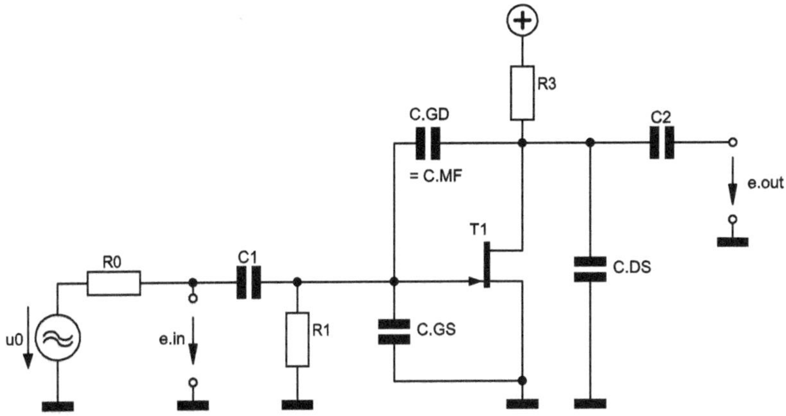


**Fig. 3.36** JFET gain stage with all relevant capacitances

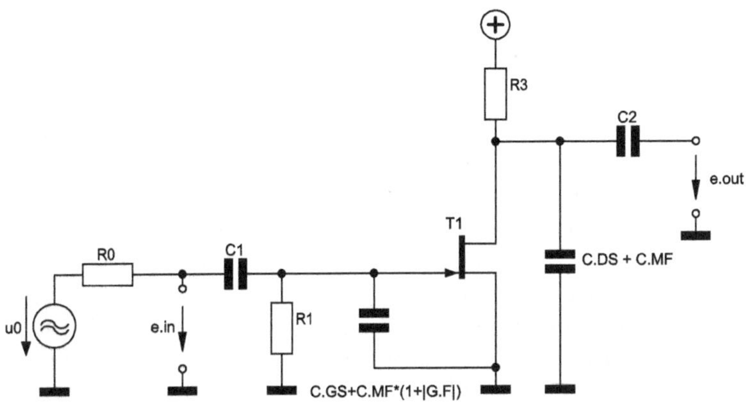


**Fig. 3.37** Equivalent circuit of Fig. 3.36 including its Miller capacitance

The input capacitance of the gain stage changes from a tiny valued $C_{GS}$ to a rather high total input capacitance $C_{in.tot.F}$

$$C_{in.tot.F} = C_{GS} + (1 + |G_F|)\, C_{M.F} \tag{3.145}$$

and the output capacitance $C_{DS}$ changes as well – but less drastically – to a total output capacitance $C_{out.tot.F}$

$$C_{out.tot.F} = C_{DS} + C_{M.F} \tag{3.146}$$

Unfortunately, the Miller-capacitance can not be minimized by a feedback process! It keeps alive in any situation.

For the gain stage input situation this means that the corner frequency of the low-pass filter formed by $R_0$ and $C_{in.tot.F}$ parallel to $R_1$ triggers the frequency and phase response of the stage. That's why for a MM cartridge the input load $C_{in.tot.F}$ plays such a significant role in the calculation of the maximal allowed capacitance load of the cartridge!

For example: a certain MM cartridge requires a max. $C_{\text{load}}$ of 300 pF. This will be built-up by the capacitances of the cable plus those of the connectors plus $C_{\text{in.tot.F}}$. If we take 100 pF for the cable-C and 25 pF for the connector-C than, the capacitance left is 175 pF, the one allowed for the JFET. A SK170 has $C_{\text{rss}} = 6\,\text{pF}$ at $-10\,\text{V}_{\text{GD}}$ and an input capacitance $C_{\text{GS}} = C_{\text{iss}} = 32\,\text{pF}$ at $5\,\text{V}_{\text{DS}}$. Hence, the maximal stage gain allowed will become:

$$G_{\text{F.a}} = \frac{175\,\text{pF} - 32\,\text{pF}}{6\,\text{pF}} - 1 = 23.83 = 27.54\,\text{dB} \tag{3.147}$$

With these results the calculations concerning the contribution allowed issue of a 2nd gain stage à la BJT chapter calculation examples have to take place as well.

### SRPP Gain Stage

Consequently, it makes no sense to try to increase a JFET gain stage the way it could be done with BJTs or valves: with a current generator as the drain load. Therefore, in principal, SRPP[29] gain stages as a 1st gain stage make no sense, except we get a source follower as the input stage in front of the 1st SRPP gain stage. But this means that the noise voltage of the JFET becomes multiplied with $\sqrt{2}$! This is exactly the same noise voltage result we would get with a long-tailed input pair of JFETs – but with less noise problems if we would use an op-amp as the 2nd stage à la Fig. 3.24 and an active RIAA transfer solution via feedback path. The noise (and overload) problems of a SRPP solution with passive RIAA transfer network between 1st stage and a following one will come from the rather low gain of the 1st stage in conjunction with this network's impedance. The network's impedance can't be configured with components of rather low values only and the result of the multiplication of the input noise voltage with the gain of the 1st stage does not swallow the noise of the passive network.

The Miller-capacitance's mechanics are the same in the BJT and valve environment. Usually, the negative effects can be ignored because the capacitances between base/grid and collector/anode are rather small compared with the equivalent ones of JFETs.

### Cascoded Gain Stage

To overcome the Miller-C problem another solution can be found in the cascoding approach: a CS configured JFET is followed by a CG configured one.

The following requirements are essential for the below shown circuit:

[29] for more details: see Sect. 3.4 (Noise in Valves)

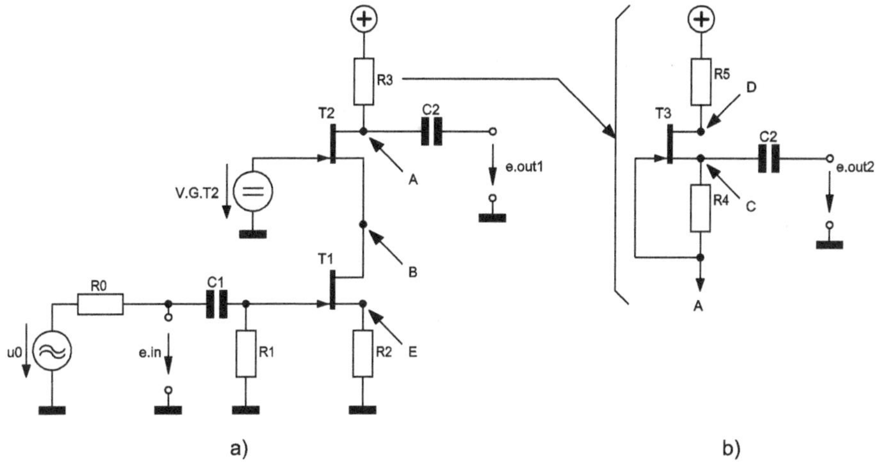


**Fig. 3.38** Cascoded JFET input stage **a** and alternative for $R_3$ **b**

- $V_{DS.T1} \geq V_{T.T1}$ (= threshold voltage of $T_1$)
- $V_{DS.T2} > V_{T.T2}$ (= " of $T_2$)
- $V_{DS.T3} \geq V_{T.T3}$ (= " of $T_3$)
- high $g_{m.Tx}$ to get high gain (e.g. $I_{D.T1\,...3} = 10\,\text{mA}$)
- $g_{m_{T1}} \geq 0.9 \times g_{m_{T2}}$
- $V_{G.T2} \geq V_{T.T1}$

To better understand what that means I guess a calculated example with a 2SK289 would help. First of all, let's start with the circuit of Fig. 3.38 without $R_2$ and with $R_3$ in place. Thus, the total gain $G_{tot}$ can be expressed as:

$$G_{tot} = \frac{\text{e.out1}}{\text{e.in}} = G_{T1} \times G_{T2} \tag{3.148}$$

$$G_{T1} = -\frac{g_{m.T1}}{g_{m.T2}} \tag{3.149}$$

$$G_{T2} = g_{m.T2} \times R_3 \tag{3.150}$$

To select $T_1$ and $T_2$ it's wise to take equal types of JFETs. With

$$g_{m.T1} = g_{m.T2} \tag{3.151}$$

the total gain becomes:

$$G_{tot} = (-1) \times g_{m.T1} \times R_3 \tag{3.152}$$

With $|G_{T1}| = 1$ the Miller-C $C_{M.T1}$ at the input got drastically reduced to only $2 \times C_M = 12\,\text{pF}$! This value plus the input capacitance of the JFET $C_{in} = 25\,\text{pF}$ is no longer a "dangerous" load for any MM cartridge – but still has to be taken into account!

With $I_D = 10$ mA the minimum threshold voltages can be found in the data sheet graphs. The gate-source voltage $V_{GS}$ becomes $-0.03$ V at $V_T = 2.5$ V. With $V_{cc} = +15$ V at point $B$ we'll get $V_B = +2.5$ V, than, the gate of $T_2$ should become a bias voltage of $V_B + V_{GS.T2} = 2.47$ V. With $V_{T2} = 2.5$ V at point A we'll get $V_A = 5$ V, hence $R_3$ must become:

$$R_3 = \frac{V_{cc} - V_A}{I_D} = 1000\,\Omega \tag{3.153}$$

From data sheet we get $g_m$ at 10 mA:

$$g_{m.2sk289} = 32\,\text{mS} \tag{3.154}$$

hence, the total gain becomes:

$$|G_{1_{tot}}| = |-1000\,\Omega \times 32 \times 10^{-3}\,\text{S}| = 32 \text{ equivalent to } 30.1\,\text{dB} \tag{3.155}$$

To get no interaction of a 2nd stage noise with the noise of the 1st stage the minimum requirements were a 37.5 dB gain (see Sect. 3.2 and Eqs. (3.82) ff). This is not met by the result of Eq. (3.155). We have to find a better solution – and this can be found in a replacement of $R_3$ by the b) circuit of Fig. 3.38 (a third 1/2 SK289 – the other 1/2 could serve the other channel of the stereo phono-amp). This is nothing else but a current generator that should have the same $g_m$ like $T_1$ and $T_2$. In addition, we can use it as the upper part of a SRPP configuration with a low-Z output at point $C$ at the source of $T_3$. I call the whole arrangement Cascoded SRPP = CSRPP. To calculate the new overall gain $G_{2_{tot}}$ we need to know the value of the resistor $r_{DS.T3}$ represented by the $T_3 - R_4 - R_5$ configuration.

To simplify things we'll take $V_{DS.T3} = V_{T3} = 2.5$ V and $V_{GS.T3} = 0.03$ V. Thus, $R_4$ becomes $0.03\,\text{V}/10\,\text{mA} = 3\,\text{R}$, we'll take $R_4 = 3\,\text{R32}$. Insertion of e.g. $R_2 = 3\,\text{R32}$ into the circuit (at point $E$) produces an advantage: reduction of distortion – but also a disadvantage: the result of this manoeuvre is the fact that the mutual conductance of $T_1$ becomes reduced to

$$g_{m.T1.red} = \frac{g_{m.T1}}{1 + g_{m.T1} \times R_2} = 28.93\,\text{mS} \tag{3.156}$$

hence, the gain of $T_1$ becomes:

$$G_{T1.red} = \frac{g_{m.T1.red}}{g_{m.T2}} = 0.904 \tag{3.157}$$

which would be in line with the above given requirements.

The $T_2$ load resistor $r_{DS.T3}$ becomes:

$$r_{DS.T3} = \frac{V_{E.T3}}{I_{D.T3}} \tag{3.158}$$

$V_{E.T3} = 30\,\text{V} \ldots 200\,\text{V}$. It's the so-called Early[30] voltage of a JFET. Usually, this type of voltage is not given in data sheets. That's why I take the lowest possible $V_E$ into the calculation course $= 30\,\text{V}$:

$$r_{DS.T3} = \frac{30\,\text{V}}{10\,\text{mA}} = 3000\,\Omega \tag{3.159}$$

To keep $V_{T.T3}$ at 2.5 V we need the right value for resistor $R_5$. It can be calculated as follows:

$$R_5 = \frac{V_{cc} - (V_{T.T1} + V_{T.T2} + V_{T.T3} + |V_{GS.T3}|)}{I_{D.T3}} = \frac{7.5332\,\text{V}}{10\,\text{mA}} = 753.32\,\Omega \tag{3.160}$$

Now, we can recalculate the total gain $G2_{tot}$:

$$G2_{tot} = G_{T1.red} \times G_{T2.adj} \tag{3.161}$$

$$G_{T2.adj} = (3000\,\Omega + 753\,\Omega) \times 32\,\text{mS} \tag{3.162}$$

$$|G2_{tot}| = |-0.904 \times 3753\,\Omega \times 32\,\text{mS}| = 108.6 \text{ equivalent to } 40.7\,\text{dB} \tag{3.163}$$

Even with the minimum Early voltage this result fulfils the above mentioned gain requirement of minimal 37.5 dB for a 1st stage!

What happens with the input referred noise voltage density $e_{N.tot}$ of this gain stage? It can be calculated as follows:

$$e_{N.in.tot} = \sqrt{e_{N.T1}^2 + e_{N.T2}^2 + 4kTB_1R_2} \tag{3.164}$$

With data sheet figures for $e_{N.T1} = e_{N.T2} = 1.12\,\text{nV/rtHz}$ $e_{N.in.tot}$ becomes[31]:

$$e_{N.in.tot} = 1.601\,\text{nV/rtHz} \tag{3.165}$$

### Weighted *SN* Calculation

With an input load $R_0$ of 1 k or 12 k and $R_1 = 47\,\text{k5}$, referenced to $B_{20\,\text{k}}$ and a nominal input voltage of $5\,\text{mV}_{rms}/1\,\text{kHz}$ we can expect the following *SN*s of an A-weighted and RIAA-equalized phono-amp for MM cartridge use (additionally see Eqs. (6.12)ff):

$$SN_{ariaa.1st.1k} = 20\log\left(\frac{\sqrt{B_{20\,\text{k}}\{e_{N.in.tot}^2 + 4kT(1\,\text{k}||47k5)\}}}{5\,\text{mV}_{rms}}\right) - 7.935\,\text{dB}$$

$$= -86.169\,\text{dBA} \tag{3.166}$$

$$SN_{ariaa.1st.12k} = -76.833\,\text{dBA} \tag{3.167}$$

[30] T/S Chap. 3

[31] remember: the noise gain is always >1, thus, $G_{T1.red} = 0.904$ doesn't change $e_{N.T1}$

Note: These result figures were calculated without $C_{load}$-effect of a MM cartridge ($C_{load}$ parallel to the input load). The results will decrease a bit! This will be demonstrated in the MM cartridge part of the book (Part II, Chaps. 4 … 5).

### Example Calculation for a Low-Noise Gain Stage

At the end of this chapter I would like to show an idea of a lowest-noise all-FET MM phono-amp. Yet, I didn't built it up but I think it's worth to study its advantages.

The gain of $T_1$ must be kept rather low to avoid negative impact of the Miller-effect on the load capacitance of the MM cartridge. Therefore I've chosen

$$G_{T1} = 10 \tag{3.168}$$

With Eqs. (3.144 … 3.146) the input capacitance (including Miller-C) becomes:

$$C_{in.tot.T1} = 10\,\text{pF} + 32\,\text{pF} + 6\,\text{pF} \times 11 = 108\,\text{pF} \tag{3.169}$$

Thus, the cable capacitance to connect the MM cartridge to the input should be <180 pF!

With Eqs. (3.141 … 3.143) $R_3$ can be determined.

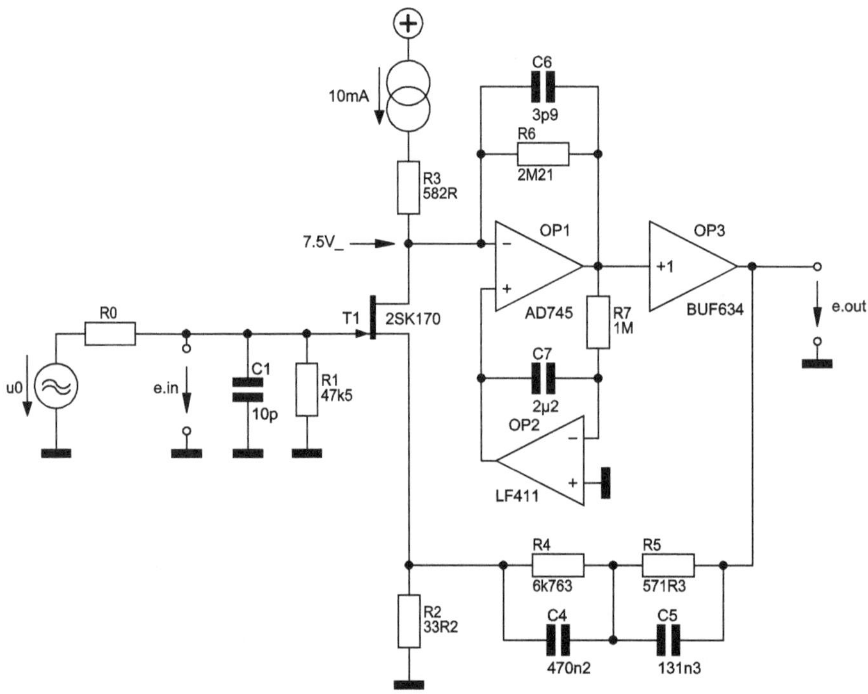


**Fig. 3.39** Draft for a lowest-noise all-FET MM phono-amp

With the data sheet values for the input noise voltage densities at 1 kHz of 2SK170 (0.85 nV/rtHz) and AD745 (3.2 nV/rtHz) the input referred noise voltage density of the draft phono-amp can be determined with a rough calculation:

$$e_{\mathrm{N.in.tot}} = \frac{\sqrt{(10 \times e_{\mathrm{N.T1}})^2 + e_{\mathrm{N.op1}}^2 + 4kTB_1R_2}}{10} = 1.167\,\mathrm{nV/rtHz} \tag{3.170}$$

With reference to a nominal input voltage of 5 $\mathrm{mV_{rms}}/1$ kHz in $B_{20\,\mathrm{k}}$ and an input load resistor $R_0$ of 1 k or 12 k without input $C$ and $1/f$-effect $SN_{\mathrm{ariaa}}$ becomes:

$$SN_{\mathrm{ariaa.amp.1k}} = 20\log\left(\frac{\sqrt{B_{20\,\mathrm{k}}\left(e_{\mathrm{N.in.tot}}^2 + 4kTB_1(1\,\mathrm{k}||47k5)\right)}}{5\,\mathrm{mV_{rms}}}\right) - 7.935\,\mathrm{dB} = -86.456\,\mathrm{dBA} \tag{3.171}$$

$$SN_{\mathrm{ariaa.amp.12k}} = -76.865\,\mathrm{dBA} \tag{3.172}$$

The $SN$s decreasing input $C$ and $1/f$-effects can easily be calculated with the mathematical course given in Part II, Chap. 5 and with Eqs. (3.4 ... 3.5).

## 3.4 Noise in Valves (US: Tubes)

Like BJTs and FETs valves can be used in three different configurations: common cathode (CC), common plate (CP) or common anode (CA) and common grid (CG). Usually, in audio amplification the CC configuration plays the major role. The types of valves used are the triode and the pentode and the pentode configured as a triode. The equivalent noise source model of a triode for the audio band looks as follows[32]:

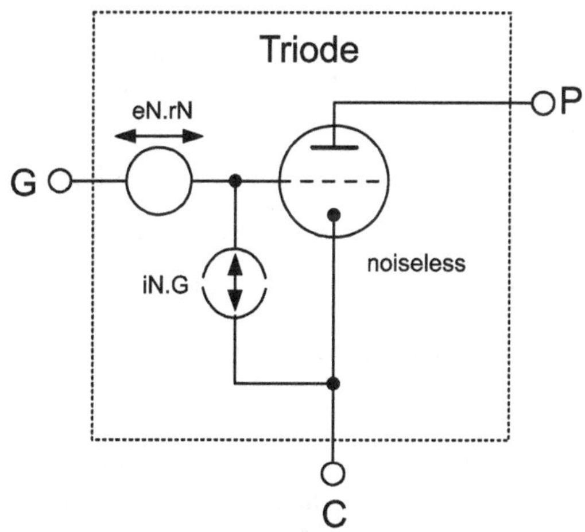


**Fig. 3.40** Simplified audio band equivalent noise source model for a triode

[32] "Telefunken Laborbuch" (laboratory handbook), Vol. 3, 2nd edition 1966

In this model $e_{N.rN}$ is the equivalent noise voltage of the triode, based on the so-called equivalent noise resistance $r_{equ}$ or $r_N$ of the valve, $i_{N.G}$ is the equivalent grid noise current. Similar to FETs this noise current is rather low ($\leq$50 fA/rtHz) and as such can be ignored in future calculations for audio purposes.

### Triode Noise Resistance

Spoilt by the enormous amount of noise data we can get for FETs and BJTs it's almost surprising to find only one important one for valves: the equivalent noise resistance $r_{N.t}$ for a triode. It is defined as[33]:

$$r_{N.t} = \frac{2}{3}\frac{T_c}{T}\frac{1}{\sigma|g_{m.t}|} \tag{3.173}$$

or

$$r_{N.t} = \frac{3.06}{|g_{m.t}|} \tag{3.174}$$

with:

- $T_c$ = temperature of the cathode (in average 1100 K)
- $T$ = room temperature (300 K)
- $\sigma$ = valve control figure for triodes (usually this value is 0.8 for all valves workable in the audio field)
- $g_{m.t}$ = mutual conductance of the triode

For example: with $g_{m.t} = 12.5$ mA/V $r_{N.e88cc}$ for a E88CC/6922/ECC88/PCC88/7DJ6 becomes:

$$r_{N.e88cc} = 244.8\,\Omega \tag{3.175}$$

### Triode Noise Voltage

For the above mentioned $g_{m.t}$ this value is rather close to the one which can be found in the data sheets: 300 R.There is only a 10% difference in the below shown noise voltage calculations:

$$e_{N.300} = \sqrt{4kT300B_1} = 2.23\,\text{nV/rtHz} \tag{3.176}$$

$$e_{N.245} = \sqrt{4kT245B_1} = 2.01\,\text{nV/rtHz} \tag{3.177}$$

### Pentode Noise Resistance

Any further grid in a valve creates additional noise. That's why pentodes with their two additional grids are less favourable for lowest-noise audio applications. For pentodes the equivalent noise resistance can be calculated as follows[34]:

[33] "Telefunken Laborbuch" Vol. 3

[34] dto.

With the pentode's

- mutual conductance of the cathode $g_{m.p.c}$
- mutual conductance of the anode $g_{m.p.a}$
- anode current $I_a$
- cathode current $I_c$
- grid 2 current $I_{g2}$

and the physical constants

- Boltzmann's constant $k$
- electron charge $q$

$g_{m.p.c}$ becomes:

$$g_{m.p.c} = g_{m.p.a}\frac{I_c}{I_a} \tag{3.178}$$

thus, the equivalent noise resistance $r_{N.p}$ for a pentode becomes:

$$r_{N.p} = \frac{2}{3}\frac{T_c}{T}\frac{1}{\sigma|g_{m.p.c}|} + \frac{q}{2kT}\frac{1}{|g_{m.p.a}|}\frac{I_a I_{g2}}{I_c} \tag{3.179}$$

or, a bit easier to handle:

$$r_{N.p} \approx \frac{3.06}{|g_{m.p.c}|} + \frac{1}{0.05\,\text{V}}\frac{1}{|g_{m.p.a}|}\frac{I_a I_{g2}}{I_c} \tag{3.180}$$

Comparison of Eq. (3.173) with Eq. (3.179) makes clear that the term after the plus sign in Eq. (3.180) is one of the noise increasing factors of the pentodes. The other one is the fact that, because of the additional currents for the grids, the pentode's $g_{m.p.c}$ must become always bigger than the $g_{m.t}$ of a triode with equal mutual conductances of the anodes.

Hence,

- if $g_{m.p.a} = g_{m.t}$
- and $I_c > I_a$
- than $g_{m.p.c} > g_{m.p.a}$
- than $r_{N.p} > r_{N.t}$

### Pentode Noise Voltage

Let's check this for the EF86/6BK8/6267, a widely used audio valve and also quite often used as a 1st stage of RIAA phono-amps. Data sheet figures for a typical amp stage are:

- $I_a = 3.0\,\text{mA}$
- $I_{g2} = 0.6\,\text{mA}$
- $I_c = 3.6\,\text{mA}\ (= I_a + I_{g2})$
- $g_{m.a} = 2.0\,\text{mA/V}$

- $g_{m.c} = 2.4\,\mathrm{mA/V}$ (see Eq. (3.178))

Thus, with Eq. (3.180) $r_{N,ef86}$ becomes:

$$r_{N.ef86} \approx 3.69\,\mathrm{k\Omega} \tag{3.181}$$

and the respective equivalent noise voltage $e_{N.ef86}$ on $B_1$ becomes:

$$e_{N.ef86} \approx 7.82\,\mathrm{nV/rtHz} \tag{3.182}$$

without taking into account the fact that a $1/f$ effect should be calculated too – if we would know the corner frequency $f_{ce}$ for the EF86. This $f_c$ is – not only for the EF86 – not given in the data sheets nor could something similar be found in my collection of WW, EW&WW and EW issues from 1968 on nor in Elector Electronics issues from 1973 on nor in the Tube CAD Journal nor in any book about valves I own. So, we have to guess it. And I bet another point will be for sure: $f_{ce}$ changes with changed anode current.

If we would guess $f_{ce} = 200\,\mathrm{Hz}$ for the valve in the anode current range from 1 mA to 2 mA, hence, with Eq. (3.4) the rms voltage of $e_{N.ef86.B}$ in $B_{20\,k}$ becomes:

$$e_{N.ef86.B} = 1.143\,\mu\mathrm{V} \tag{3.183}$$

or, expressed as spectral noise voltage density value in $B_1$:

$$e_{N.ef86} = 8.09\,\mathrm{nV/rtHz} \tag{3.184}$$

Fortunately, with A-filter based SN measurements $1/f$-noise becomes blurred because of the heavy low-frequency attenuation of that kind of filter[35].

### Selection of Low-Noise Valves

A small selection of low-noise valves – pentodes and triodes – is given in the following table.

The triodes are less noisy than the pentodes, but, configured as triodes in many cases the pentodes are much better than the triodes. This small disadvantage of the triodes can easily be surmounted by paralleling the two triodes of a dual-triode like the ones of the PCC88 family, bringing noise voltage down to 1.42 nV/rtHz in $B_1$. This value is absolutely comparable with the values of very good FETs and BJTs for MM phono amp input stages. In addition, on the market these triodes are easier to get than e.g. the EF280F/EF810F types.

Nevertheless, if we put in place only one triode as a 1st stage of a low noise amplifier for MM cartridge amplification even a noise voltage density of 2.01 nV/rtHz doesn't beat the respective ones from a FET or from a BJT, to say nothing of MC amplification! As will be shown in one of the next sections even a transformer doesn't help to improve the valve's performance compared with the one of FETs/BJTs.

[35] "Measurement filters" see Chap. 12

**Table 3.6** Selection[36,37,38] of low-noise pentodes, pentodes as triodes and triodes for phono-amp purposes (∗ equal values for E88CC)

| 1/A | B | C | D | E | F | G | H | I | J |
|---|---|---|---|---|---|---|---|---|---|
| 2 | Valve: pentodes | | Data sheet | | | Calculated | | Data sheet | |
| 3 | | | $g_{m.a}$ | $I_a$ | $f_{ce}$ | $r_{N.p}$ | $e_{N.p}$ | $r_{N.p}$ | $e_{N.p}$ |
| 4 | | | | | | | | | calc. |
| 5 | EU | US | mA/V | mA | Hz | kΩ | nV/rtHz | kΩ | nV/rtHz |
| 6 | EF86 | 6CF8, 6BK8 | 2.0 | 3.0 | ? | 3.69 | 7.82 | ? | ? |
| 7 | C3M | C3M | 6.5 | 16.0 | ? | 1.55 | 5.07 | 1.20 | 4.46 |
| 8 | EF184 | 6EJ7 | 15.6 | 14.0 | ? | 0.40 | 2.59 | 0.30 | 2.23 |
| 9 | 404A | 5847 | 12.5 | 13.5 | ? | 0.57 | 3.07 | 0.55 | 3.02 |
| 10 | ECF80 | 6BL8 | 6.2 | 10.0 | ? | 1.49 | 4.96 | 1.50 | 4.98 |
| 11 | E280F | 7722 | 26.0 | 20.0 | ? | 0.17 | 1.70 | 0.22 | 1.91 |
| 12 | EF732 | 5840 | 5.0 | 7.5 | ? | 1.87 | 5.57 | 1.60 | 5.15 |
| 13 | E810F | 7788 | 50.0 | 35.0 | ? | 0.08 | 1.13 | 0.11 | 1.35 |
| 14 | valve: pentodes as triodes | | $g_{m.a}$ | $I_a$ | $f_{ce}$ | $r_{N.p\text{-}t}$ | $e_{N.p\text{-}t}$ | $r_{N.p\text{-}t}$ | $e_{N.p\text{-}t}$ |
| 15 | | | | | | | | | calc. |
| 16 | EU | US | mA/V | mA | Hz | kΩ | nV/rtHz | kΩ | nV/rtHz |
| 17 | EF86 | 6CF8, 6BK8 | 2.0 | 3.0 | ? | 1.28 | 4.60 | ? | ? |
| 18 | C3M | C3M | 6.5 | 16.0 | ? | 0.40 | 2.60 | 0.65 | 3.28 |
| 19 | EF184 | 6EJ7 | 15.6 | 14.0 | ? | 0.15 | 1.58 | ? | ? |
| 20 | 404A | 5847 | 12.5 | 13.5 | ? | 0.19 | 1.77 | ? | ? |
| 21 | ECF80 | 6BL8 | 6.2 | 10.0 | ? | 0.39 | 2.53 | ? | ? |
| 22 | E280F | 7722 | 26.0 | 20.0 | ? | 0.10 | 1.29 | ? | ? |
| 23 | EF732 | 5840 | 5.0 | 7.5 | ? | 0.43 | 2.77 | ? | ? |
| 24 | E810F | 7788 | 50.0 | 35.0 | ? | 0.06 | 0.97 | ? | ? |
| 25 | valve: triodes | | $g_{m.a}$ | $I_a$ | $f_{ce}$ | $r_{N.t}$ | $e_{N.t}$ | $r_{N.t}$ | $e_{N.t}$ |
| 26 | | | | | | | | | calc. |
| 27 | EU | US | mA/V | mA | Hz | kΩ | nV/rtHz | kΩ | nV/rtHz |
| 28 | ECC40 | n.a. | 2.9 | 6.0 | ? | 1.10 | 4.18 | 0.28 | 2.15 |
| 29 | ECC81 | 12AT7 | 5.5 | 10.0 | ? | 0.56 | 3.03 | 0.50 | 2.88 |
| 30 | ECC83 | 12AX7 | 1.6 | 1.2 | ? | 1.91 | 5.63 | ? | ? |
| 31 | EC71 | 5718 | 6.5 | 13.0 | ? | 0.47 | 2.79 | ? | ? |
| 32 | PCC88* | 7DJ8 | 12.5 | 15.0 | ? | 0.24 | 2.01 | 0.30 | 2.23 |
| 33 | E188CC* | 7308 | 12.5 | 15.0 | ? | 0.24 | 2.01 | 0.25 | 2.04 |
| 34 | ECC808 | 6KX8 | 1.6 | 1.2 | ? | 1.91 | 5.63 | ? | ? |
| 35 | ECF80 | 6BL8 | 5.0 | 14.0 | ? | 0.61 | 3.18 | ? | ? |

In addition: having configured a valve stage with an un-bypassed cathode resistor $R_C$ (current feedback) to create a feedback path to the input of the amp via the cathode node, it must also be taken into account that, because of the input resistance $r_{c.t}$ at the cathode, a fraction of the noise of $R_C$ sums up in rms mode to the noise of the valve's equivalent noise resistance. It increases the relatively low noise of the

[36] "Valves Pocket Book", VALVO, Hamburg 1971

[37] "Röhren Taschen Tabelle" (valves pocket table), Franzis, Munich 1963

[38] "Taschenbuch Röhren, Halbleiter, Bauteile" (valves, semiconductors, components pocket book), AEG-Telefunken 1973

valve the same way like un-byassed emitter or source resistances in the solid state world.

### Triode Cathode Input Resistance

With the following data the calculation of the triode's cathode input resistance looks like:

- $\mu$ = gain factor (data sheet)
- $R_L$ = load resistance of the plate
- $r_{a.t}$ = plate or anode resistance

$r_{c.t}$ becomes[39]:

$$r_{c.t} = \frac{R_L + r_{a.t}}{\mu + 1} \tag{3.185}$$

### Example Calculation for a Low-Noise CC Gain Stage

Another negative point hits the valve's noise picture as well: the excess noise voltages of all resistances that are placed between different potentials. To get the input referred spectral noise voltage density $e_{N.t}$ for a typical triode amplification stage the following example calculation for the $1/2$ PCC88 circuit of Fig. 3.41 will give all relevant facts.

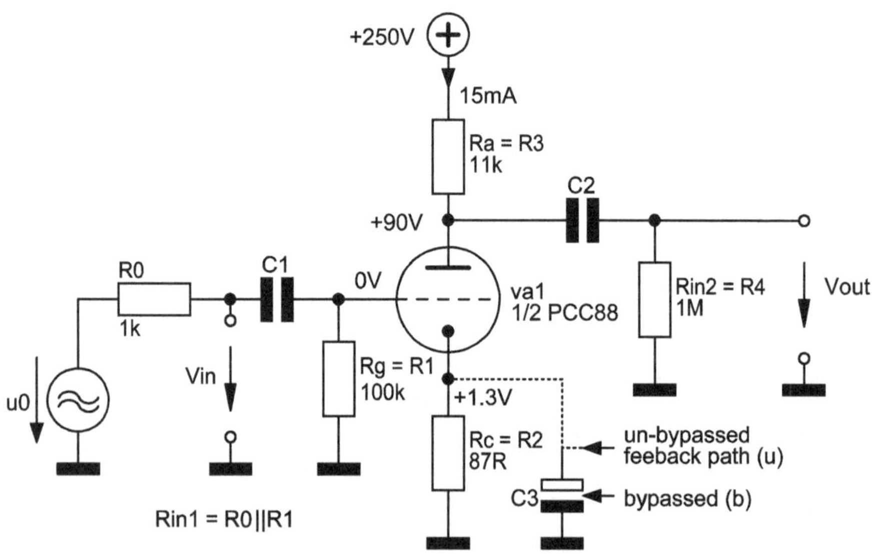


Fig. 3.41 Triode gain stage

[39] "Valve Amplifiers", Morgan Jones, Newnes, 1995

Assumed that the capacitors in Fig. 3.41 were all of a value that doesn't hurt the flat frequency response in the frequency range of interest and by ignoring the output voltage phase shift of 180°, the un-bypassed triode gain $G_{\mathrm{t.u}}$ becomes [40,41,42]:

$$G_{\mathrm{t.u}} = \frac{\mu R_{\mathrm{L}}}{r_{\mathrm{a.t}} + R_{\mathrm{L}} + (\mu + 1)R_{\mathrm{c}}} \tag{3.186}$$

the bypassed triode gain $G_{\mathrm{t.b}}$ becomes:

$$G_{\mathrm{t.b}} = g_{\mathrm{m.t}} \frac{r_{\mathrm{a.t}} R_{\mathrm{L}}}{r_{\mathrm{a.t}} + R_{\mathrm{L}}} \tag{3.187}$$

The relevant values for 1/2 PCC88 and components of the circuit of Fig. 3.41 are the following ones:

- $\mu = 33$
- $r_{\mathrm{a.t}} = 2\mathrm{k}6$
- $g_{\mathrm{m.t}} = 12.5\,\mathrm{mA/V}$
- $V_{\mathrm{a}} = 90\,\mathrm{V}$
- $V_{\mathrm{c}} = -1.3\,\mathrm{V}$
- $I_{\mathrm{a}} = 15\,\mathrm{mA}$
- $r_{\mathrm{N.t}} = 300\,\mathrm{R}$
- $e_{\mathrm{N.t}} = 2.23\,\mathrm{nV/rtHz}$
- $R_0 = 1\,\mathrm{k}$
- $R_{\mathrm{g}} = R_1 = 100\,\mathrm{k}$
- $R_{\mathrm{c}} = R_2 = 87\,\mathrm{R}$
- $R_{\mathrm{a}} = R_3 = 11\,\mathrm{k}$
- $R_{\mathrm{in2}} = R_4 = 1\,\mathrm{M}$
- $R_{\mathrm{in1}} = R_1 || R_0 = 0\mathrm{k}99$
- $R_{\mathrm{L}} = R_3 || R_4 = 10\mathrm{k}9$

the calculation course for the un-bypassed version goes as follows:

$$G_{\mathrm{t.u}} = \frac{33 \times 10\mathrm{k}9}{2.6\mathrm{k} + 10\mathrm{k}9 + (33+1) \times 87\mathrm{R}} = 21.84 \tag{3.188}$$

the spectral noise voltage density $e_{\mathrm{N.Rc}}$ of the cathode resistor $R_{\mathrm{c}} = R_2 = 87\,\mathrm{R}$ becomes

$$e_{\mathrm{N.Rc}} = \sqrt{4kTR_{\mathrm{c}}} = 1.20\,\mathrm{nV/rtHz} \tag{3.189}$$

reduced by the voltage divider formed by $r_{\mathrm{c.t}}$ and $R_{\mathrm{C}}$ effective will be only a fraction of it, with $r_{\mathrm{c.pcc88}} = 397\,\mathrm{R}$ [as of Eq. (3.185)] $e_{\mathrm{N.Rc.eff}}$ becomes

$$e_{\mathrm{N.Rc.eff}} = e_{\mathrm{N.Rc}} \frac{r_{\mathrm{c.t}}}{r_{\mathrm{c.t}} + R_{\mathrm{c}}} = 0.985\,\mathrm{nV/rtHz} \tag{3.190}$$

---

40 Additional useful valve formulae: for all valves: $r_{\mathrm{a}} \times g_{\mathrm{m}} \times \frac{1}{\mu} = 1$;

41 gain for triodes: $G_{\mathrm{t}} = g_{\mathrm{m.t}} \frac{r_{\mathrm{a.t}} R_{\mathrm{L}}}{r_{\mathrm{a.t}} + R_{\mathrm{L}}}$; $R_{\mathrm{L}} = R_{\mathrm{a}} || R_{\mathrm{in2}}$

42 gain for pentodes: $G_{\mathrm{p}} = g_{\mathrm{m.p}} \times R_{\mathrm{L}} (r_{\mathrm{a.p}} \gg R_{\mathrm{L}})$

the spectral noise voltage density $e_{N.RL}$ of the anode load resistance $R_L = R_a||R_{in2} = R_3||R_4$ becomes

$$e_{N.RL} = \sqrt{4kTR_L} = 13.43\,nV/rtHz \tag{3.191}$$

reduced by the voltage divider formed by $r_{a.t}$ and $R_L$ effective will be only a fraction of it, calculated as follows:

$$e_{N.RL.eff} = e_{N.RL}\frac{r_{a.t}}{r_{a.t}+R_L} = 2.59\,nV/rtHz \tag{3.192}$$

the spectral noise voltage density $e_{N.Rin1}$ of the grid load resistance $R_{in1} = R_S||R_g = R_S||R_1$ becomes

$$e_{N.Rin1} = \sqrt{4kTR_{in1}} = 4.05\,nV/rtHz \tag{3.193}$$

The calculations for the two excess noise voltages of the anode and cathode resistances go as follows (the grid resistance does not produce excess noise as long as the grid's potential is at $0\,V_{DC}$!):

in the three decades from 20 Hz ... 20 kHz (B) the excess noise voltages of $R_a$, $R_c$ (metal film resistors with $NI = 0.05\,\mu V/V/dec.$) become

$$e_{Nex.Ra.B} = (250\,V - 90\,V)\times 3^{1/2}\times 0.05\,\mu V/V/dec. = 13.86\,\mu V \tag{3.194}$$

$$e_{Nex.Rc.B} = (1.3\,V)\times 3^{1/2}\times 0.05\,\mu V/V/dec. = 0.113\,\mu V \tag{3.195}$$

referenced to $B_1$ the spectral noise voltage densities become

$$e_{Nex.Ra} = 98.03\,nV/rtHz \tag{3.196}$$

$$e_{Nex.Rc} = 0.8\,nV/rtHz \tag{3.197}$$

reduced by the voltage dividers formed by $R_a + (r_{a.t}||R_{in2})$ and $R_c + r_{c.t}$ effective for noise calculations will only be a fraction of them, hence,

$$e_{Nex.Ra.eff} = e_{Nex.Ra}\left(\frac{r_{a.t}||R_{in2}}{R_a + r_{a.t}||R_{in2}}\right) = 18.7\,nV/rtHz \tag{3.198}$$

$$e_{Nex.Rc.eff} = e_{Nex.Rc}\frac{r_{c.t}}{R_c + r_{c.t}} = 0.65\,nV/rtHz \tag{3.199}$$

Now, we have everything in hand to calculate the input referred spectral noise voltage density $e_{N.t}$ of the un-bypassed valve driven 1st gain stage by rms summing-up of all noise making elements of the circuit of Fig. 3.41:

$$e_{N.pcc88.u.1k} = \sqrt{\begin{array}{l} e^2_{N.pcc88} + e^2_{N.Rin1} \\ +e^2_{N.Rc.eff} + e^2_{Nex.Rc.eff} \\ +\left(\frac{e_{N.RL.eff}}{G_u}\right)^2 + \left(\frac{e_{Nex.Ra.eff}}{G_u}\right)^2 \end{array}} \tag{3.200}$$

$$e_{N.pcc88.u.1k} = 4.849\,nV/rtHz \tag{3.201}$$

## Noise Model of the CC Gain Stage

For better understanding of the above outlined calculations I add Fig. 3.42 which shows all relevant noise sources and voltage dividers of the above calculated un-by-passed triode gain stage.

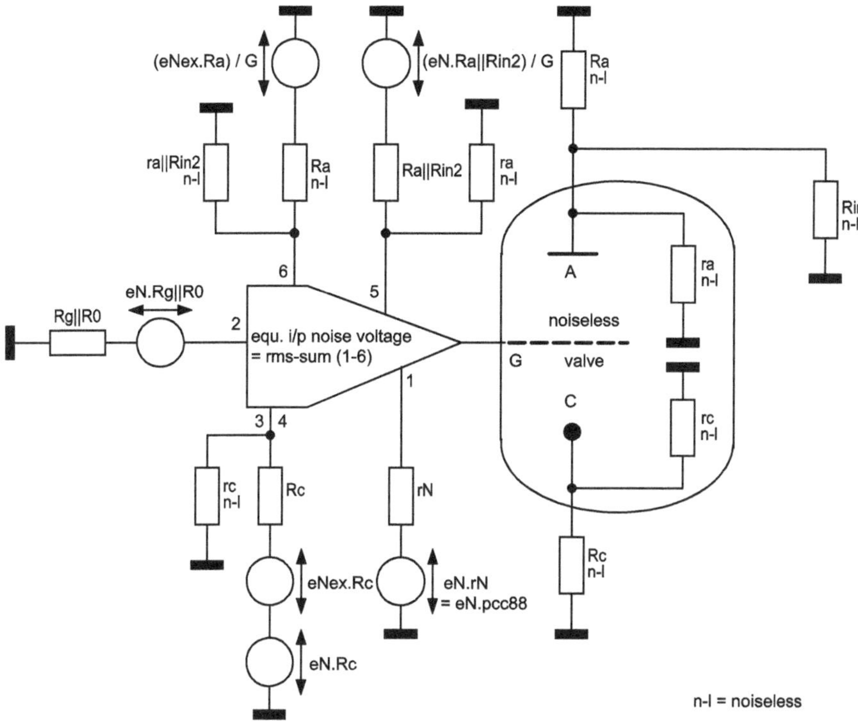


**Fig. 3.42** Triode gain stage of Fig. 3.41 with all relevant noise sources and voltage dividers

With Eqs. (3.48 ... 3.49) and the spectral noise voltage density $e_{\mathrm{N.Rin1}} = 4.05\,\mathrm{nV}/\mathrm{rtHz}$ of $R_{\mathrm{in1}} = R_1||R_0$ $NF_{\mathrm{e.pcc88.u.1k}}$ becomes:

$$NF_{\mathrm{e.pcc88.u.1k}} = 20\log\left(\frac{e_{\mathrm{N.pcc88.u.1k}}}{e_{\mathrm{N.Rin1}}}\right) = 1.564\,\mathrm{dB} \qquad (3.202)$$

With reference to a rated input voltage of $5\,\mathrm{mV_{rms}}/1\,\mathrm{kHz}$ and with the indicated source resistance of 1 k the un-weighted signal-to-noise-ratio $SN_{\mathrm{ne}}$ in a frequency band of $B_{20\,\mathrm{k}}$ of that 1st stage becomes:

$$SN_{\mathrm{ne.pcc88.u.1k}} = 20\log\left(\frac{e_{\mathrm{N.pcc88.u.1k}} \times \sqrt{B_{20\,\mathrm{k}}}}{5\,\mathrm{mV_{rms}}}\right) = -77.260\,\mathrm{dB} \qquad (3.203)$$

Reminder: the numerator in Eq. (3.203) comes from Eq. (3.48) the following way:

$$\sqrt{\int_{20\,\text{Hz}}^{20{,}000\,\text{Hz}} e^2_{\text{N.pcc88.u.1k}} df} = e_{\text{N.pcc88.u.1k}} \times \left(\sqrt{20{,}000\,\text{Hz} - 20\,\text{Hz}}\right) \tag{3.204}$$

Now, if we change the source resistance $R_0$ from 1 k to 50 R let's check what will happen with $NF_e$ and $SN_{ne}$ referenced to $0.5\,\text{mV}_{\text{rms}}/1\,\text{kHz}$ – as if we would use this gain stage for MC amplification:

$$e_{\text{N.pcc88.u.50}} = 2.82\,\text{nV/rtHz} \tag{3.205}$$

$$R_{\text{in1.50}||100\,\text{k}} = 49\,\text{R98} \tag{3.206}$$

$$e_{\text{N.50R}} = 0.91\,\text{nV/rtHz} \tag{3.207}$$

With Eq. (3.202) $NF_{\text{e.pcc88.u.50}}$ becomes

$$NF_{\text{e.pcc88.u.50}} = 9.819\,\text{dB} \tag{3.208}$$

and $SN_{\text{ne.50}}$ becomes

$$SN_{\text{ne.pcc88.u.50}} = 20\log\left(\frac{e_{\text{N.pc88.u.50}} \times \sqrt{B_{20\,\text{k}}}}{0.5\,\text{mV}_{\text{rms}}}\right) = -61.974\,\text{dB} \tag{3.209}$$

If we now repeat the whole game for a 1st gain stage with bypassed $R_c$ and taking into account that its gain $G_b$ is bigger than that of the un-bypassed stage, than, all results should improve.

$$G_{\text{t.b}} = 12.5\,\text{mA/V} \times \left(\frac{2.6\,\text{k} \times 10.9\,\text{k}}{2.6\,\text{k} + 10.9\,\text{k}}\right) = 26.23 \tag{3.210}$$

By adequately changing the above demonstrated calculations $NF_e$ and $SN_{ne}$ (ref. $5\,\text{mV}_{\text{rms}}/1\,\text{kHz}$) come up as follows:

$$e_{\text{N.pcc88.b.1k}} = 4.68\,\text{nV/rtHz} \tag{3.211}$$

$$NF_{\text{e.pcc88.b.1k}} = 1.253\,\text{dB} \tag{3.212}$$

$$SN_{\text{ne.pcc88.b.1k}} = -77.571\,\text{dB} \tag{3.213}$$

and with reference to $0.5\,\text{mV}_{\text{rms}}/1\,\text{kHz}$ and $R_0 = 50\,\text{R}$ the results look like:

$$e_{\text{N.pcc88.b.50}} = 2.51\,\text{nV/rtHz} \tag{3.214}$$

$$NF_{\text{e.pcc88.b.50}} = 8.824\,\text{dB} \tag{3.215}$$

$$SN_{\text{ne.pcc88.b.50}} = -62.969\,\text{dB} \tag{3.216}$$

It might look a bit strange calculating *NFs*, *SNs* and noise voltages with three digits after the decimal point. But, from time to time, by comparing results, there are very small differences which become better visible with the chosen description.

To sum up at this point of survey: from a noise point of view valve 1st stages are very good for MM phono-amp purposes but nothing for direct MC cartridge amplification. We'll see later on in Chap. 3.7 what will happen with *SN*s if we would take transformers to link MC cartridges to MM phono-amps. In this chapter it will satisfyingly be demonstrated with BJT input devices. But it works equally well with valves.

Further SN improvements can be found in the replacement of the anode resistor $R_a$ by an active solution or by paralleling of at least two triodes of a dual triode – or with both measures. Really, paralleling pentodes is extremely difficult, because rather seldom we find two equal ones. With the so-called SRPP[43] (Shunt Regulated Push Pull, Fig. 3.43) approach for triodes (or with any similar approach like Morgan Jones' $\mu$-follower [44]) $R_a$ nearly becomes infinite and the full gain of the valve can be used. In addition, there will be the choice between a high-Z ($= r_{a.va1}$) or low-Z ($= R_3 || r_{c.va2}$) output of this valve configuration. Excess noise at the anode of the gain producing lower valve nearly disappears as well and the noise voltage of $R_a$ changes to the noise voltage of the upper valve plus a fraction of its cathode resistor noise voltage together with its excess noise voltage. In any case, these are lower values than that of a high $R_a$. If we would take the low-Z-output of the SRPP stage the noise voltage of $R_{in2}$ no longer plays a role as well.

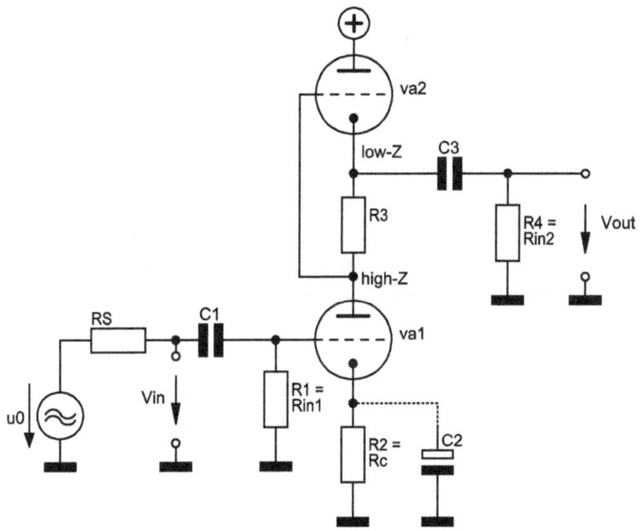


**Fig. 3.43** Principal SRPP circuit

[43] "High-end Audio Equipment", Elektor Electronics, ISBN 0-905705-40-8

[44] "Valve Amplifiers", Morgan Jones, Newnes 1995, UK

For such a SRPP gain stage detailed design proposals are given in [45]. If we would take the Fig. 3.43 gain stage model and 1/2 PCC88 as the active load (with $R_3 = R_c = 87\,\mathrm{R}$) the noise figures $NF$ and $SN$ in the $5\,\mathrm{mV_{rms}}\,/1\,\mathrm{kHz}$ – 1k-input-load-case would improve by app. +0.1 dB (un-bypassed and bypassed) and by app. +0.3 dB in the $0.5\,\mathrm{mV_{rms}}\,/1\,\mathrm{kHz}$ – 50R-input-load-case. At least, we need to know the gain of a cathode follower:

$$G_{\mathrm{t.c}} = \frac{\mu R_c}{r_{\mathrm{a.t}} + (\mu + 1)R_c} \tag{3.217}$$

For such an increase of material these noise improvement results look rather little. But we can expect additional advantages of the SRPP concept that make it worth to study it. Such as:

- high gain
- high linearity
- rather low distortion
- low output impedance
- no need of voltage feedback

This makes the SRPP an ideal choice for RIAA amps with passive equalization. Unfortunately, many offered design concepts have a resistor placed between $R_1$ and the valve's grid (I've seen some in the range from 220 R to 2 k2). From a noise point of view this is nonsense because the noise of this resistor adds in rms mode to the total noise of the circuit as well. If its noise voltage has the same value like the gain stage's noise voltage without this resistor, than, $SN$s will be worsened by 3 dB. If there is a need for such a resistor to avoid ringing of the gain stage, I guess, something is really wrong with the whole concept: circuitry as well as mechanics.

At the end of this section 2 graphs demonstrate that, not only for the usage of a PCC88 as a 1st stage, an image does tell 1000 times more than written sentences or calculations. It's nothing else but the transfer of Eqs. (3.203) and (3.204) into plots – referenced to a nominal input voltage of $5\,\mathrm{mV_{rms}}/1\,\mathrm{kHz}$. They also make clear that valves (like FETs) are very good for MM cartridge amplification – or for high output/low impedance MC cartridges. Additionally, the noise quality of a 1st MM phono-amp stage with a valve like the PCC88 can easily be checked by application of the formulae given in Chap. 4 (Noise in MM cartridges).

$SN$s referenced to a nominal phono-amp input voltage of $0.5\,\mathrm{mV_{rms}}/1\,\mathrm{kHz}$ can be taken from Fig. 3.45 by addition of 20 dB to the value of the plot at a specific $R_0$ (e.g. $R_0 = 50\,\mathrm{R}$) the following way:

$$\begin{aligned}\text{bypassed version } SN \text{ to find} &= \text{plot value of the dotted trace at 50 R plus 20 dB}\\ &= -82.969\,\mathrm{dB} + 20\,\mathrm{dB} = -62.969\,\mathrm{dB}.\end{aligned}$$

This is exactly the calculation result of Eq. (3.216).

[45] "Tube CAD Journal", May 2000, Vol. 2, Number 4

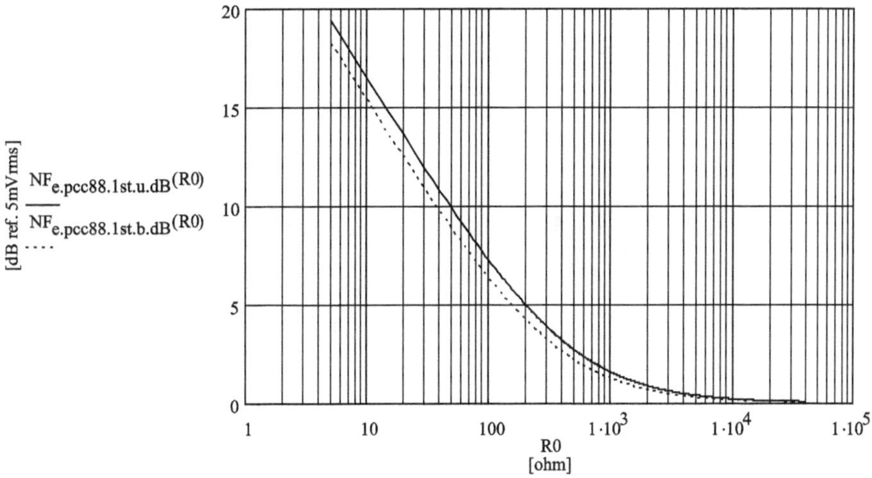


**Fig. 3.44** $NF_e$ vs. $R_0$ of the gain stage of Fig. 3.41 – un-bypassed (u) and bypassed (b) version – referenced to an input voltage of 5 $mV_{rms}$/1 kHz

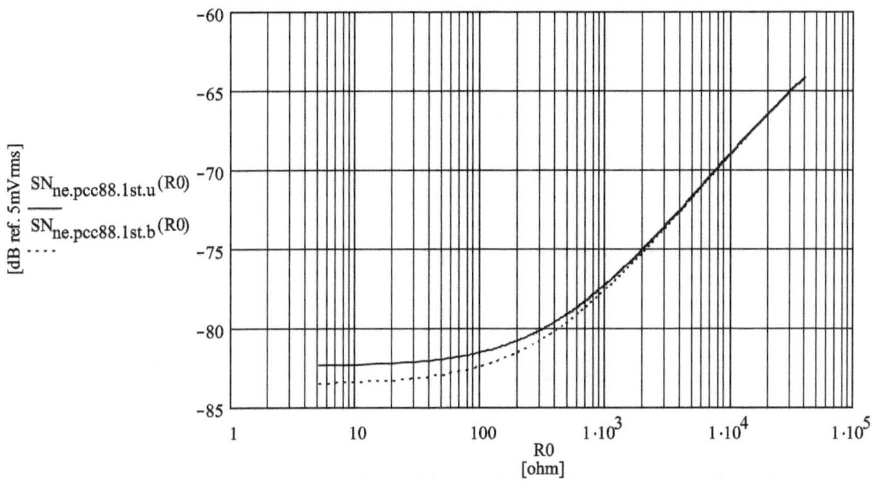


**Fig. 3.45** $SN_{ne}$ vs. $R_0$ of the gain stage of Fig. 3.41 – un-bypassed (u) and bypassed (b) version – referenced to an input voltage of 5 $mV_{rms}$/1 kHz

## Remarks on Valve Power Supplies

This is not a book that covers all aspects of the design of phono-amps. But dealing with valves is a different thing. Alone, not only the noise voltage produced by a valve in conjunction with its biasing components is the sound disturbing element of a valve circuitry. The spikes $\leq$2 kHz in Figs. 3.46… 3.47 give an idea on what I'm thinking of: it's hum that comes into the amplifying chain via heater and plate power lines as well as from mains interferences vagabonding inside the amp case

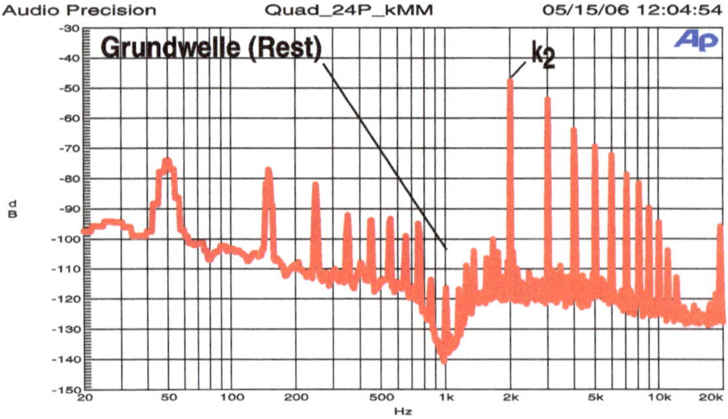


**Fig. 3.46** Quad 24P distortion measurement[46] for the MM input showing noise floor and distortion artefacts ≥2 kHz as well as heavy mains influence up to 2 kHz at 50 Hz, 150 Hz, 250 Hz, etc.

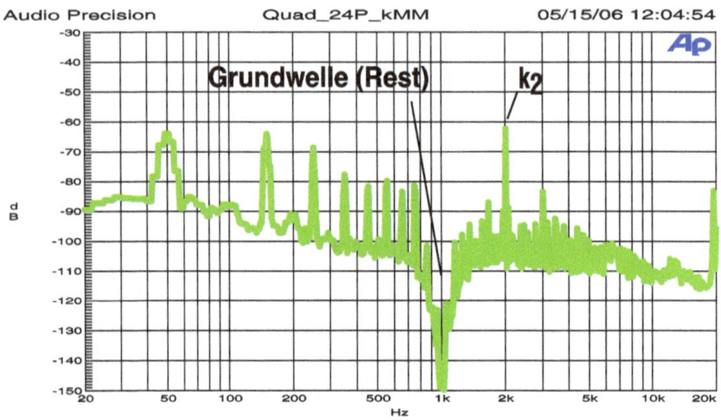


**Fig. 3.47** Like Fig. 3.46[47] for the MC input

and in the vicinity of the mains transformers. Stereoplay's Quad 24P Audio Precision MM input test results looked as follows:

with reference to $5\,\mathrm{mV_{rms}}/1\,\mathrm{kHz}$ and with an input load of 1 k $SN_{\mathrm{ariaa.1k}}$ becomes

$$SN_{\mathrm{ariaa.1k}} = -77.0\,\mathrm{dBA} \tag{3.218}$$

with an input load of a standard cartridge containing a sequence of a 500 R resistor and a 500 mH inductance $SN_{\mathrm{ariaa.sc}}$ becomes

$$SN_{\mathrm{ariaa.sc}} = -75.0\,\mathrm{dBA} \tag{3.219}$$

---

46 © stereoplay 2006-07, by kind permission of Motor-Presse International Verlagsgesellschaft, Stuttgart, Germany

47 © stereoplay 2006-07, dto.

I assume that – for the best case – $SN_{riaa.1k}$ for the 1 k loaded MM input can be calculated as:

$$SN_{riaa.1k} = SN_{ariaa.1k} + 7.935\,dB - 3.646\,dB + 3.0\,dB$$
$$= -69.7\,dB \tag{3.220}$$

The deduction comes from application of Eqs. (6.8) and (6.10) from Chap. 6 plus a guessed term of obliging and experience based 3.0 dB hum influence.

Stereoplay's Quad 24P Audio Precision MC input test results looked as follows: with reference to 0.5 $mV_{rms}/1$ kHz and with an input load of 20 R $SN_{ariaa.1k}$ becomes

$$SN_{ariaa.20R} = -72.0\,dBA \tag{3.221}$$

$SN_{riaa}$ can be calculated as:

$$SN_{riaa.20R} = SN_{ariaa.20R} + 7.935\,dB - 3.646\,dB + 3.0\,dB$$
$$= -64.7\,dB \tag{3.222}$$

The deduction is alike the MM case.

After many measurements concerning the evaluation of noise charts I found out that the following rule of thumb becomes more or less valid:

- If in a $B_{20k}$ noise spectrum à la Figs. 3.46 … 3.47 one, some or all of the mains frequency based spikes at 50 Hz, 100 Hz, 150 Hz, etc. cross an imaginary horizontal line set by $SN_{riaa}$ (= −69.7 dB resp. −64.7 dB) than, hum can be heard at normal listening loudness with volume pot setting at 11 am and no signal.

Consequently, all signals will be modulated by hum and – depending on the signal strength and on the degree of effectiveness of the loudspeakers – it can be noticed more or less. Obviously, this would be the case for the MC input of the Quad P24.

To avoid this kind of extra disturbance caused by hum, at least, the following actions should be taken into account when designing valve phono-amps:

- In any case, the valve's heaters should be supplied by carefully regulated and balanced soft-started DC voltages and currents. Tightly twisted supply wires are essential. A centre tapped transformer is a must. The 1st stage valve should be heated via separate power supply.

  Often, in data sheets of valves, the heater supply voltage is indicated as AC voltage only. We do not destroy a low-power valve – suitable for phono-amp gain stages – if we would try to supply it with DC voltage[48].

[48] Mr Brüggemann, SST Brüggemann GmbH, Frankfurt, Germany, got a lot of positive practical knowledge on that issue when designing and building-up many German valve driven radio station studios in the past

- Because of the rather low supply currents <10 mA regulated anode voltages/currents aren't necessarily a must, but supply voltage treatment with a high-value C-L-C $\Pi$-network is strongly recommended. For supply currents >10 mA I would take a regulated power supply.
- A current generator at the plate and/or cathode keeps power lines hum sufficiently off the valve as well as plate/cathode resistor excess noise off the plate/cathode.
- Mu-metal magnetic shielding attenuates most of the mains caused electromagnetic field[49].
- Complete balanced circuitry layout.
- Proper grounding.

## 3.5 Noise in Operational Amplifiers (Op-Amps)

### Intro

From a noise point of view op-amps are very much easier to handle than transistors or valves[50,51,52,53]. In this book I only refer to low-noise voltage feedback op-amps. Usually, current feedback op-amps produce much more noise.

Basically, we only need two precise figures to describe the noise situation of an op-amp: the equivalent input noise voltage $e_{N.opa}$ – in data sheets very often specified for several frequencies – and the equivalent input noise current $i_{N.opa}$ – mostly given for 1 kHz only. Quite often, additional graphs show the respective spectral noise density plots in $B_{20\,k}$. Example graphs are the ones of Figs. 3.2 ... 3.3, the principal amplifier situation is given in Fig. 3.1.

Unfortunately, despite the ease to handle them, concerning noise op-amps placed in a real life circuitry need a lot of special attention and designing a low-noise amp is still a challenging task. The following four figures show the noise situations of the two basic amplifier configurations for voltage feedback op-amps: series and shunt arrangement.

---

[49] I'm a bit surprised about the many mains induced spikes in Figs. 3.46 ... 3.47 because in the Quad 24P all valves got shielded

[50] "Noise and Operational Amplifier Circuits", L. Smith & D. H. Sheingold, Analog Dialogue 03-1969, Vol. 3, No. 1

[51] T/S Chap. 5

[52] M/C Chap. 10

[53] "Intuitive IC OP AMPS", Th. M. Frederiksen, National Semiconductor Technology Series, 1984

### Series Configuration

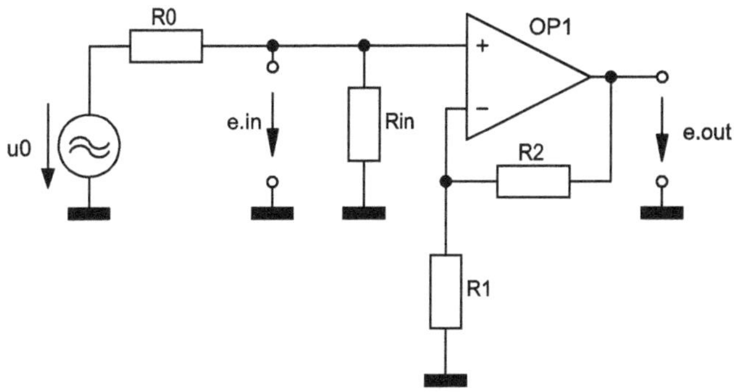


**Fig. 3.48** OP-amp in a series (non-inverting) configuration

### Shunt Configuration

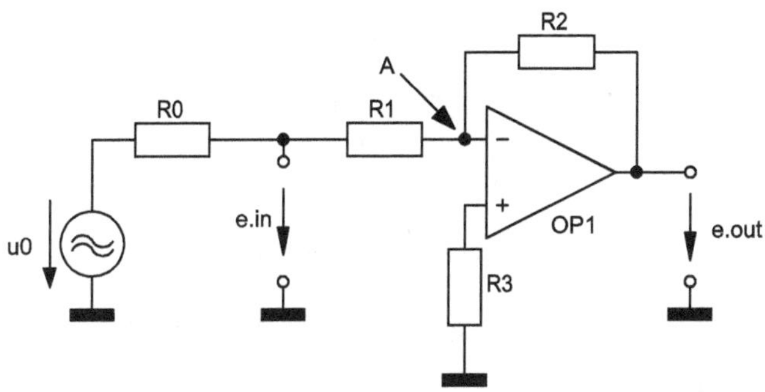


**Fig. 3.49** OP-amp in a shunt (inverting) configuration with virtual earth at point A

The main – noise relevant – differences of the two configurations are:

1. effective signal gain $G_{\text{ser.opa}}$ for the series configuration:

$$G_{\text{ser.opa}} = 1 + \frac{R_2}{R_1} \tag{3.223}$$

2. effective signal gain $G_{\text{shu.opa}}$ for the shunt configuration:

$$G_{\text{shu.opa}} = -\frac{R_2}{R_1} \tag{3.224}$$

with output phase shift of 180°! For noise calculations we only need the magnitude of that gain, because phase relationships don't play a role in a 100% uncorrelated environment.

3. op-amp input impedance $Zin_{ser.opa}$; if $Zin_{opa} \gg R_{in}$, than,

$$Zin_{ser.opa} = R_{in}||Zin_{opa} \approx R_{in} \tag{3.225}$$

4. op-amp input impedance $Zin_{shu.opa}$:

$$Zin_{shu.opa} = R_1 \tag{3.226}$$

### Series Configured Noise Model

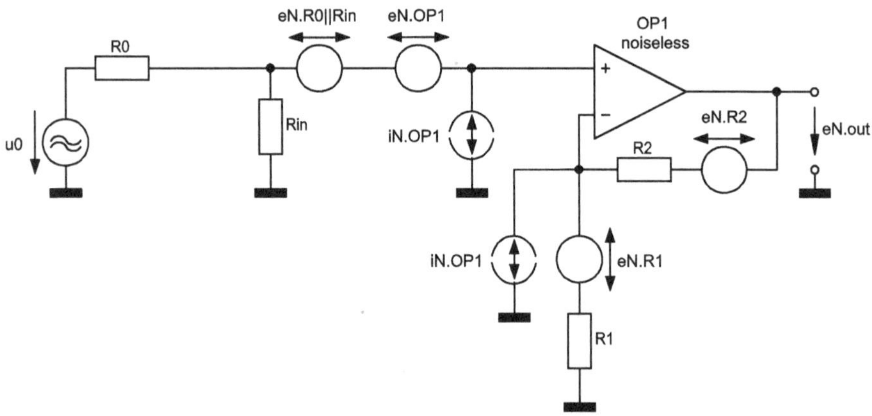


**Fig. 3.50** Op-amp series configuration with all meaningful noise sources

In the above shown noise model the noise voltage $e_{N.out.ser}$ at the output becomes:

$$e^2_{N.out.ser} = \left(1+\frac{R_2}{R_1}\right)^2 \left[e^2_{N.opa} + e^2_{N.(R0||R_{in})} + \left(i_{N.opa}(R_0||R_{in})\right)^2\right] + e^2_{N.R2} + \left(i_{N.opa}R_2\right)^2 + \left(\frac{R_2}{R_1}e_{N.R1}\right)^2 \tag{3.227}$$

Division by the gain of that configuration $G_{ser} = 1 + (R_2/R_1)$ leads to the input referred noise voltage density $e_{N.in.ser}$:

$$e_{N.in.ser} = \sqrt{e^2_{N.opa} + e^2_{N.(R0||Rin)} + \left(i_{N.opa}(R_0||R_{in})\right)^2 + e^2_{N.R1||R2} + \left(i_{N.opa}(R_1||R_2)\right)^2} \tag{3.228}$$

### Example Calculation for Noise Voltage, Noise Current, *SN*, *NF*

Example calculation at room temperature of 300 K:

- OPA = OP27
- $e_{\mathrm{N.op27}} = 3.2\,\mathrm{nV/rtHz}$ $\quad R_1 = 100\,\mathrm{R}$
- $i_{\mathrm{N.op27}} = 0.6\,\mathrm{pA/rtHz}$ $\quad R_2 = 10\,\mathrm{k}$
- $R_0 = 1\,\mathrm{k}$ $\quad R_0||R_{\mathrm{in}} = 979\,\mathrm{R2}$
- $R_{\mathrm{in}} = 47\,\mathrm{k}$ $\quad R_1||R_2 = 99\,\mathrm{k01}$
- $G_{\mathrm{ser}} = 101 = (R_1 + R_2)/R_1$ $\quad e_{\mathrm{in.nom}} = 5\,\mathrm{mV_{eff}}$

With Eq. (3.227) $e_{\mathrm{N.out.op27.ser}}$ becomes:

$$e_{\mathrm{N.out.op27.ser}} = \sqrt{\begin{array}{l} G_{\mathrm{ser}}^2\{3.2\,\mathrm{nV/rtHz}^2 + (4kT \times 979.2\,\Omega) \\ +(0.6\,\mathrm{pA/rtHz} \times 979.2\Omega)^2\} + (4kT \times 10^4\Omega) \\ +(0.6\,\mathrm{pA/rtHz} \times 10^4\Omega)^2 + 100^2(4kT \times 100\,\Omega) \end{array}} \tag{3.229}$$

Thus, the input referred noise voltage density $e_{\mathrm{N.in1.op27.ser}}$ becomes:

$$e_{\mathrm{N.in1.op27.ser}} = \frac{e_{\mathrm{N.out.op27.ser}}}{G_{\mathrm{ser}}} = 5.334\,\mathrm{nV/rtHz} \tag{3.230}$$

With Eq. (3.228) $e_{\mathrm{N.in2.op27.ser}}$ becomes:

$$e_{\mathrm{N.in2.op27.ser}} = \sqrt{\begin{array}{l} 3.2\,\mathrm{nV/rtHz}^2 + (4kT \times 979.2\,\Omega) + (0.6\,\mathrm{pA/rtHz} \times 979.2\,\Omega)^2 \\ +(4kT \times 99.01\,\Omega) + (0.6\,\mathrm{pA/rtHz} \times 99.01\,\Omega)^2 \end{array}}$$

$$e_{\mathrm{N.in2.op27.ser}} = 5.334\,\mathrm{nV/rtHz} \tag{3.231}$$

Thus,

$$e_{\mathrm{N.in1.op27.ser}} = e_{\mathrm{N.in2.op27.ser}} \tag{3.232}$$

The noise figure $NF_{\mathrm{e.op27ser}}$ and the signal-to-noise-ratio $SN_{\mathrm{ne.op27ser}}$ in $B_{20\,\mathrm{k}}$ become:

$$NF_{\mathrm{e.op27.ser}} = 20\log\left(\frac{e_{\mathrm{N.in2.op27.ser}}}{e_{N.(RS||Rin)}}\right) = 2.440\,\mathrm{dB} \tag{3.233}$$

$$SN_{\mathrm{ne.op27.ser}} = 20\log\left(\frac{\sqrt{B_{20\,\mathrm{k}}}e_{\mathrm{N.in2.op27.ser}}}{e_{\mathrm{in.nom}}}\right) = -76.432\,\mathrm{dB} \tag{3.234}$$

### Shunt Configured Noise Model

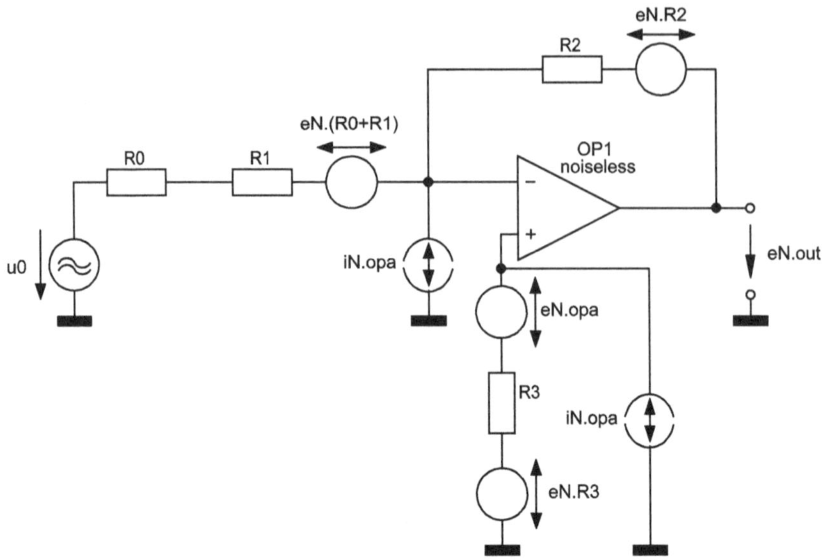


**Fig. 3.51** Op-amp shunt configuration with all meaningful noise sources

In the above shown noise model the output noise voltage $e_{N.out.shu}$ becomes:

$$e^2_{N.out.shu} = \left(\frac{R_0+R_1+R_2}{R_0+R_1}\right)^2 \left[e^2_{N.opa} + e^2_{N.(R3)} + \left(i_{N.opa} \times R_3\right)^2\right] + \left(\frac{R_2}{R_0+R_1}\right)^2 \left(e_{N.(R0+R1)}\right)^2 + e^2_{N.R2} + \left(i_{N.opa} \times R_2\right)^2 \tag{3.235}$$

Although, the magnitude of the signal gain of the shunt configured circuitry $G_{shu}$ looks like:

$$G_{shu} = \left|-\frac{e_{out.shu}}{u_0}\right| = \left|-\frac{R_2}{R_0+R_1}\right| \tag{3.236}$$

this is not the case for noise voltages. No matter which configuration we choose, the noise gain $G_{N.shu}$ always becomes (subject to be proved at the end of this section):

$$G_{N.shu} = 1 + G_{shu} = 1 + \left|-\frac{R_2}{R_0+R_1}\right| = 1 + \frac{R_2}{R_0+R_1} \tag{3.237}$$

Thus, the input referred noise voltage density $e_{N.in.shu}$ of the above shown model becomes:

$$e_{N.in.shu} = \frac{\sqrt{e^2_{N.out.shu}}}{G_{N.shu}} \tag{3.238}$$

$$e_{\mathrm{N.in.shu}} = \sqrt{e_{\mathrm{N.opa}}^2 + e_{\mathrm{N.R3}}^2 + i_{\mathrm{N.opa}}^2 R_3^2 + e_{\mathrm{N.(R0+R1)||\,R2}}^2 + i_{\mathrm{N.opa}}^2\{(R_1 + R_0)||R_2\}^2} \tag{3.239}$$

### Example Calculation for Noise Voltage, Noise Current, *SN*, *NF*

With an OP27 and:

- $R_0 = 1\,\mathrm{k}$
- $R_1 = 47\,\mathrm{k}$
- $R_2 = 4\,\mathrm{M7}$
- $R_3 = 0\,\mathrm{R}$

$e_{\mathrm{N.in.shu}}$ becomes

$$e_{\mathrm{N.in.op27.shu}} = 56.67\,\mathrm{nV/rtHz} \tag{3.240}$$

The noise figure $NF_{\mathrm{e.shu}}$ and the signal-to-noise-ratio $SN_{\mathrm{ne.shu}}$ in $B_{20\,\mathrm{k}}$ (referenced to a nominal amp input voltage of $e_{\mathrm{in.nom}} = 5\,\mathrm{mV_{rms}}/1\,\mathrm{kHz}$) become:

$$NF_{\mathrm{e.op27.shu}} = 20\log\left(\frac{e_{\mathrm{N.in.op27.shu}}}{e_{\mathrm{N.R0}}}\right) = 19.876\,\mathrm{dB} \tag{3.241}$$

$$SN_{\mathrm{ne.op27.shu}} = 20\log\left(\frac{\sqrt{B_{20\,\mathrm{k}}}\,e_{\mathrm{N.in.op27.shu}}}{e_{\mathrm{in.nom}}}\right) = -58.905\,\mathrm{dB} \tag{3.242}$$

For both examples I've chosen the input impedance for MM cartridge use (47 k) and a source resistance of 1 k. To demonstrate what happens if we change the source resistance $R_0$ from 5 R to 100 k the following two graphs (Figs. 3.52 and 3.53) will give an interesting – and expected – answer: the shunt configuration is nothing for low-noise phono-amp amplification!

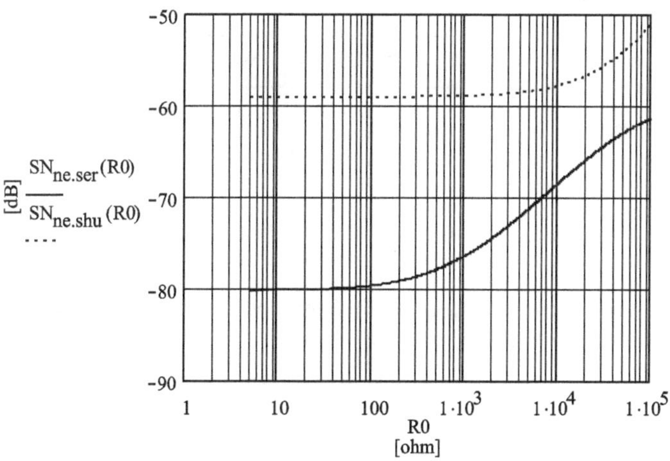


**Fig. 3.52** $NF_{\mathrm{e}}$ of series (*solid trace*) and shunt (*dotted trace*) op-amp configuration

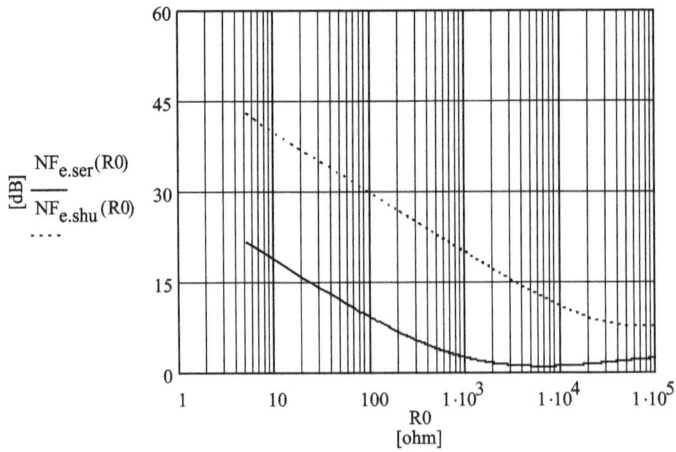


**Fig. 3.53** $SN_{ne}$ of series (*solid trace*) and shunt (*dotted trace*) op-amp configuration

### $1/f$-Noise Effects

Reminder:
If there exists a corner frequency $f_{ce}$ and/or $f_{ci}$ for noise voltage and/or noise current density of the op-amp of interest, than, in Eqs. (3.227 …3.228) and (3.239) we shoudn't forget to adapt the equivalent input noise voltage $e_{N.opa}$ and current $i_{N.opa}$ according to Eqs. (3.4) and (3.5). In addition – to get a reference value for $\sqrt{1\,\text{Hz}}$ – the results of these two Eqs. have to be divided by $\sqrt{B_{20\,\text{k}}}$ !

Hence,

$$e_{N.opa.adapt} = \frac{e_{N.opa}\sqrt{f_{ce}\ln\frac{20\,\text{kHz}}{20\,\text{Hz}} + 19{,}980\,\text{Hz}}}{\sqrt{19{,}980\,\text{Hz}}} \tag{3.243}$$

$$i_{N.opa.adapt} = \frac{i_{N.opa}\sqrt{f_{ci}\ln\frac{20\,\text{kHz}}{20\,\text{Hz}} + 19{,}980\,\text{Hz}}}{\sqrt{19{,}980\,\text{Hz}}} \tag{3.244}$$

### Noise Gain

With the help of the Figs. 3.54a, b and a step by step approach it will be easy to understand why the noise gain $G_{N.shu}$ always becomes $\geq 1$.

1. Take Eq. (3.239) and set $R_3 = 0\,\text{R}$, hence, the resulting gain for the equivalent noise voltage of the op-amp at the (+) input becomes the result shown in Eq. (3.237) and Fig. 3.54a:

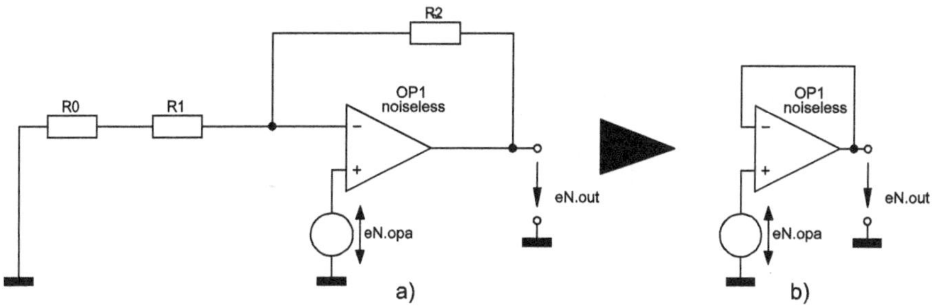


**Fig. 3.54** OP-amp equivalent noise voltage situation in a shunt configured op-amp in **a** and its evolution for $R_2 = 0\,\mathrm{R}$ in **b**

$$G_{\mathrm{N.shu}} = 1 + \frac{R_2}{R_1 + R_0} \tag{3.245}$$

2. Setting $R_2 = 0$ means that the noise from the source $(R_0 + R_1)$ becomes blocked at the $(-)$ input of the op-amp and the op-amp only acts like a voltage follower for its own noise voltage $e_{\mathrm{N.opa}}$ as of Fig. 3.54b, thus, resulting in the minimal gain $G_{\mathrm{N.shu.min}}$ of

$$G_{\mathrm{N.shu.min}} = 1 \tag{3.246}$$

3. Hence, in any configuration the noise gain $G_\mathrm{N}$ of an op-amp always equals the gain of the series configured op-amp, resulting in $G_\mathrm{N} \geq 1$[54]. It will never be $< 1$.

## 3.6 Noise in Instrumentation Amps (In-Amps)

### Intro

The difference between the op-amp and the in-amp is the fact that, usually, the input of an op-amp is configured for single ended (non-symmetrical, un-balanced) purposes and the in-amp is configured as an amp with a true differential (symmetrical, balanced) input. As well, both can be configured the opposite way – but coming along with certain disadvantages. In addition in most IC cases the in-amp's gain can be set by only one resistor whereas the op-amp's gain can be set by a range of different possibilities. The following figures show the circuitry differences as well as the gain setting possibilities for potential MM or MC amplification purposes:

[54] I remember a similar problem about the cat with 3 tails: it's a fact that no cat has 2 tails, but one cat has one tail in addition to the situation of no cat, hence, the cat has three tails! This means for op-amps: no op-amp has no noise voltage source. One op-amp has one noise voltage source more than no op-amp ... !?!

**Basic Op-Amp and In-Amp Gain Setting Options**

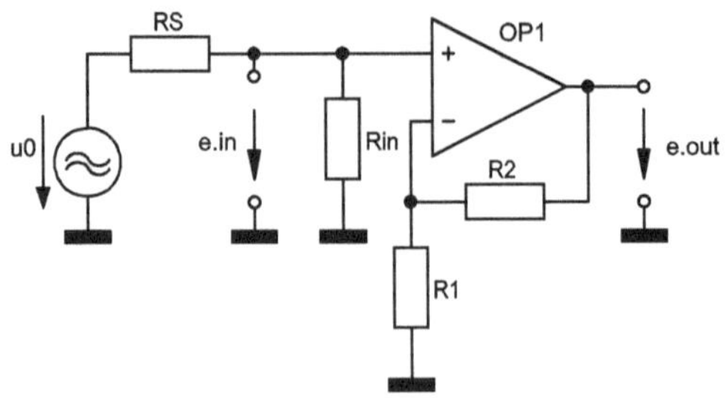


**Fig. 3.55** Non-inverting op-amp gain stage with single ended input, with gain setting impedances $R_1$ & $R_2$ and input resistance $R_{in}$

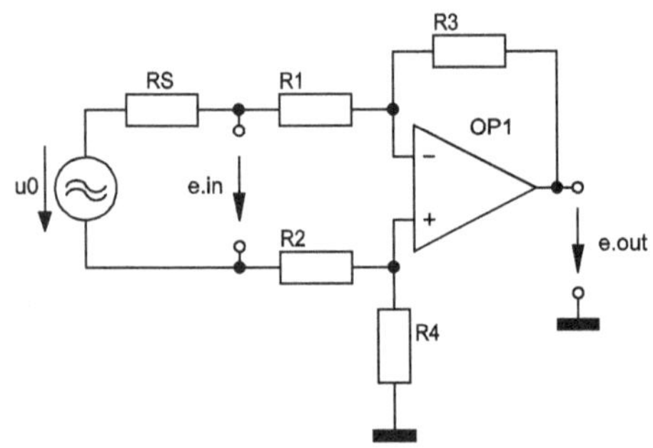


**Fig. 3.56** Op-amp configured as a in-amp in balanced mode with gain setting impedances $R_3$ & $R_4$ and input resistances $R_1$ resp. $R_2$ on each input

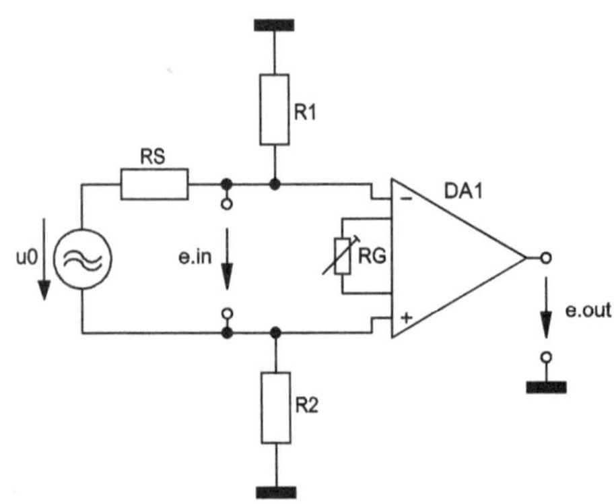


**Fig. 3.57** In-amp configured as a balanced gain stage with only one gain setting resistor $R_G$ and input resistances $R_1$ resp. $R_2$ on each input. Output phase can be set by transposing the source leads

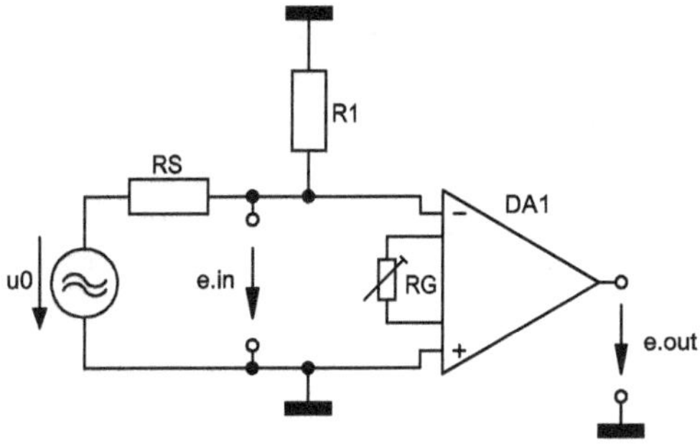


**Fig. 3.58** In-amp configured as a single ended gain stage with one gain setting resistor $R_G$ and input resistance $R_1$. Output phase is inverted. With $R_1$ at the (+) input and ground at the (−) input the output phase will be non-inverted

Before checking the advantages and disadvantages of the four amp configurations for MM and/or MC amplification purposes I'll go through the analysis of the in-amp noise model. The noise model looks as follows[55]:

## In-Amp Noise Model with 2 Different I/P Sources

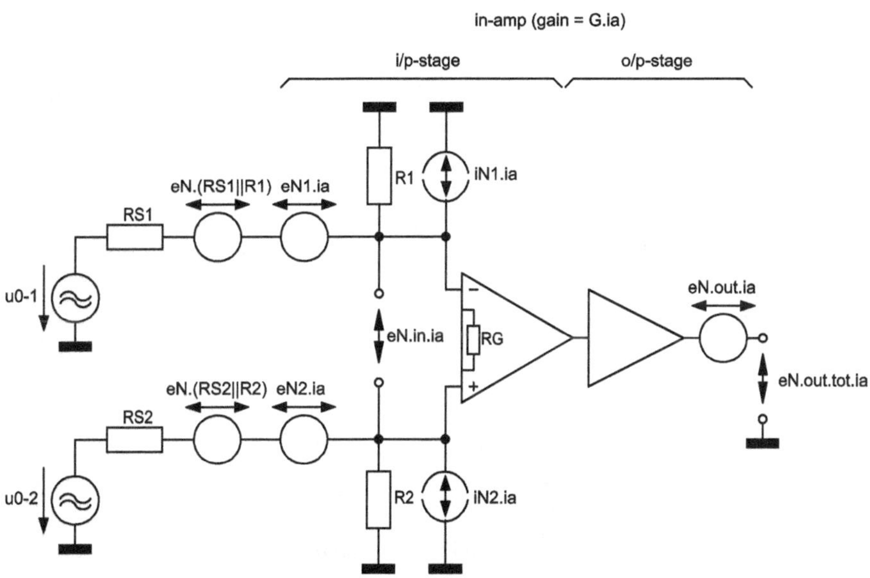


**Fig. 3.59** In-amp noise model with two different input sources

[55] M/C chapter 10

To make things less complicate the noise of the noise making components of the in-amp's input section is concentrated in the equivalent input noise voltages densities $e_{N1.ia}$ and $e_{N2.ia}$ and noise current densities $i_{N1.ia}$ and $i_{N2.ia}$. Sometimes indicated in the data sheets there exists a further noise source in the output section of the in-amp that produces the output stage noise voltage $e_{N.out.ia}$. If this noise source is not indicated in the data sheets we can expect that it's effects will be covered by $e_{N1.ia}$ and $e_{N2.ia}$.

The noise gain $G_{N.ia}$ of the in-amp equals the signal gain $G_{ia}$ set by $R_G$, thus, the total output noise voltage $e_{N.out.tot.ia}$ becomes:

$$\begin{aligned} e^2_{N.out.tot.ia} &= e^2_{N.in.tot.ia} G^2_{N.ia} + e^2_{N.out.ia} \\ &= e^2_{N.in.tot.ia} G^2_{ia} + e^2_{N.out.ia} \end{aligned} \tag{3.247}$$

The total equivalent input noise voltage density $e_{N.in.tot.ia}$ becomes:

$$\begin{aligned} e^2_{N.in.tot.ia} = {} & e^2_{N1.ia} + e^2_{N2.ia} + e^2_{N.(RS1||R1)} + e^2_{N.(RS2||R2)} \\ & + (i_{N1.ia}(RS_1||R_1))^2 + (i_{N2.ia}(RS_2||R_2))^2 + \left(\frac{e_{N.out.ia}}{G_{N.ia}}\right)^2 \end{aligned} \tag{3.248}$$

Because of the IC-internal symmetrical set-up of in-amps the following assumptions are valid. With:

$$e_{N1.ia} = e_{N2.ia} \tag{3.249}$$

$$i_{N.ia} = i_{N1.ia} = i_{N2.ia} \tag{3.250}$$

Note: $i_{N.ia}$ is the in-amp's data sheet value for the equivalent input noise current density [rtHz] at a certain frequency – mostly 1 kHz.

And with:

$$e_{N.ia} = \sqrt{e^2_{N1.ia} + e^2_{N2.ia}} \tag{3.251}$$

Note: $e_{N.ia}$ is the in-amp's data sheet value for the equivalent input noise voltage density [rtHz] at different frequencies.

And with:

$$\begin{aligned} R_1 &= R_2 \\ RS_1 &= RS_2 \\ R_0 &= R_1||RS_1 = R_2||RS_2 \\ U_0 &= u_0 - 1 = u_0 - 2 \end{aligned} \tag{3.252}$$

Eq. (3.248) for $e^2_{N.in.tot.ia}$ becomes a bit more handy:

$$e^2_{N.in.tot.ia1} = e^2_{N.ia} + 2(i_{N.ia} R_0)^2 + 2(e_{N.R0})^2 + \frac{e^2_{N.out.ia}}{G^2_{ia}} \tag{3.253}$$

thus, as well simplifying the in-amp noise model of Fig. 3.59 to:

## In-Amp Noise Model with 2 Equal I/P Sources

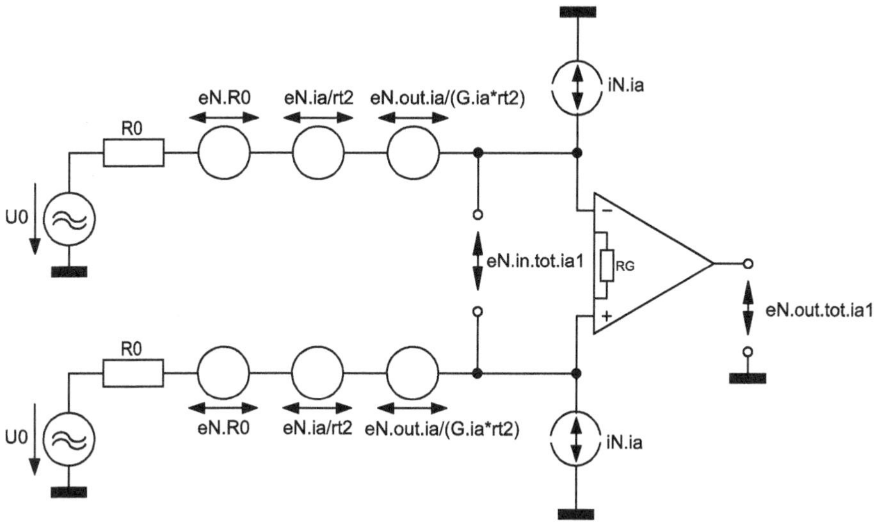


**Fig. 3.60** Simplified in-amp noise model for two equal input sources = input configuration version 1 ($rt_2 = \sqrt{2}$)

## In-Amp Noise Model with Floating I/P Source

Now, let's assume that there is no 2nd source and $R_0$ at the upper (−) input is connected to the lower (+) input of the in-amp as shown in Fig. 3.61, thus, creating a floating source connected to a balanced input. Hence, the noise model becomes modified à la Fig. 3.61.

We know from Eq. (3.25) that two equal and independent noise current sources $i_N$ working on one resistor form a sequence of noise current sources, thus, resulting in a noise current source which is $\sqrt{2}$ smaller than $i_N$. Hence, $i_{N.ia.res}$ becomes:

$$i_{N.ia.res} = \frac{i_{N.ia}}{\sqrt{2}} \tag{3.254}$$

Consequently, for the in-amp input configuration of Fig. 3.61 the total input noise voltage density $e_{N.in.tot.ia2}$ can be written like:

$$e^2_{N.in.tot.ia.2} = e^2_{N.ia} + \left(\frac{i_{N.ia}}{\sqrt{2}}\right)^2 R_0^2 + e^2_{N.RO} + \frac{e^2_{N.out.ia}}{G^2_{ia}} \tag{3.255}$$

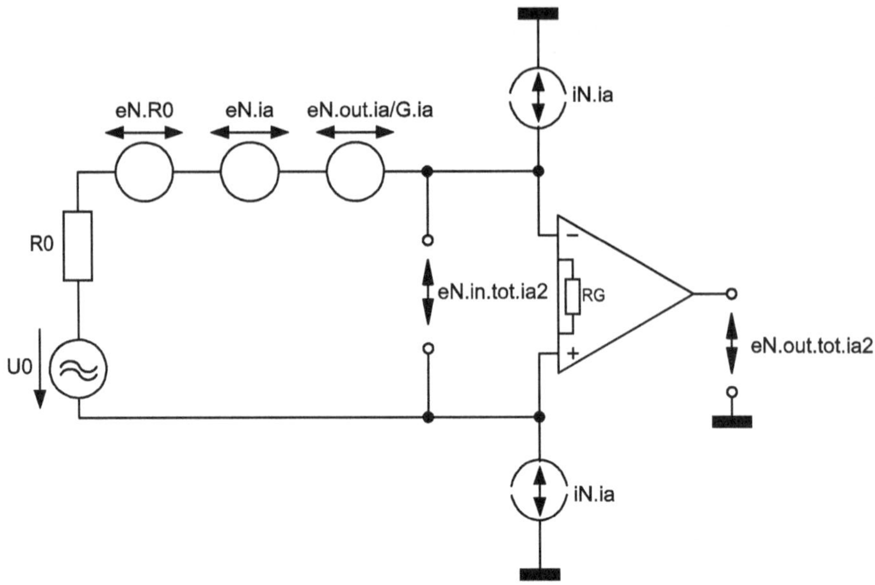


**Fig. 3.61** Noise model of a in-amp with floating (balanced) input source = input configuration version 2

## In-Amp Noise Model with Un-Bal. I/P Source

There exists a 3rd possibility to form an input configuration: in Fig. 3.61 the (+) input is connected to ground, thus forming an un-balanced input configuration.

What does this mean for the equivalent input noise voltage density $e_{\mathrm{N.in.tot.ia3}}$? The following equation and Fig. 3.62 will give the answer:

$$e^2_{\mathrm{N.in.tot.ia.3}} = e^2_{\mathrm{N.ia}} + i^2_{\mathrm{N.ia}} R_0^2 + e^2_{\mathrm{N.R0}} + \frac{e^2_{\mathrm{N.out.ia}}}{G^2_{\mathrm{ia}}} \tag{3.256}$$

Together with a 4th input configuration version (Fig. 3.60 with two times $1/2\,R_0$ instead of $R_0$) the noise voltage formulae of all versions give some interesting hints:

1st: version 2 (Fig. 3.61) equivalent input noise voltage density equals that of version 4 [this is also a practical prove of Eq. (3.21)]:

$$e_{\mathrm{N.in.tot.ia4}} = e_{\mathrm{N.in.tot.ia2}} \tag{3.257}$$

2nd: versions 2 & 4 are less noisy than version 1 (Fig. 3.60):
3rd: versions 2 & 4 are less noisy than version 3 (Fig. 3.62):
4th: version 3 is less noisy than version 1:

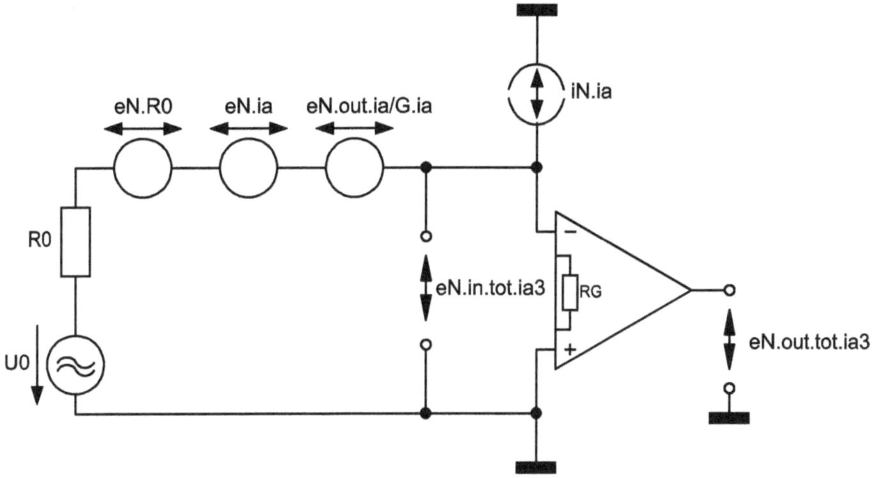


**Fig. 3.62** Noise model of a in-amp with un-balanced input source = input configuration version 3

## Example Calculation for Floating I/P Source

Two example calculations with the in-amp SSM2017 will give better understanding:

- Example case 1 – floating input source:

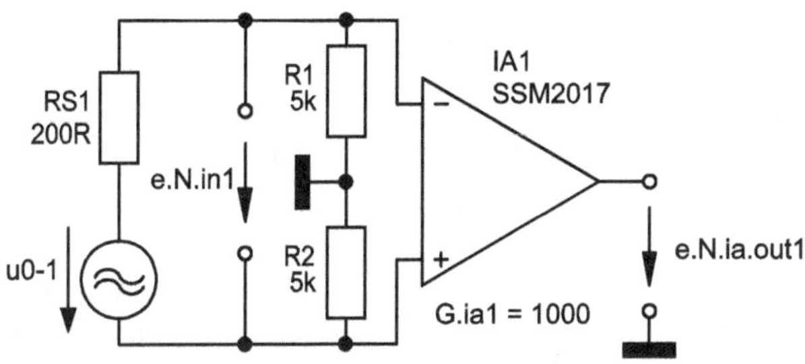


**Fig. 3.63** Typical in-amp gain stage with floating input load and grounded biasing input resistors $R_1, R_2$

With Eq. (3.255) (without $1/f$ effects and with specs from the data sheet) the calculated equivalent input noise voltage density for the whole gain stage ($G =$ 1000) with a SSM2017[56] will become:

$$e_{\mathrm{N.in1.ssm2017}} = \sqrt{\begin{array}{l}(0.95\,\mathrm{nV/rtHz})^2 + 4kT\left(\frac{200\,\mathrm{R}\times 10\,\mathrm{k}}{200\,\mathrm{R}+10\,\mathrm{k}}\right)\\ +\frac{(2\,\mathrm{pA/rtHz})^2}{2}\left(\frac{200\,\mathrm{R}\times 10\,\mathrm{k}}{200\,\mathrm{R}+10\,\mathrm{k}}\right)^2\end{array}}$$

$$e_{\mathrm{N.in1.ssm2017}} = 2.06\,\mathrm{nV/rtHz} \tag{3.258}$$

[56] Analog Devices SSM2017 data sheet, Audio Video Reference Manual 1992

$$NF_{\text{e1.ssm2017}} = 1.144\,\text{dB} \tag{3.259}$$

$$SN_{\text{ne1.ssm2017}} = -64.712\,\text{dB (ref. } 0.5\,\text{mV}_{\text{rms}}/1\,\text{kHz}, B_{20\,\text{k}}) \tag{3.260}$$

### Example Calculation for Grounded I/P Source

- Example case 2 – grounded input source:

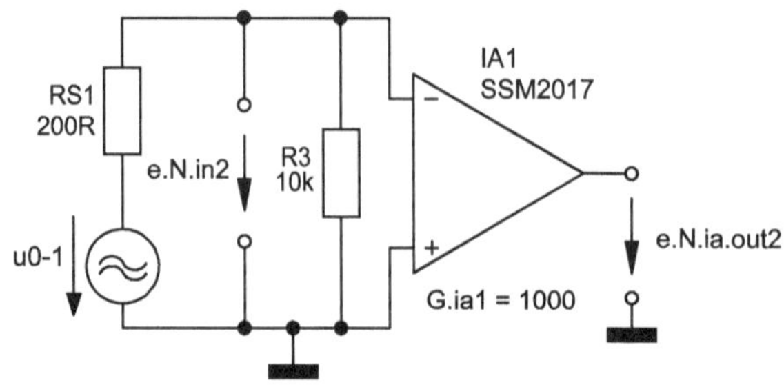


**Fig. 3.64** In-amp gain stage with grounded input source

$$e_{\text{N.in2.ssm2017}} = \sqrt{(0.95\,\text{nV/rtHz})^2 + 4kT\frac{200\,\text{R}\times 10\,\text{k}}{200\,\text{R}+10\,\text{k}} + (2\,\text{pA/rtHz})^2\left(\frac{200\,\text{R}\times 10\,\text{k}}{200\,\text{R}+10\,\text{k}}\right)^2}$$

$$e_{\text{N.in2.ssm2017}} = 2.08\,\text{nV/rtHz} \tag{3.261}$$

$$NF_{\text{e2.ssm2017}} = 1.223\,\text{dB} \tag{3.262}$$

$$SN_{\text{ne2.ssm2017}} = -64.634\,\text{dB(ref. } 0.5\,\text{mV}_{\text{rms}}/1\,\text{kHz}, B_{20\,\text{k}}) \tag{3.263}$$

The result of the 2nd example will also be the case if you replace the in-amp with an op-amp with the same noise data. The difference between Eqs. (3.260) and (3.263) becomes only 0.078 dB! But the advantage of the example case 1 concerning common mode rejection can't be beaten by example case 2.

The following four figures will show how calculation results will be influenced by changing RS from 1 R to 50 k in 5 R steps and a gain change from 1000 to 10 in one step. Changing the gain from 1000 to 10 means that the equivalent input noise voltage density jumps from 0.95 nV/rtHz to 11.83 nV/rtHz. Hence, all figures become drastically worse. There is no indication for a change in equivalent input noise current density because there is no change in collector current of the input transistors of the IC.

But, when changing the gain with $R_G$ another thing doesn't change as well: the feedback resistors of the 1st gain stage of the IC. These resistors – in conjunction with $R_G$ – are the reason for the jump of the noise voltage. In addition, it will jump much higher when changing the gain from 1000 to 1: from 0.95 nV/rtHz to 107.14 nV/rtHz. It has to do with the typical circuit topology of such ICs and it will be explained after the following figures[57]:

[57] created with Eq. (3.50) and the approach to get SN that is described by Eqs. 3.203 ... 3.204): solid traces = floating input load – dotted traces = grounded input load

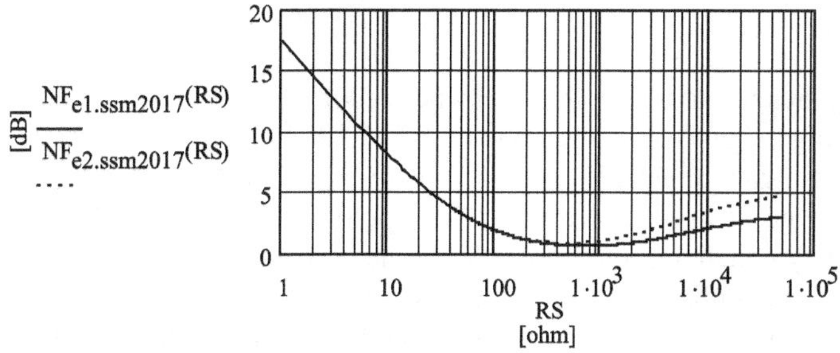


**Fig. 3.65** *NF* of circuits of Figs. 3.63 … 3.64 with gain of 1000

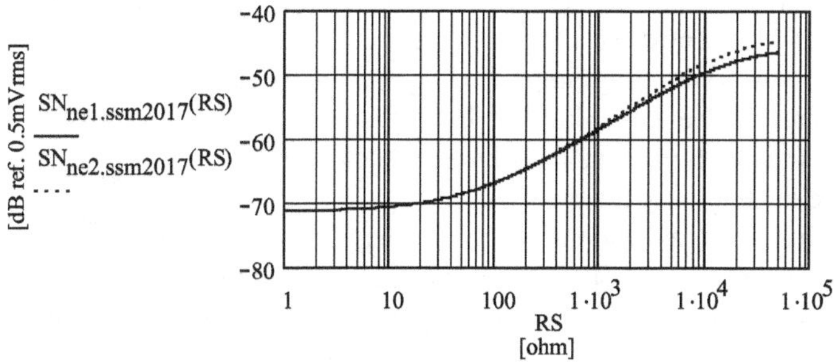


**Fig. 3.66** *SN* of circuits of Figs. 3.63 … 3.64 with gain of 1000

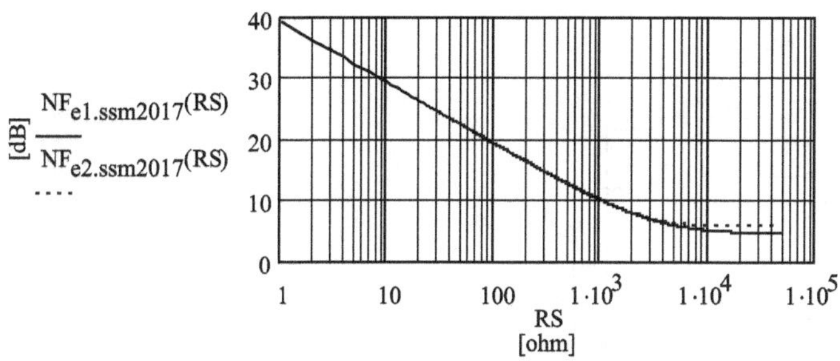


**Fig. 3.67** *NF* of circuits of Figs. 3.63 … 3.64 with gain of 10

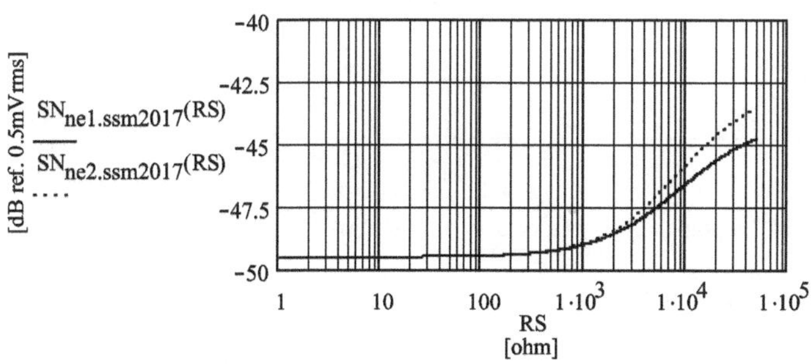


**Fig. 3.68** *SN* of circuits of Figs. 3.63 … 3.64 with gain of 10

## In-Amp IC Circuitry Topologies

The above given figures tell us a simple story: when talking about amplification of MM and MC cartridge signals in RS-regions of 1 R ... 40 R and 1 k ... 20 k we need very high in-amp gain to get very good results for NF and SN. But, it's a fact and major disadvantage of in-amps based on the topology of the SSM 2017 that we can't integrate any equalizing impedance into a negative feedback path. There is none which could be configured from the outside of the IC! And, if we would take a rather low gain to feed a passive network for the RIAA equalization we'll run into the overload trap >5 kHz. Therefore: the ideal in-amp should have access to the whole feedback circuitry – and not only to the gain setting part of it. How can this be reached? An answer will be given right after the explanations on the different IC circuit topologies that follow next with Figs. 3.69 and 3.70.

Both types consist of a summing amp at the output with gain $G_{sum} = 1$. The inputs are very different: type 1 is a kind of hybrid stage, type 2 is made of two op-amps. Characteristically, both have resistive feedback paths ($Rf_1$, $Rf_2 - R_1$, $R_2$) with access from the outside only for the INA103[59].

The equivalent noise voltages $e_n$ at the inputs are heavily influenced by the equivalent input noise current $i_{N.in\text{-}amp}$ that "flows" through these feedback resistors parallel to $R_G$. If $R_G \ll Rf_1$, $Rf_2$, $R_1$, $R_2$, thus creating a very high gain, than, as result

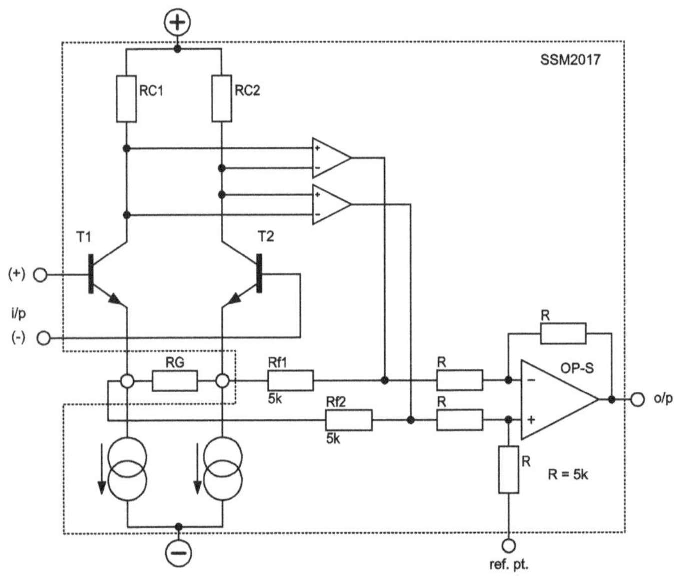


**Fig. 3.69** Basic in-amp IC topology type 1 (special audio in-amp)[58]

[58] Data sheet Analog Devices

[59] according to the INA103 data sheet comments it makes no sense to use these outputs for any equalizing purpose nor to try to decrease drastically the feedback resistors $R_1$ and $R_2$

[60] Data sheet Texas Instruments (Burr-Brown)

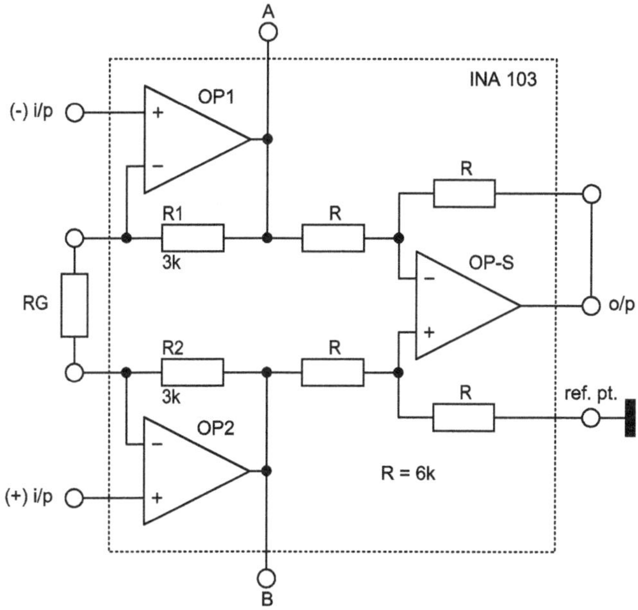


**Fig. 3.70** Basic in-amp IC topology type 2 (instrumentation amp)[60]

of the multiplication of noise current with the feedback resistors plus the noise voltage of the feedback resistors, the noise voltage $e_n$ becomes rather small. Hence, the equivalent input noise voltage density of one input transistor or input op-amp $e_{n.T1/OP1}$ becomes rather small as well. The opposite becomes true if $R_G$ is chosen bigger to get lower gain. The result of that manoeuvre: the input noise voltage grows drastically. The respective formula looks like:

$$e_{N.T1/OP1} = \sqrt{e^2_{n.T1/OP1} + e^2_{N.(Rf1\text{ or }R1||\,RG)} + i^2_{N.ia}\left(Rf_1 \text{ or } R_1||R_G\right)^2} \qquad (3.264)$$

For both gain producing input devices the data sheet spec for the equivalent input noise voltage density $e_{N.in\text{-}amp}$ becomes:

$$e_{N.ia} = \sqrt{2}e_{N.T1/OP1} \qquad (3.265)$$

A solution for the above described problem can be found in:

- we don't use IC in-amps: whether those of type 1 nor those of type 2, nor ICs of the configuration type given in Fig. 3.56 – despite their enticing input noise data!
- the development of a hybrid solution à la SSM2017 with low-noise input devices (BJTs or FETs) with adequate flexibility to choose the right feedback impedance. By attaching the cartridge in the balanced mode to the input of the amp as of Fig. 3.63 a lowest-noise design will be guaranteed. A draft design for a MC phono amp is given in Fig. 3.71[61].

[61] Draft based on an Elektor Electronics design in issue 03-1991

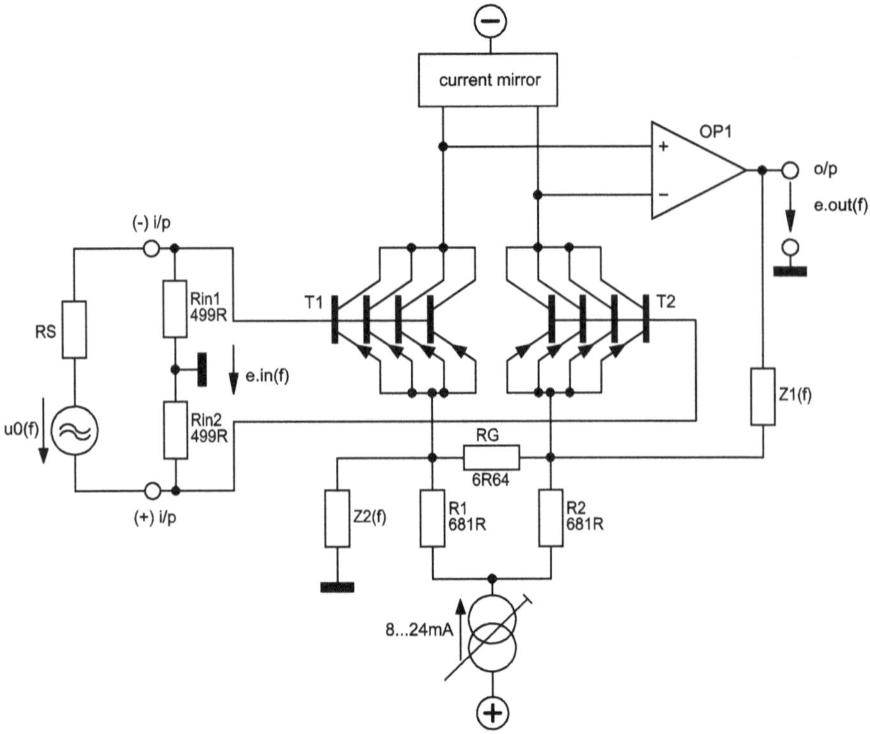


**Fig. 3.71** Draft design of a lowest-noise MC phono-amp with balanced input

## Draft Design of a Lowest-Noise In-Amp

This type of amp only works as long as both impedances $Z_1(f)$ and $Z_2(f)$ are made of identical values $(= Z(f))$ with tolerances of only $+/-1\%$, better of only $+/-$ $0.1\%$. Than, the magnitude of the gain of the draft design becomes:

$$G_{\text{dd}}(f) = \left|\frac{e_{\text{out}}(f)}{e_{\text{in}}(f)}\right| = \left|1 + \frac{2Z(f)}{R_{\text{G}}||(R_1 + R_2)}\right| \tag{3.266}$$

Equations (3.255) and (3.258) indicate what will happen to equivalent input noise current densities in use for a balanced input load: we have a sequence of two noise current sources (series connected to the input resistors). Assuming they are of identical value the resulting noise current becomes smaller by the factor of $1/\sqrt{2}$ (Eq. (3.25)). The same approach is also valid for the feedback path: two noise current sources working on the feedback resistors series connected.

If we would take for $T_1$ and $T_2$ a four transistor array like a THAT 320 (bold traces) or a THAT 300 (dotted traces) and fix $I_{\text{C}} = 1\,\text{mA}$ for each $1/4$ transistor, than, with

- the data sheet figures for these transistor arrays
- $|Z_1(f)| = |Z_2(f)| = 327\,\mathrm{R1}$
- the formulae given in the preceding sections

$NF_e$ and $SN_{ne}$ (ref. $0.5\,\mathrm{mV_{rms}}/1\,\mathrm{kHz}$, $B_{20\,k}$) vs. RS plots will look like:

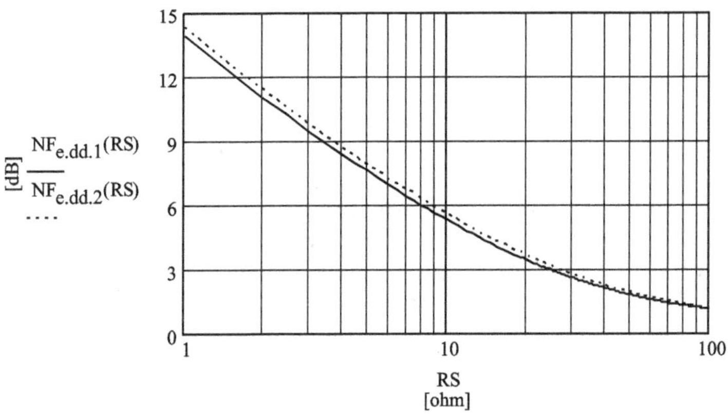


**Fig. 3.72** $NF_e$ for the in-amp draft design (*THAT 300: dotted, THAT 320: solid*)

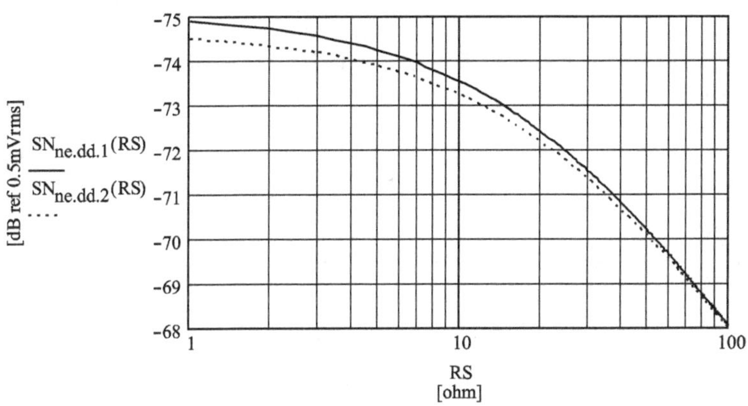


**Fig. 3.73** $SN_{ne}$ for the in-amp draft design (*THAT 300: dotted, THAT 320: solid*)

If we would take $2 \times 2$ SSM2210 NPN double transistors for $T_1$ and for $T_2$ at $I_C = 2.5\,\mathrm{mA}$ and $h_{FE} = 700$ for each single transistor, than, because of their higher $h_{FE}$, compared with the THAT 320 solution $NF_e$ and $SN_{ne}$ become slightly better: e.g. 0.02 dB at 20 R, 0.04 dB at 40 R. By halving values of $R_1$, $R_2$, $R_G$, $Z_1(f)$, $Z_2(f)$ a further SN improvement of 0.31 dB can be achieved.

Additional note for the above shown draft design:

To calculate $SN_{ariaa}$ for a given white noise producing source impedance (which is always the case for low-Z MC cartridges) simply add the respective *SN* factors of Eqs. (6.8) ff (given in Chap. 6) to the Fig. 3.73 values.

## 3.7 Noise in Transformers (Trafos)

### Intro

Transformers are a very good substitute for MC cartridge pre-pre-amps (Fig. 1.1). Without big development efforts they can be connected to a MM phono-amp's input (Fig. 3.74). Unfortunately, and despite the many claims one can read from time to time in test magazines, transformers substantially add noise to the cartridge-transformer-phono-amp-chain.

### Turns Ratio

The turns ratio *tr*[62] should be of a size that the MC cartridge's output voltage fulfils the input voltage requirements of the MM phono-amp (= nominal input voltage of the phono-amp). If the nominal output voltage of a MC cartridge is rated with e.g. 0.42 $mV_{rms}$ (at 1 kHz and 8 cm/s/s velocity) and the nominal input voltage of the MM phono-amp is rated with 5 $mV_{rms}$ at 1 kHz, than, the turns ratio of the transformer has to become 12 – according to the following Eq.:

$$e_{MM.in.nom} = e_{MC.out.nom} \times n \tag{3.267}$$

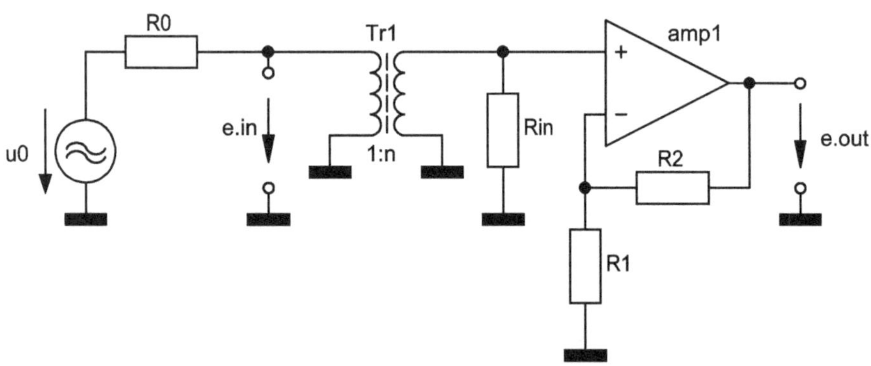


**Fig. 3.74** Step-up transformer for MC cartridge purposes connected to a MM phono-amp

[62] if primary turns have been set as "1" and the secondary turns have been set as "*n*", than the turns ratio $tr = n$

To simplify things a lot the transformers I want to talk about in this book fulfil – above all other – two basic requirements.

## Frequency and Phase Response

Within $B_{20\,\mathrm{k}}$

- their frequency response is flat ($+/-0.25$ dB max. deviation at the ends of the 3 decades) and
- their phase response is as flat as possible ($+/-5°$ max. deviation at the ends of the 3 decades)

The following figure shows typical transfer examples for a step-up transformer for use with mid-source-resistance MC cartridges with a turns ratio of $tr = 12$:

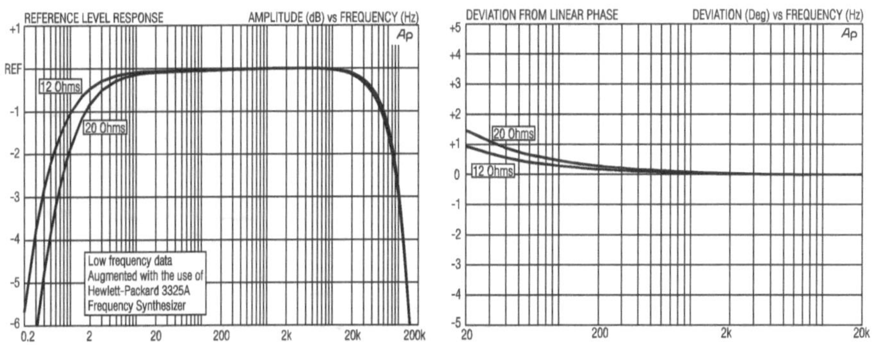


**Fig. 3.75** Frequency (*left*) and phase response (*right*) of a high-quality step-up transformer with a turns ratio of 1:12[63]

## Transformer and MC Cartridge Classification

Following their source resistances transformers and MC cartridges can be classified into three different Ω-categories:

- low-source-resistance: 1.5 Ω – 10 Ω
- mid-source-resistance: 10 Ω – 50 Ω
- high-source resistance: 50 Ω – 200 Ω

The purpose of this section will be the discussion of the respective effects on noise and gain of MC cartridge amplification via transformers.

[63] JT-347-AXT data sheet, by kind permission of Jensen Transformers, Inc., CA USA. To keep a clear overall view in this section I only refer to data of products from Jensen Transformers (JT). Of course, there exists another handful of excellent transformer manufacturers, e.g. Lundahl (Sweden), Pikatron (Germany), Sowter (U.K.), etc.

### Transformer Equations – Ideal Situation

Let's start with a look at the ideal situation of a transformer. According to the so-called transformer formulae in an equivalent circuit a step-up transformer can be replaced by the respective equivalent values for the input load resistance and/or the output load resistance.

With

- the transformer formulae for the ideal transformer (no-loss-trafo):

$$n = \frac{e_{out}}{e_{in}} = \frac{i_{in}}{i_{out}} = \sqrt{\frac{Z_{out}}{Z_{in}}} = \frac{\text{output turns}}{\text{input turns}} \tag{3.268}$$

$$P_{tr} = i_{in} \times e_{in} = i_{out} \times e_{out} \tag{3.269}$$

$$Z_{out} = n^2 \times Z_{in} \tag{3.270}$$

- and the resulting effects for the transfer of impedances from one side of the trafo to the other, demonstrated in the following figure:

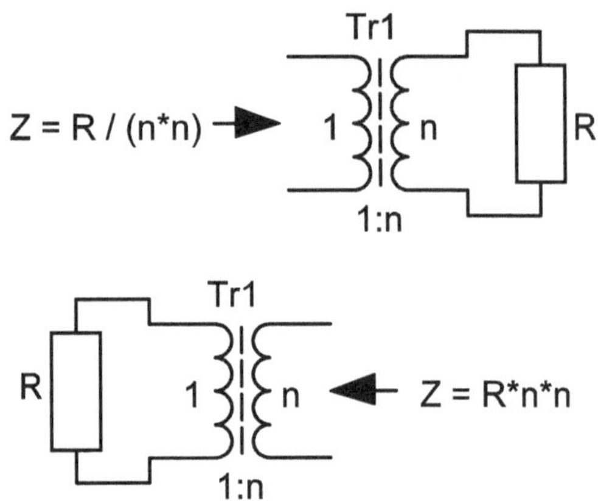


**Fig. 3.76** Impedance transfer with a transformer

- the evolution from Fig. 3.74 to an equivalent circuit to calculate any kind of voltages in an ideal environment can be described with the following figures:

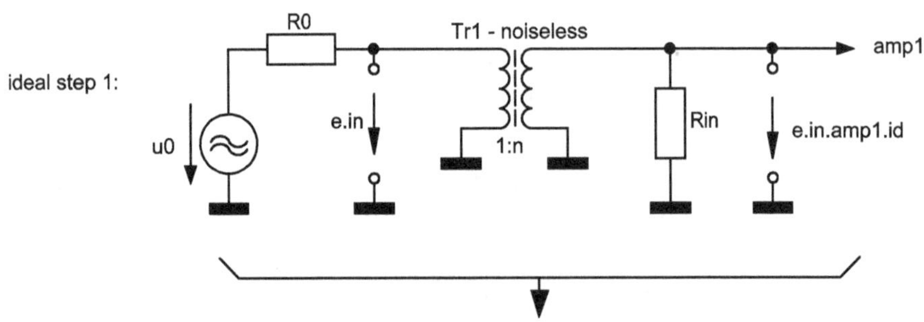


**Fig. 3.77** Ideal amp1 input situation

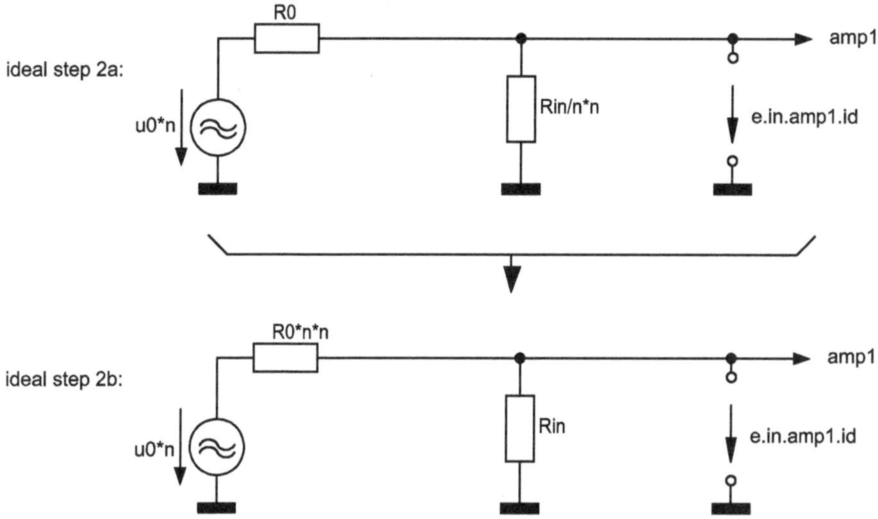


**Fig. 3.78** Equivalent circuit for amp1's input situation = re-designed Fig. 3.77
step 2a: $R_{in}$ transferred to the input side of the trafo
step 2b: $R_0$ and $u_0$ transferred to the output side of the trafo

Now, we can calculate the ideal (= id) input voltage $e_{in.amp1.id}$ of amp1 (Fig. 3.78, step 2a or 2b):

$$\begin{aligned} \text{step 2a: } e_{in.amp1.id} &= u_0 \times n \times, \left(\tfrac{R_{in}}{n^2 R_0 + R_{in}}\right) \\ \text{step 2b: } e_{in.amp1.id} &= u_0 \times n \times \left(\tfrac{R_{in}/n^2}{R_0 + R_{in}/n^2}\right) \end{aligned} \quad (3.271)$$

For $u_0$ the ideal gain $G_{id}$ of the whole arrangement in Fig. 3.74 becomes:

$$G_{id} = \frac{e_{out}}{u_0} = \frac{1}{u_0}\left(1 + \frac{R_2}{R_1}\right) e_{in.amp1.id} \quad (3.272)$$

$$G_{id} = nG_{amp1}\left(\frac{R_{in}}{n^2 R_0 + R_{in}}\right) \quad (3.273)$$

If $R_{in} \gg n^2 R_0$, than, the ideal gain $G_{id}$ for $u_0$ would become:

$$G_{id} = nG_{amp1} \quad (3.274)$$

hence, amp1' s gain must be adjusted to compensate the gain loss of the transformer connected to a lower valued $R_{in} > n^2 R_0$ by:

$$G_{loss.id} = \frac{n^2 R_0 + R_{in}}{R_{in}} \quad (3.275)$$

$$G_{amp1.adj} = G_{amp1} G_{loss} \quad (3.276)$$

**Transformer Eqs. – Real Situation**

The real (= re) situation looks a bit different, because of the facts that the coil windings of the trafo are resistances ($R_s$ and $R_p$) and there exist some other unpleasant characteristics like trafo inductances and capacitances.

The good news: as long as the transfer plots look like the ones in Fig. 3.75 and we do not leave the $B_{20\,k}$ bandwidth, than, not making a big mistake we can ignore any inductance and capacitance.

The bad news: primary and secondary coil resistance play a major role and have to be taken into any noise calculation course. The following figure and Eqs. will show how to handle this.

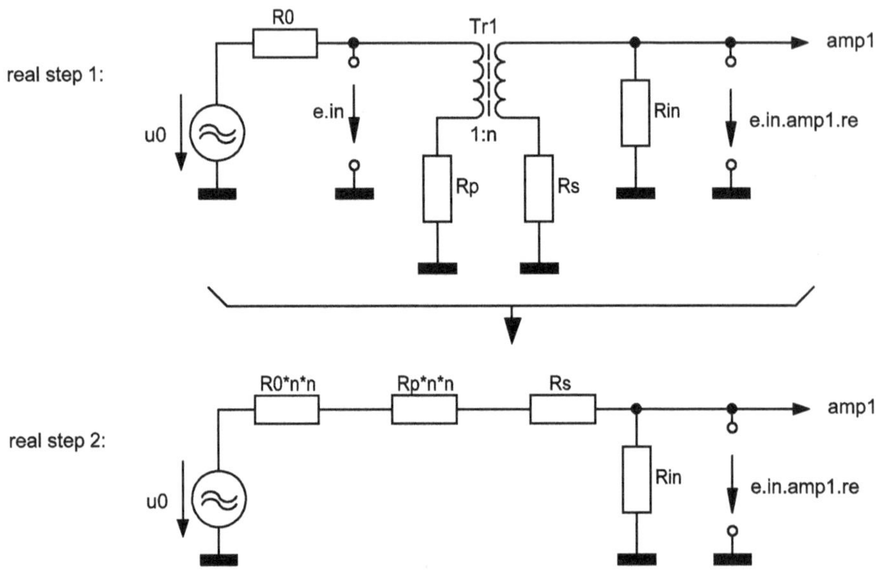


**Fig. 3.79** Real situation of a source $u_0$ connected to amp1 via transformer $Tr_1$; step 1: situation with ideal transformer $Tr_1$ plus it's coil resistances $R_p$ and $R_s$; step 2: transfer into an equivalent circuit

Derived from Eqs. (3.271 ... 3.276) and with $R_p$ = primary coil resistance and $R_s$ = secondary coil resistance we can formulate the equations for the real situation:

$$e_{\text{in.amp1.re}} = n \times u_0 \times \left( \frac{R_{\text{in}}}{R_{\text{in}} + R_s + n^2 R_p + n^2 R_0} \right) \tag{3.277}$$

$$G_{\text{loss.re}} = \frac{R_{\text{in}} + R_s + n^2 R_p + n^2 R_0}{R_{\text{in}}} \tag{3.278}$$

$$G_{\text{amp1.adj.re}} = G_{\text{amp1}} G_{\text{loss.re}} \tag{3.279}$$

and the overall gain $G_{re}$ for $u_0$ becomes:

$$G_{re} = \frac{e_{out}}{u_0} = nG_{ampl.adj.re} \tag{3.280}$$

### *SN* Example Calculations

To calculate noise effects and *SN*s of Fig. 3.74 we need the following data:

- MC cartridge: $u_{0nom}$, $R_0$, $R_{L.nom}$
- transformer: $n$, $R_s$, $R_p$, $e_{in.nom.tr1}$
- ampl: $R_{in}$, $R_1$, $R_2$, $e_N$, $i_N$, $e_{in.nom.ampl}$

Basically, there are two different approaches possible to calculate noise voltage and *SN*s. Both will lead to the same results.

- **approach 1:** "noise voltage approach", useful for any kind of resistive input load
- **approach 2:** "noise figure approach", useful for the calculation for a specific MC cartridge

To demonstrate the two approaches I calculate one real **Example** with a Jensen transformer JT-44-AX [64]. The input situation looks as follows:

- MC cartridge[65]:
  $u_{0nom} = 0.56\,\text{mV}_{rms}$
  $R_{L.nom} = 1\,\text{k}$ — nominal load resistance (min.: 100 R)
  $R_0 = 40\,\text{R}$
- transformer:
  $R_s = 950\,\text{R}$
  $R_p = 3\,\text{R}$
  $n = 10$
- ampl (topology of Fig. 3.24a or b):
  $R_{in} = 47\,\text{k5}$
  $R_1 = 130\,\text{R}$
  $R_2 = 24 \times 130\,\text{R}$
  $e_{N.ampl} = \sqrt{2} \times 1.54\,\text{nV/rtHz}$ — long tailed pair of 2SC2546
  $i_{N.ampl} = 0.23\,\text{pA/rtHz}$ — $I_C = 100\,\mu\text{A}, h_{FE} = 600$
- trafo-amp-chain:
  $e_{in.nom.tr1} = 0.5\,\text{mV}_{rms}$ — at the input of $Tr_1$, generator source resistance $R_{gen} = 0\,\text{R}$

### The Noise Voltage Approach

**Approach 1 goes like follows:**
Derived from Fig. 3.79 the total resistance at the input of ampl becomes:

$$R_{in.tot.40} = \left(n^2(R_0 + R_p) + R_s\right) || R_{in} = 4727\,\Omega \tag{3.281}$$

[64] JT-44-AX data sheet

thus, with $R_f = R_1||R_2$ the total input referred equivalent noise voltage density $e_{N.tot.40}$ at the input of $Tr_1$ becomes:

$$e_{N.tot.40} = \frac{\sqrt{2e_N^2 + i_N^2 R_{in.tot}^2 + 4kTR_{in.tot}B_1 + 4kTR_f B_1 + i_N^2 R_f^2}}{n} \tag{3.282}$$

$$e_{N.tot.40} = 0.93\,\mathrm{nV/rtHz} \tag{3.283}$$

with the calculation of $G_{loss}$

$$G_{loss.40} = \frac{R_{in} + R_s + n^2 R_p + n^2 R_0}{R_{in}} = 1.111 \tag{3.284}$$

$$G_{e.loss.40} = 20\log(G_{loss}) = 0.911\,\mathrm{dB} \tag{3.285}$$

we can calculate $SN_{ne.40}$ for the 40 R source:

$$SN1_{ne.40} = 20\log\left(\frac{\sqrt{\int\limits_{20\mathrm{Hz}}^{20{,}000\,\mathrm{Hz}} e_{N.tot.40}^2\,\mathrm{d}f}}{e_{in.nom.tr1}}\right) + G_{e.loss.40} \tag{3.286}$$

$$SN1_{ne.40} = -70.702\,\mathrm{dB} \tag{3.287}$$

the RIAA-equalized and A-weighted $SN_{ariaa}$ becomes:

$$SN1_{ariaa.40} = -SN1_{ne.40} - 7.935\,\mathrm{dB} = -78.637\,\mathrm{dB} \tag{3.288}$$

The derivation of the RIAA-equalizing and A-weighting term "7.935 dB" is based on calculations by application of Eqs. (3.56 … 3.58) and will be explained later on in Chap. 6 "Noise in MC Phono-Amps".

If we would change $R_0$ from 1.0 R to 200 R the results for $e_{N.tot}(R_0)$ and $SN_{ariaa}(R_0)$ of the above given example would look as demonstrated in Figs. 3.80 and 3.81.

It must be noticed that the chosen transformer JT-44-AX is a special device for mid-source-resistances. That's why for low-source-resistances another transformer type would create even better SN results (e.g. JT-347-AXT with $n = 12$).

Example values for $R_0 = 10\,\mathrm{R}$:

- $SN_{ariaa.10}$(JT-44-AX) $= -82.200\,\mathrm{dB}$
- $SN_{ariaa.10}$(JT-347-AXT) $= -82.987\,\mathrm{dB}$
- improvement $= +0.787\,\mathrm{dB}$

### The Noise Figure Approach

**Approach 2 goes like follows:**
This approach starts with the signal-to-noise ratio calculation of $R_0$ in $B_{20\,k}$ with reference to $e_{in.nom.tr1}$, followed by a noise figure calculation given in the Jensen

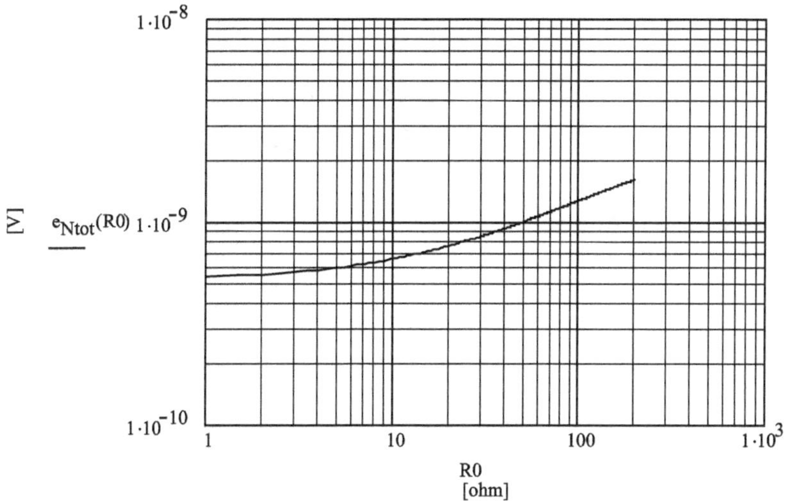


**Fig. 3.80** Total equivalent noise voltage density vs. $R_0$ at the input of $Tr_1$ (created with Eq. (3.282))

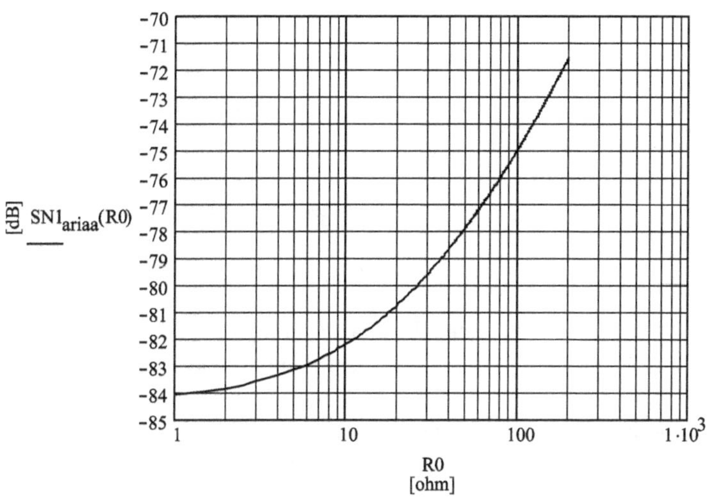


**Fig. 3.81** Signal-to-Noise-ratio $SN_{\text{ariaa}}(R_0)$ vs. $R_0$ in $B_{20\,\text{k}}$ with reference to the nominal input voltage $e_{\text{in.nom}} = 0.5\,\text{mV}_{\text{rms}}/1\,\text{kHz}$ of the transformer-amp1-chain (created with Eq. (3.288))

application note AS040[66] and a noise figure calculation of amp1 and, of course, a $G_{\text{loss}}$ $(= G_{\text{red}})$ calculation to meet the requirements which result from the specific AS040-NF-calculation results vs. the open-circuit voltage $u_{00}$ of $u_{0\text{nom}}$.

The calculation course starts with the input source resistance of the MC cartridge $R_0$:

---

[66] AS040, Jensen Transformers, Inc., Ca. USA

- $SN_{\text{ne}}$ calculation in $B_{20\,\text{k}}$:

$$e_{\text{N.20k.40}} = \sqrt{4kT40RB_{20\,\text{k}}} \tag{3.289}$$

$$SN2_{\text{ne.40}} = 20\log\left(\frac{e_{\text{N.20k.40}}}{e_{\text{in.nom.tr1}}}\right) = -72.760\,\text{dB} \tag{3.290}$$

- NF calculations a là AS040:
  $Tr_1$ impedance noise effects (re = real, id = ideal):

$$Z_{\text{out.re.40}} = \left(n^2(R_0 + R_\text{p}) + R_\text{s}\right)||R_\text{in} = 4727\,\Omega \tag{3.291}$$

$$Z_{\text{out.id.40}} = R_\text{in}||n^2R_0 = 3689\,\Omega \tag{3.292}$$

$$NF1_{\text{e.40}} = 20\log\left(\sqrt{\frac{Z_{\text{out.re.40}}}{Z_{\text{out.id.40}}}}\right) = 1.077\,\text{dB} \tag{3.293}$$

  $Tr_1$ signal level effects:

$$f_{\text{out.id.40}} = \frac{R_\text{in}}{R_\text{in} + n^2R_0} = 0.922 \tag{3.294}$$

$$f_{\text{out.re.40}} = \frac{R_\text{in}}{R_\text{in} + n^2(R_\text{p} + R_0) + R_\text{s}} = 0.900 \tag{3.295}$$

$$NF2_{\text{e.40}} = 20\log\left(\frac{f_{\text{out.id.40}}}{f_{\text{out.re.40}}}\right) = 0.208\,\text{dB} \tag{3.296}$$

  Total NF of $Tr_1$:

$$NF3_{\text{e.40}} = NF1_{\text{e.40}} + NF2_{\text{e.40}} = 1.285\,\text{dB} \tag{3.297}$$

- By the way: Calculation of $NF_3$ is the ideal tool to compare different trafo models – when in search for the right one for a specific input load situation of amp1 as well as of $Tr_1$!
- calculation of the loss $G_{\text{e.red.40}}$ that occurs when connecting a MC cartridge to a load resistance that is bigger or smaller than the rated nominal load resistance $R_{\text{L.nom}}$:
  based on the impedance transfer of $R_\text{in}$ to the input of $Tr_1$ (Fig. 3.76) the load resistance $R_{\text{L.red}}$of the MC cartridge becomes:

$$R_{\text{L.red}} = \frac{R_\text{in} + R_\text{s} + n^2R_\text{p}}{n^2} = 487.5\,\Omega \tag{3.298}$$

  with $R_{\text{L.red}}$ (instead of $R_{\text{L.nom}}$) as load the nominal generator voltage $u_{0\text{nom}}$ of the MC cartridge changes as well and becomes – in this case – the reduced value $u_{0\text{red}}$:

$$u_{0\text{red}} = u_{0\text{nom}}\frac{(R_{\text{L.nom}} + R_0)R_{\text{L.red}}}{R_{\text{L.nom}}(R_{\text{L.red}} + R_0)} = 0.54\,\text{mV}_\text{rms} \tag{3.299}$$

thus, this extra loss $G_{e.red.40}$ becomes:

$$G_{e.red.40} = 20\log\left(\frac{R_{L.nom}(R_{L.red}+R_0)}{R_{L.red}(R_{L.nom}+R_0)}\right) = 0.344\,\text{dB} \tag{3.300}$$

- calculation of $NF_e$ of amp1:

$$NF_{e.amp1.40} = 20\log\left(\sqrt{\frac{2e_N^2 + i_N^2 Z_{out.re.40}^2 + 4kT(Z_{out.re.40})B_1 + 4kTR_fB_1 + i_N^2 R_f^2}{4kT(Z_{out.re.40})B_1}}\right) \tag{3.301}$$

$$NF_{e.amp1.40} = 0.422 \tag{3.302}$$

- calculation of $\text{SN2}_{ariaa}$:

$$SN2_{ariaa.40} = SN2_{ne.40} - 7.935\,\text{dB} + NF3_{e.40} + G_{e.red.40} + NF_{e.amp1.40} \tag{3.303}$$

$$SN2_{ariaa.40} = -78.643\,\text{dB} \tag{3.304}$$

- Remember Section 3.1: NFs are a measure that describes the additional noise contribution of an amp stage – whereas G's describe voltage divider situations which have to be balanced to the nominal input and/or output voltage values by additional amp gain somewhere in the whole amp-chain!

### Delta of Approach 1 and 2

$$\text{Delta}_{SN} = |SN1_{ariaa.40} - SN2_{ariaa.40}| = 0.00685\,\text{dB} \tag{3.305}$$

The following figure shows the evolution of $\text{Delta}_{SN}$ from $R_0 = 1\,\text{R} \ldots\ 200\,\text{R}$:

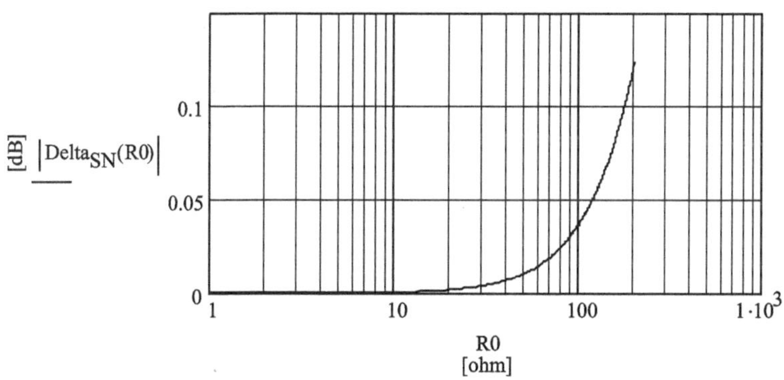


**Fig. 3.82** *SN*-result deltas of the two approaches 1&2 for transformer coupled amp inputs with changing input load $R_0$

## Questions & Answers

In conjunction with trafos other questions might be:

### Q1

- Q1: What' the role of the optimal source resistance $R_{opt}$ for amp1?

Lowest *SN*s can only be achieved as long as $R_{opt}$ of amp1 equals the amps source resistance at the input of amp1, thus, the sum of $[n^2(R_0+R_p)+R_s]$ parallel to $R_{in}$ should be near $R_{opt}$ of that amp:

$$R_{opt.amp1} = \left(n^2(R_0+R_p)+R_s\right)||R_{in} = \frac{e_{N.amp1}}{i_{N.amp1}} \tag{3.306}$$

With the given values of the Example 1 $R_{opt.amp1}$ becomes:

$$R_{opt.amp1} = \frac{\sqrt{2}\times e_N}{i_N} = 9.469\,\mathrm{k}\Omega \tag{3.307}$$

According to the following Fig. 3.83 and Eq. (3.281) this result is far away from the required value for $R_{in.amp1.40} = 4\,\mathrm{k}727$. The evolution of $R_{in.tot}$ vs. $R_0$ looks as follows (created with Eq. (3.281)):

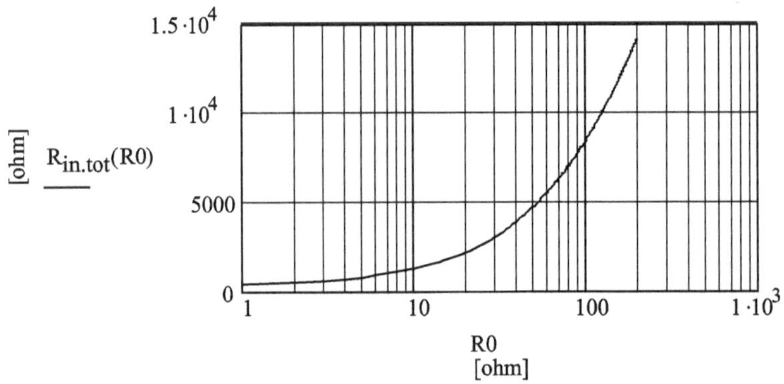


**Fig. 3.83** Input load $R_{in.tot}$ of amp1 vs. $R_0$

There exist several ways to overcome the above shown miss-match between $R_{opt.amp1}$ and $R_{in.amp1.40}$. An easy way to perform is given in the following graph and is based on a change of the collector current $I_C$ of the input devices.

This graph demonstrates several aspects. Amp1 works with $I_{C.amp1} = 100\,\mu\mathrm{A}$. $R_{opt}$ for this case would be $9\,\mathrm{k}469$. The adequate $I_C$ for $R_{opt} = 4\,\mathrm{k}797$ would be $205\,\mu\mathrm{A}$. If we would change $I_C$ from $100\,\mu\mathrm{A}$ to $205\,\mu\mathrm{A}$ the respective value for

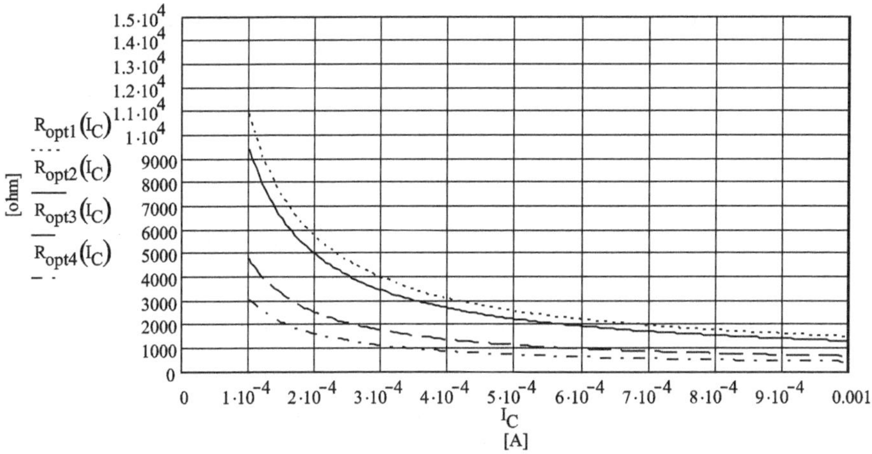


**Fig. 3.84** $R_{opt}$ vs. $I_C$ for various $h_{FE}$ values:
- $R_{opt1}(I_C)$ for $h_{FE} = 800$ (2SC2546F, $rbb' = 13$ R74)
- $R_{opt2}(I_C)$ for $h_{FE} = 600$ (2SC2546E, $rbb' = 13$ R74)
- $R_{opt3}(I_C)$ for $h_{FE} = 154$ (2SC2546D, $rbb' = 13$ R74)
- $R_{opt4}(I_C)$ for $h_{FE} = 63$ (1/4 THAT 320, $rbb' + ree' = 27$ R)

(created with Eqs. (3.69), (3.70), (3.75))

$SN_{ariaa.40}$ would become 79.530 dB. This can be picked from the solid trace in the following graph at 205 μA. Compared with Eq. (3.288) this new $SN_{ariaa.40}$ value represents an improvement of 0.893 dB!

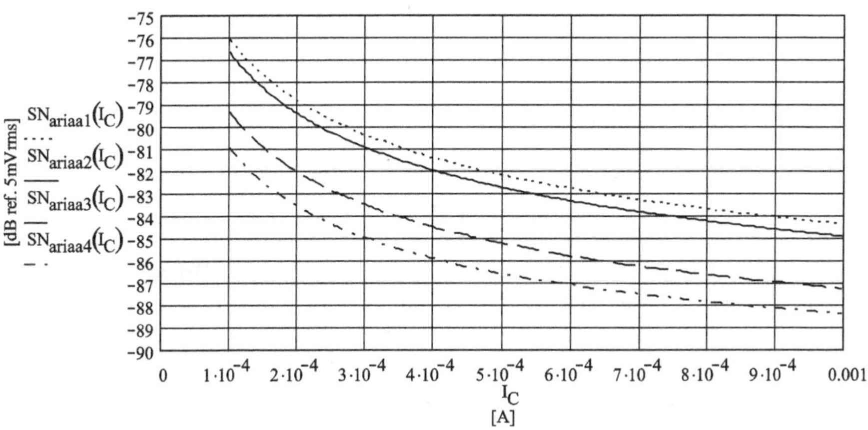


**Fig. 3.85** $SN_{ariaa.n}$ vs. $I_C$ for various $h_{FE}$ values; $n$ follows the values for $h_{FE}$ in Fig. 3.84 (created with Eq. (3.56))

I've included three other plots into the two graphs above. The reasons for that:

- to demonstrate the influence of BJT-$I_C$ and $-h_{FE}$ on $SN$s
- the plots with subscript 3 indicate a type of BJT that would fulfil the needs for $R_{opt.amp1.40} = 4\,k797$

- the plots with subscript 4 indicate that low-noise BJTs with low $h_{FE}$ don't manage situations with rather high $R_{opt}$ (e.g. BFW16A or THAT 300 family) very well.

## Q2

- Q2: What happens with SN when changing $R_1$ ($R_2$ accordingly) of Fig. 3.74 to lower or higher values than that of the **Example**?

  Figure 3.86 will give the answer.

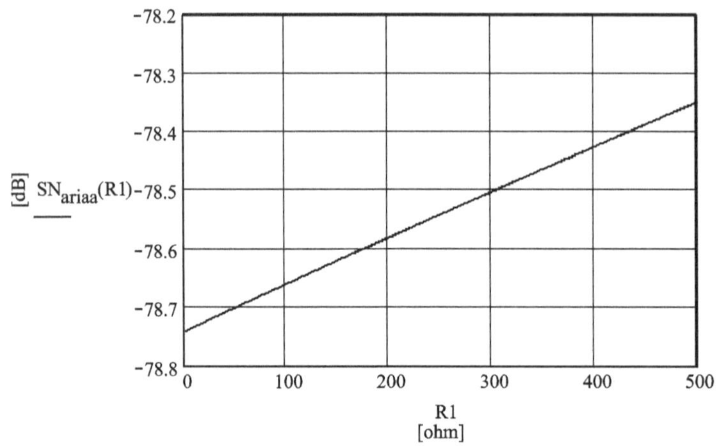


**Fig. 3.86** $SN_{ariaa.40}$ vs. $R_1$ of Fig. 3.74
- $SN_{ariaa.40}(10\,\mathrm{R}) = -78.734\,\mathrm{dB}$
- $SN_{ariaa.40}(130\,\mathrm{R}) = -78.637\,\mathrm{dB}$
- $SN_{ariaa.40}(499\,\mathrm{R}) = -78.350\,\mathrm{dB}$

10 R is app. 0.1 dB better than 130 R and this is app. 0.3 dB better than 499 R! Because of the fact that with decreasing values of $R_1$ the output of amp1 becomes more and more loaded the need to integrate a buffer at the output of amp1 will occur (with $R_1 < 25\,\mathrm{R}$ at least)! It must be underlined that these findings are only valid for purely white input noise voltages. In the case of frequency dependent noise voltages of impedances like those of MM cartridges the differences become bigger (more infos on these facts will be presented in Chap. 4 "Noise in MM Cartridges").

## Q3

- Q3: What happens if $R_2$ in Fig. 3.74 is an impedance that consists of resistors and capacitors?

In most RIAA-amplification cases the feedback path of the MM phono-amp is made of an impedance network to perform the RIAA transfer in a whole or in a certain part of it.

For noise calculation purposes you only need to calculate the magnitude of the impedance at 1 kHz and take the result of it as the value for $R_2$. Any RIAA effect on the calculated *SN*s will be taken into consideration later on as shown in Chap. 6 "Noise in MC Phono-Amps".

## Q4

- Q4: What about JFETs in a Fig. 3.24 topology? Compared with BJTs do they improve *SN*s very much?

This question will be answered by an example calculation. We take the following new **Example**[67] input devices figures for a JFET:

- $e_{N.amp1} = \sqrt{2} \times 1.25\,\mathrm{nV/rtHz}$ long tailed pair of two 1/2 2SK389
- $i_{N.amp1} = 50\,\mathrm{fA/rtHz}$ $I_D = 1\,\mathrm{mA}$

Insertion of these new values into Eqs. (3.282) ... (3.287) leads to

$$SN3_{ariaa.40} = -78.777\,\mathrm{dB} \tag{3.308}$$

which is a **0.140 dB improvement** on the result of the BJT case of Eq. (3.288).

## Q5

- Q5: What about valves? How do they influence *SN*s?

With valves we can't take the amp topology of Fig. 3.24. Instead, we'll take that of Fig. 2.7 (top or bottom) and replace OP1 ... OP2 by SRPP stages à la Fig. 3.43 and OP3 by a cathode follower.

A low-noise SRPP input stage can be built with e.g. two E188CC parallel at the bottom and two E188CC (or a pentode triode configured) at the top, hence, to answer this question with an **Example**[68] calculation we have to take the following figures for paralleled triodes as input devices:

- $e_{N.amp1} = 1.44\,\mathrm{nV/rtHz} = 2.04\,\mathrm{nV}/\sqrt{2}$
- $i_{N.amp1} = 50\,\mathrm{fA/rtHz}$ $I_a = 2 \times 2\,\mathrm{mA}$
- $R_f = 0\,\mathrm{R}$ no feedback necessary, only passive RIAA equalization useful.

Insertion of these new values into Eqs. (3.282) ... (3.287) will lead to

[67] see paragraph on "*SN* Example Calculations" on page 111

[68] see paragraph on "*SN* Example Calculations" on page 111

$$SN4_{\text{ariaa.40}} = -78.942\,\text{dB} \tag{3.309}$$

which is a **0.305 dB improvement** on the result of the BJT case of Eq. (3.288) and a **0.165 dB improvement** on the result of the JFET case of Eq. (3.308).

## Q6

- Q6: What about JFETs parallel connected?

Like in the valves case a similar input stage can be configured by putting e.g. two 1/2 2SK389 parallel, hence, for an **Example**[69] calculation we have to take the following figures for a JFET input with paralleled devices:

- $e_{\text{N.amp1}} = 0.88\,\text{nV/rtHz} = 1.25\,\text{nV}/\sqrt{2}$
- $i_{\text{N.amp1}} = 50\,\text{fA/rtHz}\ I_{\text{D}} = 1\,\text{mA}$
- $R_{\text{f}} = 0\,\text{R}$ no feedback necessary, only passive RIAA equalization

Insertion of these new values into Eqs. (3.282) … (3.287) will lead to

$$SN4_{\text{ariaa.40}} = 79.013\,\text{dB} \tag{3.310}$$

which is a **0.376 dB** improvement on the result of the BJT case of Eq. (3.288), a **0.236 dB** improvement on the result of the JFET case of Eq. (3.308) and a **0.071 dB** improvement on the result of the valve case of Eq. (3.309).

## Q7

- Q7: How does the resistor-capacity network at the output of a trafo influences *SN*s?

To reduce resonances that might occur when connecting a trafo to $R_{\text{in}} = 47\,\text{k5}$ of a MM phono-amp manufacturers recommend to include this kind of impedance into the circuit. This measure also keeps the frequency response flat – as long as the Miller-C can be kept rather low. Otherwise, the right values for the R-C sequence will become a question of trial and error.

For the noise calculations presented in this chapter it was a major prerequisite at the beginning of this section that this R-C network should keep the frequency response as flat as possible, hence, no influence on *SN*s!

## Q8

- Q8: What's the influence of the inductance of a MC cartridge on *SN*s?

---

69 see paragraph on "*SN* Example Calculations" on page 111

I decided not to take the inductance into the *SN* calculation course because it's influence is marginal. In addition, it would make calculation formulae much more complex – see the following **Example**[70] adaptation and approach 1 calculation:

**Example** with additional inductance $L_0 = 56\,\mu\text{H}$ for the DL-103:

Derived from Eqs. (3.281)... (3.288) $SN4_{\text{ariaa.40}}$ becomes:

$$Z_0(f) = |R_0 + j2\pi f L_0| \tag{3.311}$$

$$Z_{\text{in.tot.40}}(f) = \left(n^2(Z_0(f) + R_p) + R_s\right) || R_{\text{in}} \tag{3.312}$$

$$e_{\text{N.tot.40}}(f) = \frac{\sqrt{2e_N^2 + i_N^2[Z_{\text{in.tot.40}}(f)^2 + R_f^2] + 4kTB_1[Z_{\text{in.tot.40}}(f) + R_f]}}{n} \tag{3.313}$$

$$G_{\text{loss.40}}(f) = \frac{R_{\text{in}} + R_s + n^2(R_p + Z_0(f))}{R_{\text{in}}} \tag{3.314}$$

$$SN4_{\text{ne.40}} = 20\log\left(\frac{\sqrt{\int\limits_{20\text{Hz}}^{20\text{ kHz}} [e_{\text{N.tot.40}}(f) \times G_{\text{loss.40}}(f)]^2 df}}{e_{\text{in.nom.tr1}}}\right) \tag{3.315}$$

$$SN4_{\text{aiaa.40}} = SN4_{\text{ne}} - 7.935\,\text{dB} \tag{3.316}$$

$$= -78.619\,\text{dB} \tag{3.317}$$

Thus, the difference between results from Eqs. (3.288) and (3.317) is only 0.018 dB!

But I must point out that this calculation is not correct because I've treated the inductance $L$ like a resistance. But with that manoeuvre I've simplified the whole matter a bit instead of taking a rather complex mathematical path which will end up in an even smaller difference between the above mentioned equations. With much bigger inductances this path is demonstrated in more detail for the MM cartridge amp input section in the respective chapter of this book.

## Q9

- Q9: Is it allowed to change the phono amp's $R_{\text{in}}$ from the manufacturer specified load resistance to another value, e.g. from 47 k5 to 100 k or to 10 k? What happens with *SN*s?

The idea behind this question comes from the fact that – for example – the Denon DL-103 MC cartridge has a nominal load resistance of 1 k, but it can also be used with a minimum load of 100 R (according to the data sheet of the manufacturer).

[70] see paragraph on "*SN* Example Calculations" on page 111

**Specified for 47 k**

A1: Generally trafos are specified for specific load resistances. To get a flat frequency response an additional correction R-C network at the trafo's output is required as well. It is assumed that any Miller-C of the input stage does only play a marginal role. Thus, a change of the output load automatically means a change of that network. If not obtainable from the manufacturer it becomes a question of trial and error. In any case these networks do not worsen *SN*s essentially (see Q6).

Assumed we could solve the network question the influence on *SN*s for an output load resistance change from 100 k to 10 k looks as follows (see figures below). By taking the **Example**[71] calculation (with two different $R_0$s: $R_{01} = 40\,\mathrm{R}$ and $R_{02} = 10\,\mathrm{R}$) together with the respective and adapted formulae the A-weighted and RIAA-equalized $SN\mathrm{s}(R_{in})$ for the BJT input stage of the **Example** become:

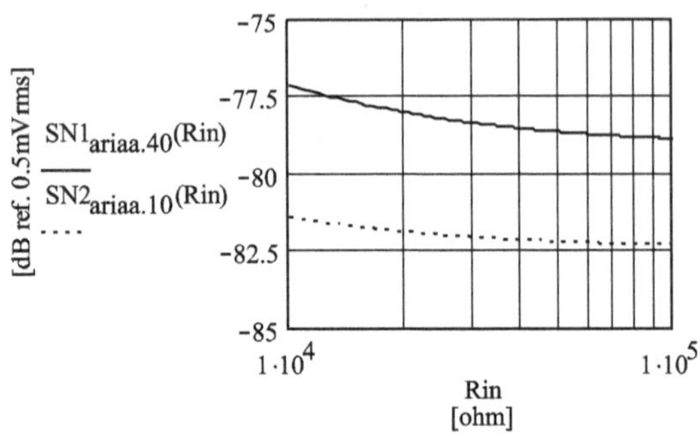


**Fig. 3.87** $SN_{ariaa.40}(R_{in})$ and $SN_{ariaa.10}(R_{in})$ vs. $R_{in}$

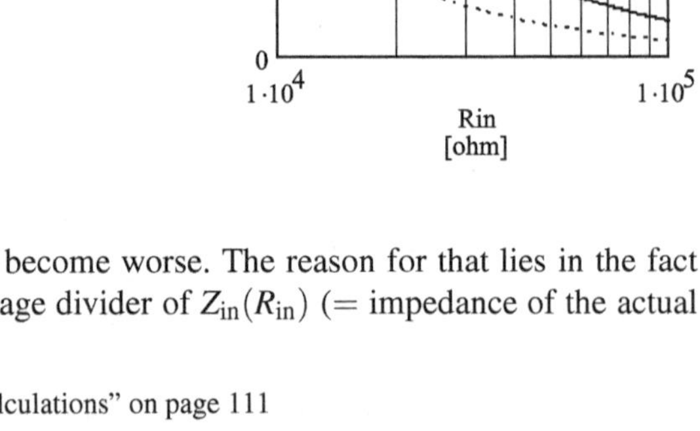


**Fig. 3.88** $G_{e.loss.40}(R_{in})$ and $G_{loss.10}(R_{in})$ vs. $R_{in}$

With decreasing $R_{in}$ all *SN*s become worse. The reason for that lies in the fact that the loss created by the voltage divider of $Z_{in}(R_{in})$ (= impedance of the actual

[71] see paragraph on "*SN* Example Calculations" on page 111

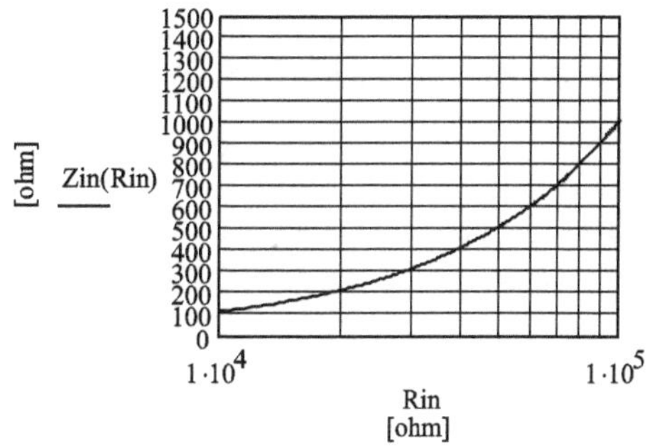


**Fig. 3.89** $Z_{in}(R_{in})$ vs. $R_{in}$

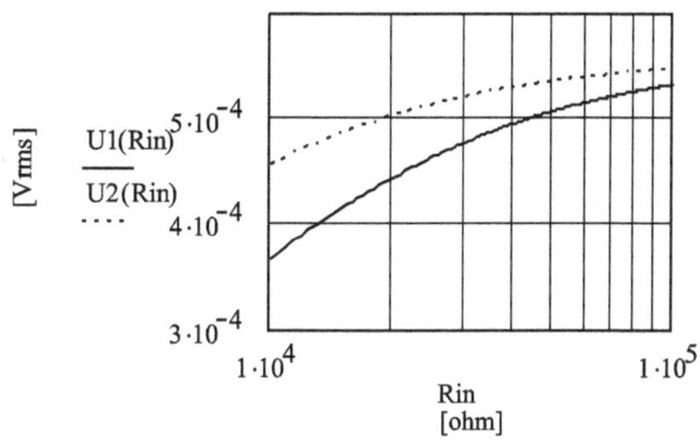


**Fig. 3.90** MC cartridge output voltage $U_1(R_{in})$ vs. $R_{in}$ with $Z_{in}$ as load

load for the MC cartridge at the input of the trafo) and $R_0$ is increasing, thus, all $G_{e.loss.xy}$ are increasing as well.

All figures are derived from Eqs. (3.281) ... (3.288) with variable $R_{in}$. It's obvious that a small $R_0$ triggers better *SN* results. But it has also to be taken into account that a small $R_0$ might not be the right input load of the chosen trafo. To find the right balance between what is recommended by a manufacturer and what is a designer's wish is – in most cases – a question of listening tests with subjective valuation and not a question of crotchety stick on mathematics.

## Specified for $\ll$ 47 k

A2: Manufacturers also offer trafos specified for load resistances smaller than 47 k5, e.g. the JT-346-AX[72] with $R_{L.nom} = 6\,k81$ and a maximal $R_{0max} = 5\,R$. In addition, this trafo has 3 separate coils at the input and 1 at the output (an improved T-version has 2 separate coils at the output), thus, creating a broad range of turns ratios between 1:4 and 1:12 (1:2/1:12).

With an op-amp like the AD797 ($e_{N.ad797} = 0.9\,nV/rtHz$, $i_{N.ad797} = 2\,pA/rtHz$, configured as of Fig. 3.74 with $R_1 = 47\,R5$, $R_2 = 24 \times R_1$) plus a MC cartridge like the Ortofon Samba[73] ($u_{0samba} = 0.57\,mV_{rms}/1\,kHz/8\,cm/s$, $R_{0samba} = 3\,R8$,

[72] Jensen Transformers data sheet

[73] stereoplay 04-2007

$L_{0samba} = 7.96\,\mu H$, recommended load resistance: 100 R) and the turns ratio set to $n = 12$ with reference to $0.5\,mV_{rms}/1\,kHz$ and in $B_{20\,k}$ $SN1_{ariaa}$ becomes:

$$SN1_{ariaa.samba} = -86.667\,dB \tag{3.318}$$

The same cartridge connected to a trafo specified for a load resistance of 47 k5 (e.g. JT-347-AXT[74]) leads to a SN2 of:

$$SN2_{ariaa.samba} = -85.277\,dB \tag{3.319}$$

Thus, the improvement of the low-$R_{load}$ trafo is 1.390 dB. This is accompanied by a change of the load resistance for the cartridge: 49 R (JT-346) and 336 R (JT-347)! Only listening tests will demonstrate what's right or wrong with the 49 R input load of the cartridge. Normally, a load of $10 \times R_0$ should work quite well.

A further change of *SN*s can be achieved by replacing the AD797 with a circuit configuration à la Fig. 3.24 with 2SC2546E input devices.

With $e_{N.2sc2546} = 0.5\,nV/rtHz$ and $i_{N.2sc2546} = 2.31\,pA/rtHz$ at $I_C = 10\,mA$ (Table 3.3) and $R_{1,2}$ chosen as above SN1 will change to SN3:

$$SN3_{ariaa.samba} = -87.564\,dB \tag{3.320}$$

*SN*2 will change to *SN*4:

$$SN4_{ariaa.samba} = -84.990\,dB \tag{3.321}$$

We should compare the *SN*1...4 results with those of a CE configured BJT stage with 2 × MAT02 (SN5...) or 1 × BFW16A (*SN*6...) as input devices that could replace the above described trafo-amp-chain:

with Fig. 3.26, Eqs. (3.93 ... 3.94), $R_E = 3\,R32$ and for

- MAT02: $R_1 = 47\,k5$, $R_2 = 10\,k$, $I_C = 4 \times 2.5\,mA$

and

- BFW16A: $R_1 = 10\,k$, $R_2 = 4.75\,k$, $I_C = 25\,mA$

*SN*5... for MAT02 becomes:

$$SN5_{ariaa.samba} = -84.726\,dB \tag{3.322}$$

*SN*6... for BFW16A becomes:

$$SN6_{ariaa.samba} = -85.982\,dB \tag{3.323}$$

Conclusions:

1. A lowest-noise trafo driven phono-amp for MC cartridges with low-source resistances <10 R should have an input stage with a trafo for low load resistances ≪47 k5 – followed by a hybrid op-amp with a BJT input stage with rather high collector current (see *SN*3... and Eq. (3.320)).

[74] Jensen Transformers data sheet

2. A lowest-noise trafo driven phono-amp for MC cartridges with mid-source-resistances >20 R should have an input stage with a trafo for a load resistance of 47 k5 – followed by a low-noise valve or FET driven phono-amp (a reconfigured BJT-stage will work as well – reconfigured means: change of collector current to a value that fits into the $R_{opt}$ needs – but it should be taken into account that this adapted collector current changes the phono-amps noise behaviour towards MM cartridges as well!!!).

## 3.8 Noise of Vinyl Records (VRs) – On how much Phono-Amp SN is Needed?

### Intro

It's hard to get precise figures on the headline's topic. Several sources[75] claim that *SN*s range from -60 ... −70 dBA for 33 1/3 LP records and -63 ... −73 dBA for 45 Single or Maxi records, both with reference to a peak velocity of 8 cm/s/1 kHz. This is equal to a rms velocity of 5.66 cm/s/1kHz. In addition, *SN*s of records heavily depend on the manufacturing process as well as on the vinyl material itself. Today, only three vinyl production companies exist world-wide in USA, in France and in Japan. The chemical formulae of the materials are the big secrets and they are the basic responsible factors for the noise of the records on the market – besides noisy phono-amps, of course.

### Cutting Technologies: Lacquer and DMM

Today, there are existing two different manufacturing processes for 33 1/3 and 45 records: the traditional one – and today's standard industry process – called lacquer technology (half- and full-speed) and the most advanced one: DMM = Direct Metal Mastering[76]. From a noise point of view the DMM technology is more fertile than the other one. The noise of DMM produced records is less strong, thus, it challenges much more the noise performance of a phono-amp.

### DMM Process

The DMM manufacturing process starts with the direct cut of the signal groove in a 0.1 mm metal layer of micro-crystalline copper. The copper is fixed on a high-grade steel base. The result of this process is a Cu master called MOTHER. It can be

[75] Inter alia: Van den Hul Phono FAQ: Q77 at www.vandenhul.com

[76] The internet is a very good source for more and deeper infos on vinyl cutting issues – inter alia: www.firstcask.com/varsity/cutting.htm

scanned with a MM or MC cartridge that is fixed on a tonearm, feeding an appropriate phono-amp. To get sell-ready vinyl records any further manufacturing process via so-called SONS creates more noise on the vinyl record. Two different DMM manufacturing processes became the favourite ones for the production companies: a very fast and short process for very low-numbered record volumes (lv) and a second one for the big volumes (bv).

The lv production starts right after the MOTHER got made. A stamper made from that MOTHER presses the desired volume of records. The bv production needs a further copy step called SON. These SONs are the basis of several stampers with which the records finally got pressed. Approximately 1000 records per stamper.

From a noise point of view the best case would be a 100% exact copy of the MOTHER – without any further production process influences. It's obvious that this case is nearly impossible because any additional manufacturing step increases the noise level by bringing in further lots of chemistry and manufacturing imponderables. Vinyl record producers[77,78] told me that, between MOTHER and final record, a noise floor increase of at least 2 dBA in $B_{20\,\mathrm{k}}$(= best case) must be taken into account. Because of additional manufacturing processes for the lacquer technology average cases lie in the range of 4 … 8 dBA. In other words: if the 33 1/3 DMM MOTHER has a noise level of e.g. −72 dBA, the best case final record will have a noise floor of −70 dBA, "lousy" – non-DMM – cases will have noise floors of −64 dBA or even less.

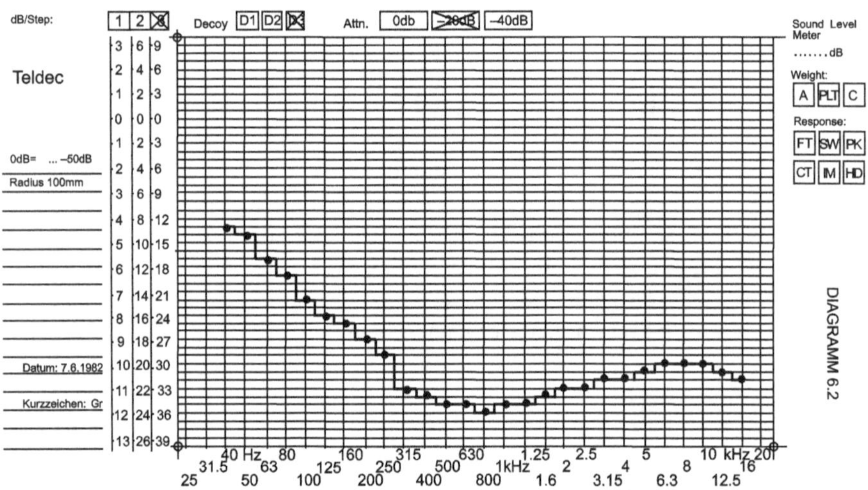


**Fig. 3.91** Typical 3rd octave scanned spectral noise distribution of a signal-free groove of a Cu master = MOTHER (created 7th of June, 1982 by Teldec, see also Figs. 3.96 … 3.97)

[77] Günter Pauler, Stockfisch Records and Pauler Acoustics, Northeim, Germany

[78] SST Brüggemann GmbH, Frankfurt, Germany

## 3rd Octave DMM Spectral Noise Density

With the above given informations in mind I tried to get more precision on the MOTHER's noise production. The service manual of the Neumann VMS-80/DMM cutting lathe became the most important source. As a part of it the authors gave informations about the noise of the MOTHER and the respective SN measurement process and standards. Figure 3.91 shows diagram 6.2[79], the third octave plot of a typical noise distribution when scanning a MOTHER's signal-free groove. The graph was made by Teldec[80], the inventor of the DMM process.

The following table gives the requirements for (the above shown) SN measurement:

**Table 3.7** SN measurement requirements for Cu MOTHERS

| Groove width and equipment | Reference level = 0 dB at 1 kHz at 8 cm/s peak velocity | SN minimal [dBA] | SN typical [dBA] |
|---|---|---|---|
| 40 μm | | −70 | −72 |
| 80 μm | | −60 | −63 |
| MM cartridge: | Shure V15V | | |
| Tracking force: | 1 gram or 10 mN | | |
| Tonearm: | SME 3009 Series II | | |
| Level Meter: | Sennheiser UPM 550 or equivalent | | |
| Phono-amp: | Neumann PUE 74 | | |

The above mentioned tonearm/cartridge combination is not a fixed one at all. The frequency response calibration flexibility of the Neumann PUE 74 phono-amp (Figs. 3.96 … 3.100) allows to make use of a variety of combinations, such as: SME 3012 + V15II[81] or Ortofon RMA-297 + TSD15 (EMT) + EMT-transformer + special phono-amp[82]. The Fig. 3.91 plot includes – besides the pure MOTHER noise – several additional noise sources: rumble, tonearm resonances, noise of the cartridge, noise of the phono-amp.

## Vinyl Record SN Calculation Step by Step

To get the pure MOTHER noise we have to eliminate all these other noise sources and – for comparison reasons – we have to enlarge the plot's frequency band from

[79] Service Manual Neumann VMS-80/DMM; since 1989 Georg Neumann GmbH, Berlin, became a subsidiary of Sennheiser electronic GmbH & Co. KG, Wedemark, Germany

[80] By kind permission of Warner Music Group Holding GmbH, Hamburg, Germany, Teldec = Telefunken + Decca, since 1988 part of WEA International, Inc.

[81] Günter Pauler, Stockfisch Records and Pauler Acoustics, Northeim, Germany

[82] SST Brüggemann GmbH, Frankfurt, Germany

**Table 3.8a** Calculation of Cu master *SN*s for Fig. 3.91

| | | | | | | | | | | | |
|---|---|---|---|---|---|---|---|---|---|---|---|
| 1 | 1/3-Octave-Band | 1/2 1 | 2 | 3 | 4 | 5 | 6 | 7 | 8 | 9 | 10 |
| 2 | Bandcenter, $f$ (Hz) | 21 | 25 | 31.5 | 40 | 50 | 63 | 80 | 100 | 125 | 160 |
| 3 | Bandcenter, $f$ (Hz) calculated | 21.2 | 25.3 | 31.9 | 40.1 | 50.5 | 63.5 | 80.0 | 100.6 | 126.5 | 159.5 |
| 4 | Frequency Band Start (Hz) | 20.0 | 22.4 | 28.2 | 35.5 | 44.7 | 56.2 | 70.8 | 89.1 | 112 | 141 |
| 5 | Frequency Band Stop (Hz) | 22.4 | 28.2 | 35.5 | 44.7 | 56.2 | 70.8 | 89.1 | 112 | 141 | 178 |
| 6 | Bandwidth, $B$ (Hz) | 2 | 6 | 7 | 9 | 12 | 15 | 18 | 23 | 29 | 37 |
| 7 | Ref. Level 5.12 $mV_{rms}/1\ kHz/8\ cm/s$ | 5.120E-03 | 5.120E-03 | 5.120E-03 | 5.120E-03 | 5.120E-03 | 5.120E-03 | 5.120E-03 | 5.120E-03 | 5.120E-03 | 5.120E-03 |
| 8 | Teldec Gain $A_{RIAA}$ (RIAA) [dB] | -10.50 | -11.00 | -12.00 | -13.00 | -14.00 | -17.00 | -19.00 | -22.00 | -24.00 | -25.00 |
| 9 | 0-dB-Level [dB] | -50.00 | -50.00 | -50.00 | -50.00 | -50.00 | -50.00 | -50.00 | -50.00 | -50.00 | -50.00 |
| 10 | Rumble & tonearm effect [dB] | -23.00 | -23.00 | -22.00 | -21.00 | -19.00 | -16.00 | -13.00 | -10.00 | -10.00 | -9.00 |
| 11 | Total $SN_{RIAA.ex}/1/3$ Octave [dB] | -83.50 | -84.00 | -84.00 | -84.00 | -83.00 | -83.00 | -82.00 | -82.00 | -84.00 | -84.00 |
| 12 | $en_{RIAA.ex}$(in B) [V/Hz] | 3.42E-07 | 3.23E-07 | 3.23E-07 | 3.23E-07 | 3.62E-07 | 3.62E-07 | 4.07E-07 | 4.07E-07 | 3.23E-07 | 3.23E-07 |
| 13 | $en_{RIAA.ex}/B^2$ [V/rtHz] | 1.53E-07 | 1.34E-07 | 1.2E-07 | 1.07E-07 | 1.07E-07 | 9.49E-08 | 9.51E-08 | 8.5E-08 | 6E-08 | 5.31E-08 |
| 14 | $en^2_{RIAA.ex}$ [$V^2$/Hz] | 1.17E-13 | 1.04E-13 | 1.04E-13 | 1.04E-13 | 1.31E-13 | 1.31E-13 | 1.65E-13 | 1.65E-13 | 1.04E-13 | 1.04E-13 |
| 15 | $\Sigma$ (14) [$V^2$] | 4.11E-12 | | | | | | | | | |
| 16 | $(15)^{0.5}$ [V] | 2.03E-06 | | | | | | | | | |
| 17 | 5.12 $mV_{rms}$ at 1 kHz/8 cm/s vel. | 5.12E-03 | | | | | $SN_{\mathbf{riaa.20k.ex}} = \mathbf{-68.049}$ | | | | |
| 18 | Gain $R(f)$ (RIAA) | 9.122 | 8.844 | 8.393 | 7.748 | 7.004 | 6.203 | 5.315 | 4.513 | 3.786 | 3.102 |
| 19 | $R(f)$ [dB] | 19.202 | 18.933 | 18.478 | 17.784 | 16.907 | 15.852 | 14.510 | 13.088 | 11.563 | 9.831 |
| 20 | Gain AR($f$) (ARIAA) | 0.110 | 0.113 | 0.119 | 0.129 | 0.143 | 0.161 | 0.188 | 0.222 | 0.264 | 0.322 |
| 21 | $en_{AR(f)}$ (in $B$) [V] | 3.75E-08 | 3.65E-08 | 3.85E-08 | 4.17E-08 | 5.17E-08 | 5.84E-08 | 7.65E-08 | 9.01E-08 | 8.53E-08 | 1.04E-07 |
| 22 | $SN_{ne.ex}$ per 1/3 Octave | -102.702 | -102.933 | -102.478 | -101.784 | -99.907 | -98.852 | -96.510 | -95.088 | -95.563 | -93.831 |
| 23 | $Nf_{e.amp.2}$ [dB] | -1.182 | -1.182 | -1.182 | -1.182 | -1.182 | -1.182 | -1.182 | -1.182 | -1.182 | -1.182 |
| 24 | $SN_{ne.V15V.ex}$ [dB] | -103.884 | -104.115 | -103.660 | -102.966 | -101.089 | -100.034 | -97.692 | -96.270 | -96.745 | -95.013 |
| 25 | $en_{ne.v15v.ex}$ (in $B$) [V] | 3.27E-08 | 3.19E-08 | 3.36E-08 | 3.64E-08 | 4.52E-08 | 5.1E-08 | 6.68E-08 | 7.87E-08 | 7.45E-08 | 9.09E-08 |
| 26 | $en_{ne.v15v.ex}$[V/rtHz] | 2.11E-08 | 1.32E-08 | 1.24E-08 | 1.2E-08 | 1.33E-08 | 1.33E-08 | 1.56E-08 | 1.64E-08 | 1.38E-08 | 1.49E-08 |
| 27 | $en_{v15v}$ at band center freq. (MCD-plot) | 3.595E-09 | 3.595E-09 | 3.595E-09 | 3.595E-09 | 3.595E-09 | 3.596E-09 | 3.596E-09 | 3.597E-09 | 3.598E-09 | 3.600E-09 |
| 28 | $en_{v15v} \times B^{0.5}$ [V] | 5.569E-09 | 8.658E-09 | 9.713E-09 | 1.090E-08 | 1.219E-08 | 1.374E-08 | 1.538E-08 | 1.721E-08 | 1.938E-08 | 2.190E-08 |
| 29 | $en_{ne.Cu.ex} = (e^2_{ne.v15v.ex} - en^2_{v15v})^{0.5}$ [V/rtHz] | 2.501E-08 | 2.900E-08 | 3.121E-08 | 3.436E-08 | 4.316E-08 | 4.922E-08 | 6.493E-08 | 7.692E-08 | 7.318E-08 | 8.967E-08 |

**Table 3.8a** (continued)

| 1 | 1/3-Octave-Band | 1/2 1 | 2 | 3 | 4 | 5 | 6 | 7 | 8 | 9 | 10 |
|---|---|---|---|---|---|---|---|---|---|---|---|
| 2 | Bandcenter, $f$ (Hz) | 21 | 25 | 31.5 | 40 | 50 | 63 | 80 | 100 | 125 | 160 |
| 30 | $SN_{ne.Cu.ex}$ per 1/3 Octave [dB] | -106.224 | -104.937 | -104.300 | -103.465 | -101.484 | -100.342 | -97.936 | -96.464 | -96.897 | -95.132 |
| 31 | $en_{ne.Cu.ex}/B^2$ [V/rtHz] | 1.12E-08 | 1.2E-08 | 1.16E-08 | 1.13E-08 | 1.27E-08 | 1.29E-08 | 1.52E-08 | 1.61E-08 | 1.36E-08 | 1.47E-08 |
| 32 | $en^2_{ne.Cu.ex}$ | 6.253E-16 | 8.411E-16 | 9.740E-16 | 1.180E-15 | 1.863E-15 | 2.423E-15 | 4.217E-15 | 5.917E-15 | 5.356E-15 | 8.041E-15 |
| 33 | Σ (32) [$V^2$] | 3.41E-11 | | | | | | | | | |
| 34 | $(33)^{0.5}$ [V] | 5.84E-06 | | | | $SN_{\mathbf{ne.Cu.20k.ex}} = \mathbf{-58.861}$ | | | | | |
| 35 | $en_{riaa.CU.20k.ex}$ [V/rtHz] | 2.281E-07 | 2.565E-07 | 2.619E-07 | 2.662E-07 | 3.023E-07 | 3.053E-07 | 3.451E-07 | 3.471E-07 | 2.770E-07 | 2.781E-07 |
| 36 | $SN_{riaa.Cu.20k.ex}$ [dB] | -87.023 | -86.004 | -85.822 | -85.682 | -84.577 | -84.490 | -83.426 | -83.376 | -85.334 | -85.301 |
| 37 | $en^2_{riaa.CU.20k.ex}$ [$V^2$/Hz] | 5.203E-14 | 6.579E-14 | 6.861E-14 | 7.085E-14 | 9.138E-14 | 9.322E-14 | 1.191E-13 | 1.205E-13 | 7.675E-14 | 7.735E-14 |
| 38 | Σ (37) [$V^2$] | 3.02E-12 | | | | | | | | | |
| 39 | $(38)^{0.5}$ [V] | 1.738E-06 | | | | $SN_{\mathbf{riaa.Cu.20k.ex}} = \mathbf{-69.387}$ | | | | | |
| 40 | Gain $A(f)$ | 0.003 | 0.006 | 0.011 | 0.019 | 0.031 | 0.049 | 0.076 | 0.110 | 0.155 | 0.217 |
| 41 | $en_{ariaa.CU.20k.ex}$ [V/rtHz] | 7E-10 | 1.52E-09 | 2.77E-09 | 5.02E-09 | 9.44E-09 | 1.49E-08 | 2.62E-08 | 3.83E-08 | 4.3E-08 | 6.03E-08 |
| 42 | $SN_{ariaa.Cu.20k.ex}$ [dB] | -137.288 | -130.540 | -125.351 | -120.171 | -114.688 | -110.713 | -105.833 | -102.521 | -101.524 | -98.581 |
| 43 | $en^2_{ariaa.CU.20k.ex}$ [$V^2$/Hz] | 4.895E-19 | 2.315E-18 | 7.646E-18 | 2.520E-17 | 8.906E-17 | 2.224E-16 | 6.842E-16 | 1.467E-15 | 1.846E-15 | 3.635E-15 |
| 44 | Σ (43) [$V^2$] | 1.71E-12 | | | | | | | | | |
| 45 | $(44)^{0.5}$ [V] | 1.31E-06 | | | | $SN_{\mathbf{ariaa.Cu.20k.ex}} = \mathbf{-71.855}$ | | | | | |

**Table 3.8b** Calculation of Cu master *SN*s for Fig. 3.91

| 1 | 1/3-Octave-Band | 11 | 12 | 13 | 14 | 15 | 16 | 17 | 18 | 19 | 20 |
|---|---|---|---|---|---|---|---|---|---|---|---|
| 2 | Bandcenter, $f$ (Hz) | 200 | 250 | 315 | 400 | 500 | 630 | 800 | 1000 | 1250 | 1600 |
| 3 | Bandcenter, $f$ (Hz) calc. | 201.0 | 253.0 | 318.5 | 401.0 | 504.5 | 635.0 | 799.5 | 1.005.5 | 1.265.0 | 1.595.0 |
| 4 | Frequency Band Start | 178 | 224 | 282 | 355 | 447 | 562 | 708 | 891 | 1120 | 1410 |
| 5 | Frequency Band Stop | 224 | 282 | 355 | 447 | 562 | 708 | 891 | 1120 | 1410 | 1780 |
| 6 | Bandwidth, $B$ (Hz) | 46 | 58 | 73 | 92 | 115 | 146 | 183 | 229 | 290 | 370 |
| 7 | Ref. Level 5.12 $mV_{rms}$ | 5.120E-03 | 5.120E-03 | 5.120E-03 | 5.120E-03 | 5.120E-03 | 5.120E-03 | 5.120E-03 | 5.120E-03 | 5.120E-03 | 5.120E-03 |
| 8 | Teldec Gain $A_{RIAA}$ (RIAA) | -27.00 | -29.00 | -33.00 | -34.00 | -35.00 | -35.00 | -36.00 | -35.00 | -35.00 | -34.00 |
| 9 | 0-dB-Level | -50.00 | -50.00 | -50.00 | -50.00 | -50.00 | -50.00 | -50.00 | -50.00 | -50.00 | -50.00 |
| 10 | Rumble & tonearm | -8.00 | -6.00 | -2.00 | -2.00 | -1.00 | -1.00 | 0.00 | 0.00 | 0.00 | 0.00 |
| 11 | Total $SN_{RIAA.ex}/1/3$ Oct. | -85.00 | -85.00 | -85.00 | -86.00 | -86.00 | -86.00 | -86.00 | -85.00 | -85.00 | -84.00 |
| 12 | $en_{RIAA.ex}$ (in $B$) | 2.88E-07 | 2.88E-07 | 2.88E-07 | 2.57E-07 | 2.57E-07 | 2.57E-07 | 2.57E-07 | 2.88E-07 | 2.88E-07 | 3.23E-07 |
| 13 | $en_{RIAA.ex}/B^2$ | 4.25E-08 | 3.78E-08 | 3.37E-08 | 2.68E-08 | 2.39E-08 | 2.12E-08 | 1.9E-08 | 1.9E-08 | 1.69E-08 | 1.68E-08 |
| 14 | $en^2_{RIAA.ex}$ | 8.29E-14 | 8.29E-14 | 8.29E-14 | 6.58E-14 | 6.58E-14 | 6.58E-14 | 6.58E-14 | 8.29E-14 | 8.29E-14 | 1.04E-13 |
| 15 | | | | | | | | | | | |
| 16 | | | | | | | | | | | |
| 17 | | | | | | | | | | | |
| 18 | Gain R($f$) (RIAA) | 2.576 | 2.157 | 1.815 | 1.546 | 1.356 | 1.208 | 1.091 | 1.000 | 0.918 | 0.828 |
| 19 | R($f$) | 8.219 | 6.677 | 5.179 | 3.784 | 2.648 | 1.642 | 0.754 | 0.000 | -0.744 | -1.643 |
| 20 | Gain AR($f$) (ARIAA) | 0.388 | 0.464 | 0.551 | 0.647 | 0.737 | 0.828 | 0.917 | 1.000 | 1.089 | 1.208 |
| 21 | $en_{AR(f)}$ in $B$ | 1.12E-07 | 1.33E-07 | 1.59E-07 | 1.66E-07 | 1.89E-07 | 2.12E-07 | 2.35E-07 | 2.88E-07 | 3.14E-07 | 3.9E-07 |
| 22 | $SN_{ne.ex}$ per 1/3 Octave | -93.219 | -91.677 | -90.179 | -89.784 | -88.648 | -87.642 | -86.754 | -85.000 | -84.256 | -82.357 |
| 23 | $Nf_{e.amp.2}$ | -1.182 | -1.182 | -1.182 | -1.182 | -1.182 | -1.182 | -1.182 | -1.182 | -1.182 | -1.182 |
| 24 | $SN_{ne.V15V.ex}$ | -94.401 | -92.859 | -91.361 | -90.966 | -89.830 | -88.824 | -87.936 | -86.182 | -85.438 | -83.539 |
| 25 | $en_{ne.v15v.ex}$ in $B$ | 9.75E-08 | 1.16E-07 | 1.38E-07 | 1.45E-07 | 1.65E-07 | 1.85E-07 | 2.05E-07 | 2.51E-07 | 2.74E-07 | 3.41E-07 |
| 26 | $en_{ne.v15v.ex}$ | 1.44E-08 | 1.53E-08 | 1.62E-08 | 1.51E-08 | 1.54E-08 | 1.53E-08 | 1.52E-08 | 1.66E-08 | 1.61E-08 | 1.77E-08 |
| 27 | $en_{v15v}$ | 3.603E-09 | 3.608E-09 | 3.616E-09 | 3.629E-09 | 3.648E-09 | 3.678E-09 | 3.728E-09 | 3.802E-09 | 3.914E-09 | 4.107E-09 |
| 28 | $en_{v15v} \times B^{0.5}$ | 2.444E-08 | 2.748E-08 | 3.089E-08 | 3.480E-08 | 3.912E-08 | 4.444E-08 | 5.044E-08 | 5.753E-08 | 6.666E-08 | 7.900E-08 |
| 29 | $en_{ne.Cu.ex}$ | 9.648E-08 | 1.155E-07 | 1.375E-07 | 1.441E-07 | 1.644E-07 | 1.848E-07 | 2.048E-07 | 2.507E-07 | 2.733E-07 | 3.402E-07 |

**Table 3.8b** (continued)

| 1 | 1/3-Octave-Band | 11 | 12 | 13 | 14 | 15 | 16 | 17 | 18 | 19 | 20 |
|---|---|---|---|---|---|---|---|---|---|---|---|
| 2 | Bandcenter, $f$ (Hz) | 200 | 250 | 315 | 400 | 500 | 630 | 800 | 1000 | 1250 | 1600 |
| 30 | $SN_{\mathrm{ne.Cu.ex}}$ per 1/3 Octave | -94.497 | -92.934 | -91.421 | -91.013 | -89.868 | -88.854 | -87.959 | -86.201 | -85.453 | -83.551 |
| 31 | $en_{\mathrm{ne.Cu.ex}}/B^2$ | 1.42E-08 | 1.52E-08 | 1.61E-08 | 1.5E-08 | 1.53E-08 | 1.53E-08 | 1.51E-08 | 1.66E-08 | 1.6E-08 | 1.77E-08 |
| 32 | $en^2_{\mathrm{ne.Cu.ex}}$ | 9.308E-15 | 1.334E-14 | 1.890E-14 | 2.076E-14 | 2.703E-14 | 3.413E-14 | 4.194E-14 | 6.287E-14 | 7.468E-14 | 1.157E-13 |
| 33 | | | | | | | | | | | |
| 34 | | | | | | | | | | | |
| 35 | $en_{\mathrm{riaa.CU.20k.ex}}$ | 2.485E-07 | 2.491E-07 | 2.496E-07 | 2.227E-07 | 2.230E-07 | 2.232E-07 | 2.233E-07 | 2.507E-07 | 2.509E-07 | 2.816E-07 |
| 36 | $SN_{\mathrm{riaa.Cu.20k.ex}}$ | -86.277 | -86.258 | -86.242 | -87.229 | -87.220 | -87.212 | -87.206 | -86.201 | -86.197 | -85.194 |
| 37 | $en^2_{\mathrm{riaa.CU.20k.ex}}$ | 6.177E-14 | 6.206E-14 | 6.228E-14 | 4.961E-14 | 4.972E-14 | 4.981E-14 | 4.988E-14 | 6.287E-14 | 6.293E-14 | 7.928E-14 |
| 38 | | | | | | | | | | | |
| 39 | | | | | | | | | | | |
| 40 | Gain $A(f)$ | 0.287 | 0.368 | 0.465 | 0.577 | 0.688 | 0.803 | 0.912 | 1.000 | 1.069 | 1.121 |
| 41 | $en_{\mathrm{ariaa.CU.20k.ex}}$ | 7.13E-08 | 9.18E-08 | 1.16E-07 | 1.29E-07 | 1.53E-07 | 1.79E-07 | 2.04E-07 | 2.51E-07 | 2.68E-07 | 3.16E-07 |
| 42 | $SN_{\mathrm{ariaa.Cu.20k.ex}}$ | -97.125 | -94.933 | -92.886 | -92.004 | -90.468 | -89.120 | -88.003 | -86.201 | -85.621 | -84.201 |
| 43 | $en^2_{\mathrm{ariaa.CU.20k.ex}}$ | 5.082E-15 | 8.419E-15 | 1.349E-14 | 1.653E-14 | 2.354E-14 | 3.210E-14 | 4.152E-14 | 6.287E-14 | 7.186E-14 | 9.965E-14 |
| 44 | | | | | | | | | | | |
| 45 | | | | | | | | | | | |

**Table 3.8c** Calculation of Cu master *SN*s for Fig. 3.91

| 1 | 1/3-Octave-Band | 21 | 22 | 23 | 24 | 25 | 26 | 27 | 28 | 29 | 30 | 1/2 31 |
|---|---|---|---|---|---|---|---|---|---|---|---|---|
| 2 | Bandcenter, $f$ (Hz) | 2000 | 2500 | 3150 | 4000 | 5000 | 6300 | 8000 | 10,000 | 12,500 | 16,000 | 18,900 |
| 3 | Bandcenter, $f$ (Hz) calc. | 2010.0 | 2530.0 | 3185.0 | 4010.0 | 5045.0 | 6350.0 | 7995.0 | 10,055.0 | 12,650.0 | 15,950.0 | 18,900.0 |
| 4 | Frequency Band Start | 1780 | 2240 | 2820 | 3550 | 4470 | 5620 | 7080 | 8910 | 11,200 | 14,100 | 17,800 |
| 5 | Frequency Band Stop | 2240 | 2820 | 3550 | 4470 | 5620 | 7080 | 8910 | 11,200 | 14,100 | 17,800 | 20,000 |
| 6 | Bandwidth, $B$ (Hz) | 460 | 580 | 730 | 920 | 1150 | 1460 | 1830 | 2290 | 2900 | 3700 | 2200 |
| 7 | Ref. Level 5.12 $mV_{rms}$ | 5.120E-03 | 5.120E-03 | 5.120E-03 | 5.120E-03 | 5.120E-03 | 5.120E-03 | 5.120E-03 | 5.120E-03 | 5.120E-03 | 5.120E-03 | 5.120E-03 |
| 8 | Teldec Gain $A_{RIAA}$ (RIAA) | -33.00 | -33.00 | -32.00 | -32.00 | -31.00 | -30.00 | -30.00 | -30.00 | -31.00 | -32.00 | -33.00 |
| 9 | 0-dB-Level | -50.00 | -50.00 | -50.00 | -50.00 | -50.00 | -50.00 | -50.00 | -50.00 | -50.00 | -50.00 | -50.00 |
| 10 | Rumble & tonearm | 0.00 | 0.00 | 0.00 | 0.00 | 0.00 | 0.00 | 0.00 | 0.00 | 0.00 | 0.00 | 0.00 |
| 11 | Total $SN_{RIAA.ex}/1/3$ Oct. | -83.00 | -83.00 | -82.00 | -82.00 | -81.00 | -80.00 | -80.00 | -80.00 | -81.00 | -82.00 | -83.00 |
| 12 | $en_{RIAA.ex}$ (in $B$) | 3.62E-07 | 3.62E-07 | 4.07E-07 | 4.07E-07 | 4.56E-07 | 5.12E-07 | 5.12E-07 | 5.12E-07 | 4.56E-07 | 4.07E-07 | 3.62E-07 |
| 13 | $en_{RIAA.ex}/B^2$ | 1.69E-08 | 1.51E-08 | 1.51E-08 | 1.34E-08 | 1.35E-08 | 1.34E-08 | 1.2E-08 | 1.07E-08 | 8.47E-09 | 6.69E-09 | 5.34E-09 |
| 14 | $en^2_{RIAA.ex}$ | 1.31E-13 | 1.31E-13 | 1.65E-13 | 1.65E-13 | 2.08E-13 | 2.62E-13 | 2.62E-13 | 2.62E-13 | 2.08E-13 | 1.65E-13 | 1.31E-13 |
| 15 | | | | | | | | | | | | |
| 16 | | | | | | | | | | | | |
| 17 | | | | | | | | | | | | |
| 18 | Gain R($f$) (RIAA) | 0.742 | 0.653 | 0.560 | 0.467 | 0.389 | 0.317 | 0.254 | 0.206 | 0.166 | 0.130 | 0.110 |
| 19 | R($f$) | -2.589 | -3.700 | -5.038 | -6.605 | -8.210 | -9.980 | -11.894 | -13.734 | -15.609 | -17.708 | -19.134 |
| 20 | Gain AR($f$) (ARIAA) | 1.347 | 1.531 | 1.786 | 2.139 | 2.573 | 3.155 | 3.933 | 4.861 | 6.032 | 7.680 | 9.052 |
| 21 | $en_{AR(f)}$ in $B$ | 4.88E-07 | 5.55E-07 | 7.26E-07 | 8.7E-07 | 1.17E-06 | 1.62E-06 | 2.01E-06 | 2.49E-06 | 2.75E-06 | 3.12E-06 | 3.28E-06 |
| 22 | $SN_{ne.ex}$ per 1/3 Octave | -80.411 | -79.300 | -76.962 | -75.395 | -72.790 | -70.020 | -68.106 | -66.266 | -65.391 | -64.292 | -63.866 |
| 23 | $Nf_{e.amp.2}$ | -1.182 | -1.182 | -1.182 | -1.182 | -1.182 | -1.182 | -1.182 | -1.182 | -1.182 | -1.182 | -1.182 |
| 24 | $SN_{ne.V15V.ex}$ | -81.593 | -80.482 | -78.144 | -76.577 | -73.972 | -71.202 | -69.288 | -67.448 | -66.573 | -65.474 | -65.048 |
| 25 | $en_{ne.v15v.ex}$ in $B$ | 4.26E-07 | 4.84E-07 | 6.34E-07 | 7.59E-07 | 1.02E-06 | 1.41E-06 | 1.76E-06 | 2.17E-06 | 2.4E-06 | 2.73E-06 | 2.86E-06 |
| 26 | $en_{ne.v15v.ex}$ | 1.99E-08 | 2.01E-08 | 2.35E-08 | 2.5E-08 | 3.02E-08 | 3.69E-08 | 4.11E-08 | 4.54E-08 | 4.46E-08 | 4.48E-08 | 4.71E-08 |
| 27 | $en_{v15v}$ | 4.374E-09 | 4.766E-09 | 5.359E-09 | 6.251E-09 | 7.445E-09 | 9.208E-09 | 1.188E-08 | 1.559E-08 | 2.089E-08 | 2.694E-08 | 2.718E-05 |
| 28 | $en_{v15v} \times B^{0.5}$ | 9.380E-08 | 1.148E-07 | 1.448E-07 | 1.896E-07 | 2.525E-07 | 3.518E-07 | 5.083E-07 | 7.460E-07 | 1.125E-06 | 1.638E-06 | 1.843E-03 |
| 29 | $en_{ne.Cu.ex}$ | 4.257E-07 | 4.840E-07 | 6.335E-07 | 7.589E-07 | 1.024E-06 | 1.409E-06 | 1.757E-06 | 2.172E-06 | 2.402E-06 | 2.726E-06 | 2.863E-06 |

**Table 3.8c** (continued)

| 1 | 1/3-Octave-Band | 21 | 22 | 23 | 24 | 25 | 26 | 27 | 28 | 29 | 30 | 1/2 31 |
|---|---|---|---|---|---|---|---|---|---|---|---|---|
| 2 | Bandcenter, $f$ (Hz) | 2.000 | 2.500 | 3.150 | 4.000 | 5.000 | 6.300 | 8.000 | 10.000 | 12.500 | 16.000 | 18.900 |
| 30 | $SN_{ne.Cu.ex}$ per 1/3 Octave | -81.603 | -80.489 | -78.150 | -76.581 | -73.976 | -71.205 | -69.290 | -67.450 | -66.575 | -65.475 | -65.049 |
| 31 | $en_{ne.Cu.ex}/B^2$ | 1.98E-08 | 2.01E-08 | 2.34E-08 | 2.5E-08 | 3.02E-08 | 3.69E-08 | 4.11E-08 | 4.54E-08 | 4.46E-08 | 4.48E-08 | 4.22E-08 |
| 32 | $en^2_{ne.Cu.ex}$ | 1.812E-13 | 2.342E-13 | 4.014E-13 | 5.760E-13 | 1.049E-12 | 1.986E-12 | 3.087E-12 | 4.716E-12 | 5.768E-12 | 7.430E-12 | 8.197E-12 |
| 33 | | | | | | | | | | | | |
| 34 | | | | | | | | | | | | |
| 35 | $en_{riaa.CU.20k.ex}$ | 3.160E-07 | 3.161E-07 | 3.547E-07 | 3.548E-07 | 3.981E-07 | 4.467E-07 | 4.467E-07 | 4.468E-07 | 3.982E-07 | 3.549E-07 | 3.163E-07 |
| 36 | $SN_{riaa.Cu.20k.ex}$ | -84.191 | -84.189 | -83.188 | -83.187 | -82.186 | -81.185 | -81.184 | -81.184 | -82.183 | -83.183 | -84.183 |
| 37 | $en^2_{riaa.CU.20k.ex}$ | 9.986E-14 | 9.991E-14 | 1.258E-13 | 1.259E-13 | 1.585E-13 | 1.995E-13 | 1.996E-13 | 1.996E-13 | 1.586E-13 | 1.260E-13 | 1.001E-13 |
| 38 | | | | | | | | | | | | |
| 39 | | | | | | | | | | | | |
| 40 | Gain $A(f)$ | 1.148 | 1.158 | 1.148 | 1.117 | 1.066 | 0.987 | 0.877 | 0.751 | 0.613 | 0.462 | 0.341 |
| 41 | $en_{ariaa.CU.20k.ex}$ | 3.63E-07 | 3.66E-07 | 4.07E-07 | 3.96E-07 | 4.24E-07 | 4.41E-07 | 3.92E-07 | 3.35E-07 | 2.44E-07 | 1.64E-07 | 1.08E-07 |
| 42 | $SN_{ariaa.Cu.20k.ex}$ | -82.990 | -82.918 | -81.986 | -82.222 | -81.630 | -81.299 | -82.329 | -83.672 | -86.433 | -89.884 | -93.524 |
| 43 | $en^2_{ariaa.CU.20k.ex}$ | 1.317E-13 | 1.339E-13 | 1.659E-13 | 1.571E-13 | 1.801E-13 | 1.944E-13 | 1.533E-13 | 1.125E-13 | 5.959E-14 | 2.692E-14 | 1.165E-14 |
| 44 | | | | | | | | | | | | |
| 45 | | | | | | | | | | | | |

**Table 3.9** *SN* results for various bandwidths

| 1/A | B | C | D | E | F | G | H | I |
|---|---|---|---|---|---|---|---|---|
| 2 | source | frequency range | *SN* | Diagram 6.2<br>incl. rumble | excl. rumble | Cu.ex min ($NF_{amp2}$ = 1.182 dB) | Cu.ex max ($NF_{amp1}$ = 1.332 dB) | Remarks |
| 3 | | | | | | | | |
| 4 | | | [± 0.5 dB tol.] | [dB] | [dB] | [dB] | [dB] | |
| 5 | **Teldec graph** | **40 Hz ... 16 kHz** | $\boldsymbol{SN_{riaa.16k}}$ | **−58.6** | **−68.6** | −69.9 | −70.0 | **bold values** taken from |
| 6 | converted to: | 20 Hz ... 20 kHz | $SN_{riaa.20k}$ | −54.3 | −68.1 | −69.4 | −69.5 | Teldec diagram 6.2 for |
| 7 | converted to: | 17.8 Hz ... 22.4 kHz | $SN_{riaa.22k}$ | −54.1 | −68.0 | −69.3 | −69.5 | Neumann VMS-80/DMM |
| 8 | converted to: | 40 Hz ... 16 kHz | $SN_{ne.16k}$ | −59.0 | −59.1 | −60.4 | −60.6 | Cu figures calculated |
| 9 | converted to: | 20 Hz ... 20 kHz | $SN_{ne.20k}$ | −57.7 | −57.9 | −58.9 | −59.0 | |
| 10 | converted to: | 17.8 Hz ... 22.4 kHz | $SN_{ne.22k}$ | −57.4 | −57.6 | −58.7 | −58.9 | |
| 11 | converted to: | 40 Hz ... 16 kHz | $SN_{ariaa.16k}$ | −70.3 | −70.7 | −71.9 | −72.0 | |
| 12 | converted to: | 20 Hz ... 20 kHz | $SN_{ariaa.20k}$ | −70.2 | −70.7 | −71.9 | −72.0 | |
| 13 | converted to: | 17.8 Hz ... 22.4 kHz | $SN_{ariaa.22k}$ | −70.2 | −70.7 | −71.9 | −72.0 | |

40 Hz … 16 kHz to 20 Hz … 20 kHz. With the findings of the MM chapter of this book it's a rather easy task to proceed accordingly.

The following steps have to be taken in $B_{20\,\mathrm{k}}$ (see Table 3.8a … c):

1. Transfer of the Fig. 3.91 $SN_{\mathrm{riaa}}$ values into a third octave spread sheet à la National Semiconductor AN 104[83] with a enlarged frequency band of $B_{20\,\mathrm{k}}$. Having done this we'll get $SN_{\mathrm{riaa.20k.in}}$ (in = includes rumble). The nominal output voltage of the V15V cartridge is $e_{\mathrm{nom.V15V}} = 5.12\,\mathrm{mV_{rms}}$. This is also the reference level for all following SN calculations.
2. Take out rumble and other noise making artefacts <1 kHz to get $SN_{\mathrm{riaa.20k.ex}}$ (see rumble- and resonance-free V15V plot in Fig. 4.5, trace (2), ex = excluding rumble etc. and lines 10, 11 of table).
3. Take out the RIAA transfer to get the non-equalized $SN_{\mathrm{ne.20k.ex}}$ per 3rd octave (lines 18 … 22). With Eq. (1.2.6) calculate $\mathrm{R}(f)$ per each 3rd octave's center frequency, than, the Anti-RIAA transfer function $\mathrm{AR}(f)$ becomes:

$$\mathrm{AR}(f) = \frac{1}{R(f)} \tag{3.324}$$

### *NF* of Neumann Phono-Amp PUE 74

4. Take out $NF$ of amp = $NF_{\mathrm{e.amp}}$ (by subtraction in line 23) and you'll get $SN_{\mathrm{ne.V15V.ex}}$, the $SN$ of Cu and of V15V. $NF_{\mathrm{e.amp}}$ of the Neumann phono-amp PUE 74 could be calculated with:

   - $NF_{\mathrm{e.amp.1}} = 1.332\,\mathrm{dB}$ ($r_{\mathrm{bb'1}} = 1040\,\mathrm{R}$ – maximal)
   - $NF_{\mathrm{e.amp.2}} = 1.182\,\mathrm{dB}$ ($r_{\mathrm{bb'2}} = 120\,\mathrm{R}$ – minimal)

   The NF-difference is marginal (0.15 dB) and it does not influence the calculated $SN$s very much (Table 3.9).

   The input transistor of the PUE 74 is a BC212B. Assumed that Neumann has selected for the lowest noise producing devices there are still questions open about the value of $r_{\mathrm{bb'}}$. The Telefunken data sheet[84] gives a typical NF of 2.5 dB at 1 kHz and a source resistance $R_{\mathrm{G}} = 2\,\mathrm{k}$ with $I_{\mathrm{C}} = 200\,\mu\mathrm{A}$, $h_{\mathrm{FE}} = 300$, $B =$ 1 rtHz. Let's assume a best case NF of 2.0 dB. Than, with Eqs. (3.50), (3.69) and (3.70) MCD calculated $r_{bb'1}$ becomes 1040 R:

$$\left[10^{\frac{2.0}{20}} = \frac{\sqrt{4kTBR_{\mathrm{G}} + i_{\mathrm{N.T}}^2 \times (R_{\mathrm{G}}^2 + r_{bb'1}^2) + e_{\mathrm{n.T}}^2 + 4kTBr_{bb'1}}}{4kTBR_{\mathrm{G}}}\right] \tag{3.325}$$
$$\text{solve}, r_{\mathrm{bb'1}} \rightarrow (1040\,\Omega, -78{,}595\,\Omega)$$

[83] "Noise Specs Confusing?" Jim Sherwin, Application Note AN 104, National Semiconductor Linear Applications Databook 1986

[84] Telefunken 1972/1973 Semiconductors Handbook (Consumer Devices)

For the BC212B Motorola data sheet Fig. 6 gives $r_{bb'2} = 120\,\mathrm{R}$. This figure is rather surprising because in the same data sheet a table with the Dynamic Characteristics gives noise figure values like the ones of Telefunken. And they lead to the $r_{bb'}$ of Eq. (3.325).

### *SN* of Shure V15V

5. Take out the noise voltage of the V15V cartridge $en_{V15V} \times B^{0.5}$ per each 3rd octave (lines 27, 28) and you'll get the noise voltage of the Cu layer $en_{ne.Cu}$ per 3rd octave (line 29). This can be calculated with the following equation:

$$en_{ne.Cu.ex} = \sqrt{en^2_{ne.V15V} - en^2_{V15V}} \tag{3.326}$$

$en_{V15V}$ values per 3rd octave mid frequencies can be taken from the respective Eq. (4.20) of Chap. 4 via Figs. 3.92 and 3.97. To generate the plot all amp related figures in that equation have to be set to "0" ($e_{N.T}$, $i_{N.T}$, $Z_4(f)$).

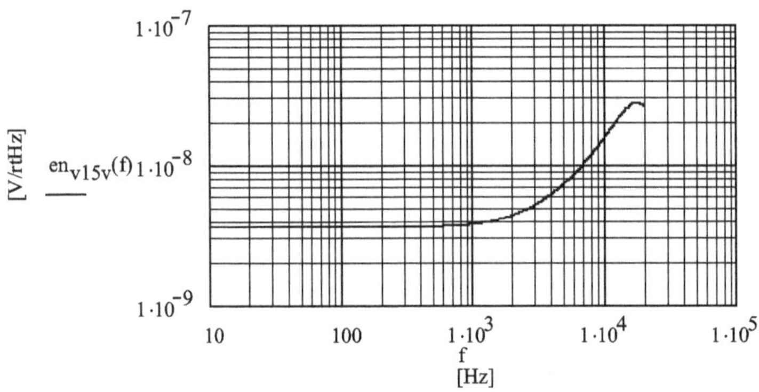


**Fig. 3.92** Noise voltage density of a V15V cartridge attached to the input impedance network of a phono-amp (Fig. 3.97)

6. Take the root of the sum of the squares of the noise voltages in each third octave of step 5 (lines 32 ... 34):

$$en_{ne.Cu.tot} = \sqrt{\sum_{\text{1st 3rd. oct.}}^{\text{31th 3rd oct.}} en_{ne.Cu.ex}(\text{3rd oct.})^2} \tag{3.327}$$

### *SN* of VR MOTHER

7. Calculation of $SN_{ne}$ of Cu layer in $B_{20\,k}$ – excluding rumble, etc.:

$$SN_{\text{ne.Cu.20k.ex}} = 20 \times \log\left(\frac{en_{\text{ne.Cu.tot}}}{e_{\text{nom.V15V}}}\right) \tag{3.328}$$

$$SN_{\text{ne.Cu.20k.ex.min}} = -58.861\,\text{dB} \tag{3.329}$$

$$SN_{\text{ne.Cu.20k.ex.max}} = -59.011\,\text{dB} \tag{3.330}$$

The respective 3rd octave plot is shown in Fig. 3.93 (there is no major difference between the $NF_{\text{e.amp1}}$ and $NF_{\text{e.amp2}}$ plots).

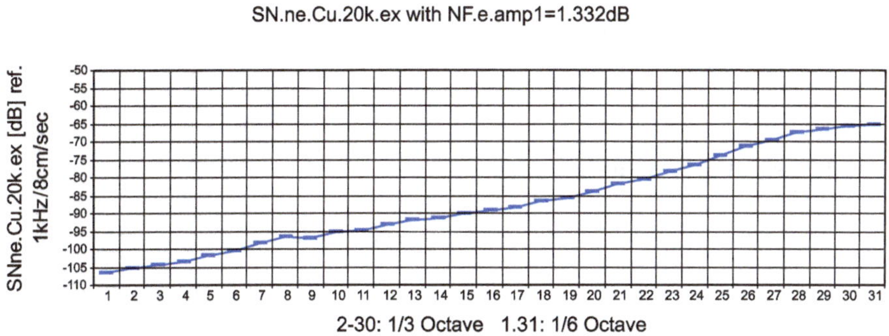


**Fig. 3.93** Final 3rd octave SN of the Fig. 3.91 MOTHER Cu layer $SN_{\text{ne.Cu.20k.ex}}$ with noise figure $NF_{\text{e.amp.1}} = 1.332$ dB

## Noise Voltage of VR MOTHER

8. After division by the square of the bandwidth of each 3rd octave (line 31) we'll get the noise voltage density of the Cu layer, given in Fig. 3.94.
   Note: in reality there exists no noise of Cu. In this case it's a kind of virtual noise created by the grain size of the Cu material in conjunction with the movement of the cartridges stylus in the groove. The respective (tiny) peak velocity multiplied with the transfer factor $TF_{V15V}$ of the cartridge creates the below shown trace.

   - Both, points 7. and 8. results, indicate that the noise created by the Cu layer via the V15V has a bit of a white noise character. I guess the plot's slope increase >2 kHz has to do with the R-L impedance of the cartridge. The impedance noise production is not white at all and its 3rd octave slope is growing from +3 dB (purely white) to > +6 dB at frequencies >2 kHz (Fig. 4.5, trace (4)). I think, a MC cartridge like the TSD15 from EMT would create a clean white noise plot. But I' didn't test it yet.

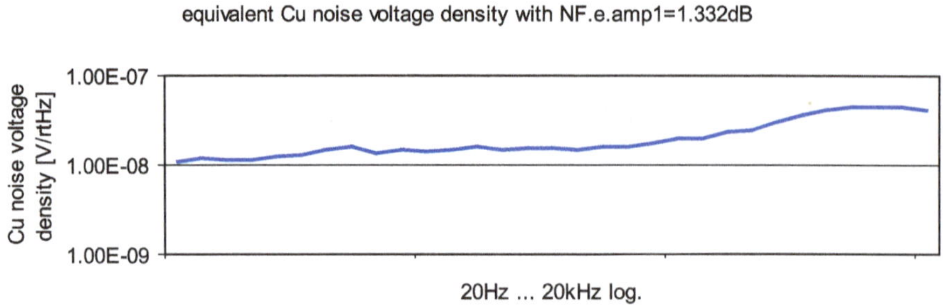


**Fig. 3.94** Equivalent noise voltage density of the Fig. 3.91 MOTHER Cu layer (root of line 32). The plot for $NF_{\text{e.amp2}}$ looks the same

## Transfer Factor of a Cartridge

- The transfer factor TF of any cartridge is defined as:

$$\text{TF} = \text{rms output voltage of cartridge at 1 cm/s/1 kHz peak velocity} \tag{3.331}$$

Thus,

$$\begin{aligned} &TF_{\text{cartridge}}\left[\frac{\text{mV}_{\text{rms}}\,\text{s}}{\text{cm}}\right] \times \text{velocity}\left[\tfrac{\text{cm}}{\text{s}}\text{ at 1 kHz}\right] \\ &= \text{rms output voltage of cartridge at 1 kHz at a certain velocity} \end{aligned} \tag{3.332}$$

TF for the V15V is given as:

$$TF_{\text{V15V}} = 0.64\,\text{mV}_{\text{ms}}\,\text{s/cm} \tag{3.333}$$

hence, the nominal rms output voltage at 0 dB becomes:

$$e_{\text{V15V.nom}} = TF_{\text{V15V}} \times 8\,\text{cm/s/1kHz} = 5.12\,\text{mV}_{\text{rms}}/1\,\text{kHz} \tag{3.334}$$

9. To get the RIAA equalized noise of the MOTHER Cu layer multiplication with the RIAA transfer per third octave (line 18 × line 29) will lead to $SN_{\text{riaa.Cu.20k.ex}}$ (lines 35 … 39):

$$SN_{\text{riaa.Cu.20k.ex.min}} = -69.387\,\text{dB} \tag{3.335}$$

$$SN_{\text{riaa.Cu.20k.ex.max}} = -69.537\,\text{dB} \tag{3.336}$$

10. Further multiplication with the respective A-weighting factors per each 3rd octave (line 40) will lead to the A-weighted and RIAA-equalized $SN_{\text{ariaa.Cu.20k.ex}}$ of the Cu layer – without all other noise making effects (lines 40 … 45):

$$SN_{\text{ariaa.Cu.20k.ex.min}} = -71.855\,\text{dBA} \tag{3.337}$$

$$SN_{\text{ariaa.Cu.20k.ex.max}} = -72.005\,\text{dBA} \tag{3.338}$$

- Because of the 1 dB minimal steps of the plot in Fig. 3.91 all calculated *SN*s have a tolerance of ±0.5 dB!

### Worst Case *SN*s for a VR

Thus, the worst case Cu-layer SN for a phono-amp (with a V15V at the input) that has to check the quality of a 33 1/3 Cu MOTHER in a DMM manufacturing process will become (rounded to 1 digit after the decimal point):

$$\begin{aligned} SN_{\text{ariaa.Cu.20k.ex.wc}} &= SN_{\text{ariaa.Cu.20k.ex.max}} - 0.5\,\text{dB} \\ &= -72.5\,\text{dBA} \end{aligned} \tag{3.339}$$

Based on the van den Hul answers and the comments of the two producers at the beginning of this chapter the different record SN figures for the phono-amp's worst case[85] (wc) become:

**Table 3.10** Maximal *SN*s for various types of records

| | | |
|---|---|---|
| vinyl 33 1/3 LP record: | $SN_{\text{ariaa.33.V.wc}}$ | $= -70.5$ dBA |
| Cu 33 1/3: | $SN_{\text{ariaa.33.Cu.wc}}$ | $= -72.5$ dBA |
| vinyl Single or Maxi 45 record: | $SN_{\text{ariaa.45.V.wc}}$ | $= -73.5$ dBA |
| Cu 45: | $SN_{\text{ariaa.45.Cu.wc}}$ | $= -75.5$ dBA |

How much the phono-amp's noise level got challenged will be demonstrated with Fig. 3.95 traces. Usually, in test magazines and data sheets of MM phono-amps the input reference (nominal) voltage $e_{\text{in.nom}}$ is 5 $\text{mV}_{\text{rms}}$ at 1kHz. All x-ordinate *SN*s of the following figure are referenced to that.

### Sum of Two *SN*s

- Example: If we have a phono-amp with a rated $SN_{\text{ariaa}} = -77$ dBA ref. 5 $\text{mV}_{\text{rms}}$/1 kHz in $B_{20\,\text{k}}$, than, the solid trace in Fig. 3.95 indicates that the noise contribution of the phono-amp is still significant and it reduces the 45 Maxi records noise in the worst case scenario from −73.5 dBA to −71.9 dBA (loss = 1.6 dB).

The formulae to calculate the traces in Fig. 3.95 look as follows (example with 45 record and Cu layer):

$$\begin{aligned} & SN_{\text{ariaa.res.45.Cu.wc}}(SN_{\text{ariaa.amp.nom}}) \\ &= 20\log\left(\sqrt{10^{\left(\frac{SN_{\text{ariaa.45.Cu.wc}}}{10}\right)} + 10^{\left(\frac{SN_{\text{ariaa.amp.cor}}(SN_{\text{ariaa.amp.nom}})}{10}\right)}}\right) \end{aligned} \tag{3.340}$$

---

[85] wc: because these rather low noise levels challenge any phono-amp's noise characteristic most

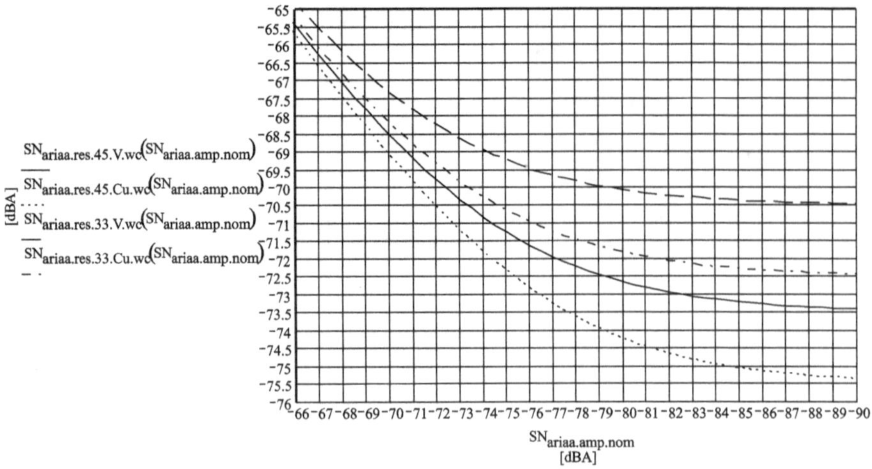


**Fig. 3.95** Traces of the sum of two A-weighted *SN*s for Cu and vinyl records – worst case scenario for Neumann's measurement phono-amp PUE 74 with $NF_{e.amp1} = 1.332$ dB and with different A-weighted *SN*s ($SN_{ariaa.amp.nom}$ at the x-ordinate) referenced to a nominal 5 $mV_{rms}$ input voltage at 1 kHz. Left ordinate gives resulting *SN*s of the sum of a SN of the x-ordinate plus one of the dBA-values given in Table 3.10

## Cartridge O/P Voltage Correction Factor

In case of a nominal MM cartridge output voltage other than $5\,mV_{rms}/1\,kHz$ a correction factor $SN_{cor}$ has to be included into the calculation course. With $e_{out.nom.v15v} = 5.12\,mV_{rms}/1\,kHz$ and $e_{in.nom.amp} = 5\,mV_{rms}/1\,kHz$ for a V15V cartridge this correction factor becomes:

$$SN_{cor} = 20\log\left(\frac{e_{out.nom.v15v}}{e_{in.nom.amp}}\right) = 0.206\,dB \tag{3.341}$$

Thus, making:

$$SN_{ariaa.amp.cor}(SN_{ariaa.amp.nom}) = SN_{ariaa.amp.nom} - SN_{cor} \tag{3.342}$$

Concerning MC cartridges and with reference to a nominal MC phono-amp's input voltage of $0.5\,mV_{rms}/1\,kHz$ the calculation process looks the same.

## *SN*s of a Selection of MM and MC Phono-Amps

To further reduce the noise of the vinyl record in the 80-ies of last century Teldec improved its DMM process by adding a 80 kHz signal that overlays the signals in $B_{20\,k}$. It is said[86] that this process should have improved the records surface by "polishing" it with that high frequency tone, thus, reducing significantly the

[86] "Highlights 15 … 18", cover remarks on the respective 1984 stereoplay 180 g LP records

noise level as well. I couldn't get any information that Fig. 3.91 includes this effects.

As a result of the above shown findings it's obvious that it makes sense to further investigate in the search and design for lowest-noise phono-amps! Tables 3.11 and 3.12 give some interesting informations about the today's and yesterday's situation on that chase for *SN* improvement.

## DMM *SN* Measurement Set-Up

Teldec's DMM measurement set-up might look as follows (I couldn't get any other detailed infos about that):

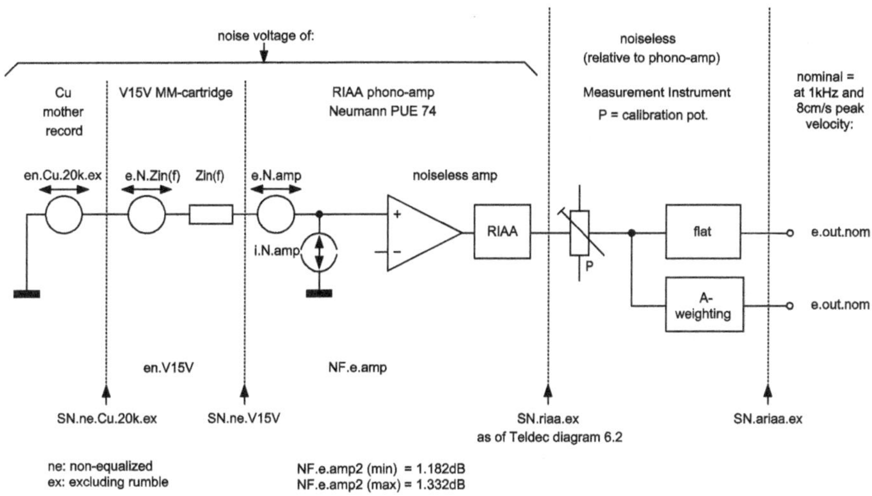


**Fig. 3.96** Teldec DMM measurement set-up with all meaningful noise sources

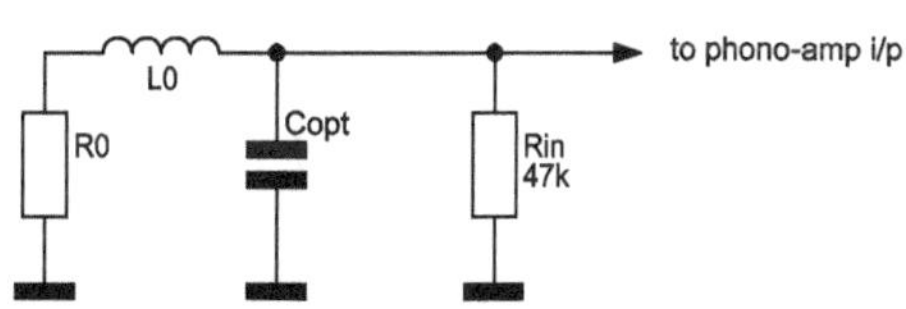


**Fig. 3.97** $Z_{\text{in}}(f)$ of the above shown figure ($R_0 + L_0 =$ V15V)

[87] For the BUVO MM all calculations and measurements where carried out with 250p instead of 125p (for further details see Chap. 4: the influence of the cartridge loading capacitor)

[88] No. 15 and 16: "Ultra-Low-Noise Preamplifier for Moving-Coil Phono Cartridges", E. H. Nordholt and R. M. van Vierzen, JAES 04-1980, Vol. 28, No. 4

**Table 3.11** Selection of MM phono-amps and their *SN*s (* = [87])

| Nr. | Manufacturer + Type | $SN_{ariaa}$ [dBA] – measured w. ref. to 5 $mV_{rms}/B_{20\,k}$ | | | | w. ref. to $e_{nom.v15v}$ | $SN_{ariaa}$ [dBA] – calculated w. ref. to 5 $mV_{rms}/B_{20\,k}$ | | | | w. ref. to $e_{nom.v15v}$ | Input devices | Source + remarks |
|---|---|---|---|---|---|---|---|---|---|---|---|---|---|
| | | 0 R (+ 125 pF* + 47 k5) | 1 k (+ 125 pF* + 47 k5) | Standard (= 1 k + 500 mH + 125 pF + 47 k5) | | V15V (+ 250 p + 47 k5) | 0 R (+ 125 pF* + 47 k5) | 1 k (+ 125 pF* + 47 k5) | Standard (= 1 k + 500 mH + 125 pF + 47 k5) | | V15V (+ 250 p + 47 k5) | | |
| | | Input load | | | | | Input load | | | | | | |
| 1 | Pass X-One | | −84.5 | −80.0 | | | | | | | | JFET | stereoplay 10-01 |
| 2 | Clearaudio Balanced Refer. | | −84.0 | −76.0 | | | | | | | | ? | stereoplay 03-05 |
| 3 | QUAD 42P | | −77.0 | −75.0 | | | | | | | | JFET + valve (6111) | stereoplay 07-06 |
| 4 | Linn Linto ($R_{in}$ = 150 R) | | | | | | | | | | | ? | stereoplay 04-98 |
| 5 | Lehmann Audio Black Cube | | −83.0 | −75.0 | | | | | | | | SSM2017 improved | stereoplay 07-06 |
| 6 | Pro-Ject Tube Box II | | −84.0 | −74.0 | | | | | | | | JRC2068D | stereoplay 07-06 |
| 7 | McLaren PPA20 | | – | −78.5 | | | | | | | | ? | stereoplay 01-02 |
| 8 | NAD PP1 | | | −79.0 | | | | | | | | 5532 | stereoplay 04-98 |
| 9 | NAD PP2 | | −83.0 | −77.0 | | | | | | | | BJTs | stereoplay 11-03 |
| 10 | Music Fidelity XP20 | | | −79.0 | | | | | | | | 1 × LM394 | stereoplay 04-98 |
| 11 | Esoteric Audio Res. 834P | | | −75.0 | | | | | | | | double triode | stereoplay 04-98 |
| 12 | Luxman E03 | | | −74.5 | | | | | | | | ? | stereoplay 04-98 |
| 13 | Burmester 838 MC | | | | | | | | | | | ? | stereoplay 08-84 |
| 14 | Marantz SC-1000 | −85.0 | | | | | | | | | | ? | stereoplay 07-84 |
| 15 | Luxman C-05 | −86.0 | | | | | | | | | | ? | stereoplay 07-84 |
| 16 | Walker-MM (input load: 600 mH‖50 k) | | | | | | | | −80.0 | | | 1 × BC184L | WW 05-72 |
| 17 | Walker-MM (new calc. with Standard) | | | | | | | | −79.8 | | | 1 × BC184L | WW 05-72 |
| 18 | BUVO-MM | | −85.1 | −80.1 | −81.4 | −81.7 | −90.6 | −85.3 | −80.3 | −81.8 | −82.1 | 2 × 2SC2546E | * 250 pF |
| 19 | Neumann PUE 74 | | | | | | −83.9 | −82.1 | −78.8 | −79.9 | −80.1 | 1 × BC212B | $r_{bb}$ = 1040 R |
| 20 | Neumann PUE 74 | | | | | | −86.9 | −83.8 | −79.5 | −80.9 | −81.1 | 1 × BC212B | $r_{bb}$ = 120 R |

**Table 3.12** Selection of MC phono-amps and their $SN$s[88]

| Nr. | Manufacturer + Type | $SN_{ariaa}$ [dBA] – measured with reference to 0.5 $mV_{rms}/B_{20\,k}$ | | | | | $SN_{ariaa}$ [dBA] – calculated | | | | | Input devices | Source + remarks |
|---|---|---|---|---|---|---|---|---|---|---|---|---|---|
| | | 0R | 2R5 | 20R | 25R | 43R | 0R | 2R5 | 20R | 25R | 43R | | |
| | | Input load | | | | | Input load | | | | | | |
| 1 | Pass X-One | | | | −79.5 | | | | | | | JFET | stereoplay 10-01 |
| 2 | Clearaudio Balanced Refer. | | | −75.0 | | | | | | | | ? | stereoplay 03-05 |
| 3 | QUAD 42P | | | −72.0 | | | | | | | | trafo | stereoplay 07-06 |
| 4 | Linn Linto ($R_{in} = 150\,R$) | | | | −81.0 | | | | | | −79.2 | special BJTs | stereoplay 04-98 |
| 5 | Lehmann Audio Black Cube | | | −76.0 | | | | | | | | ? | stereoplay 07-06 |
| 6 | Pro-Ject Tube Box II | | | −65.0 | | | | | | | | JRC2068D | stereoplay 07-06 |
| 7 | McLaren PPA20 | | | −75.0 | | | | | | | | ? | stereoplay 01-02 |
| 8 | Audio Physik Strada | | | −66.0 | | | | | | | | ? | stereoplay 04-02 |
| 9 | NAD PP2 | | | −76.0 | | | | | | | | ? | stereoplay 11-03 |
| 10 | Music Fidelity XP20 | | | | −74.1 | | | | | | −73.7 | 2 × 2SC550 | stereoplay 04-98 |
| 11 | Esoteric Audio Res. 834P | | | | −74.0 | | | | | | | trafo | stereoplay 04-98 |
| 12 | Luxman E03 | | | | −75.5 | | | | | | | ? | stereoplay 04-98 |
| 13 | Burmester 838 MC | | | | −81.0 | | | | | | | 4 BJTs par. | stereoplay 08-84 |
| 14 | Luxman C-05 | −83.0 | | | | | | | | | | ? | stereoplay 07-84 |
| 15 | Nordholt/Vierzen special-device-MC | | −83.1 | | | | | | | | | 1 × spec. made dev. | JAES 04-80 |
| 16 | Nordholt/Vierzen BFY16A-MC | | −81.0 | | | | | | | | −78.2 | 1 × BFY16A | JAES 04-80 |
| 17 | BUVO-MC-solid-state I | −84.9 | | −81.4 | −80.6 | −79.2 | −85.6 | −84.9 | −81.7 | −81.1 | −79.5 | 2 × SSM2210 par. | $R_{in} = 480R$ |
| 18 | BUVO-MC-solid-state II | | | | | | −85.6 | −84.9 | −81.9 | −81.4 | −80.0 | 2 × SSM2210 par. | $R_{in} = 150R$ |
| 19 | BUVO-MC-trafo | | | −80.7 | | −78.4 | | | | | −78.4 | trafo: JT-44-AX | $R_{in} = 475R$ |

Neumann's MM phono-amp diagram looks as follows[89]:

## Neumann PUE 74 Details

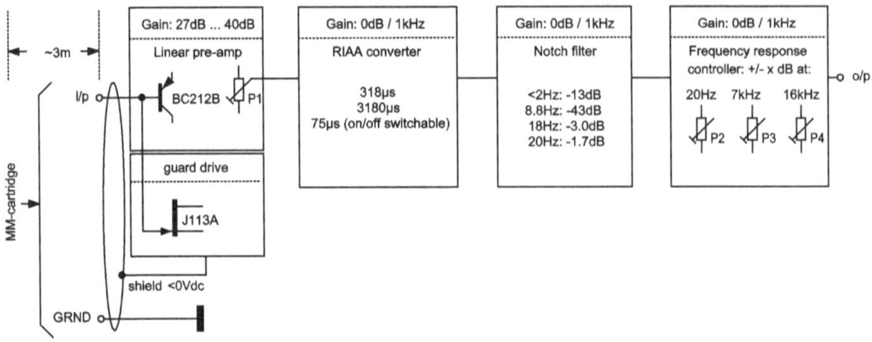


**Fig. 3.98** Neumann PUE 74 Pick-Up Equalizer

This MM phono-amp consists of a special active guard driver input section (around JFET J113A) that enables the insertion of a relatively long low-capacitance shielded cable between cartridge and phono-amp (app. 3m). The guard driver is sensed at the input of the phono-amp[90] and it keeps the input's common mode suppression high. By change of feedback $P_1$ controls the input sensitivity of the amplifying 1st stage. This stage has a flat frequency response. The input device is a BC212B, working with $I_C = 30\,\mu A$ (Fig. 3.99).

The 2nd stage is a conventional RIAA converter built around a LF356 op-amp, configured in series mode.

The 3rd stage double-T notch filter cuts away any noise and/or resonances below 20Hz (op-amp LF356).

In the 4th stage several frequency control potentiometers ($P_2 \ldots P_4$ around a 3rd LF356) allow an exact calibration of the frequency response of the whole cutting arrangement.

The overall frequency response is given in Fig. 3.100 – with potentiometers $P_2 \ldots P_4$ in middle position. It was simulated with pSpice and the input was fed via a precision Anti-RIAA transfer producing measurement instrument (Fig. 12.5b).

Although the Fig. 3.100 plot doesn't look very flat with the pot's mid positions a perfectly balanced frequency response in $B_{20\,k}$ can be trimmed.

The whole calibration process is rather complex and it's description is not part of this book. But one thing should not be lost of sight: the calibration process starts with a special lowest tolerance measurement and calibration record (DIN,

[89] By kind permission of Georg Neumann GmbH, Berlin, Germany

[90] "The handbook of linear applications", Burr Brown 1987 (TI)

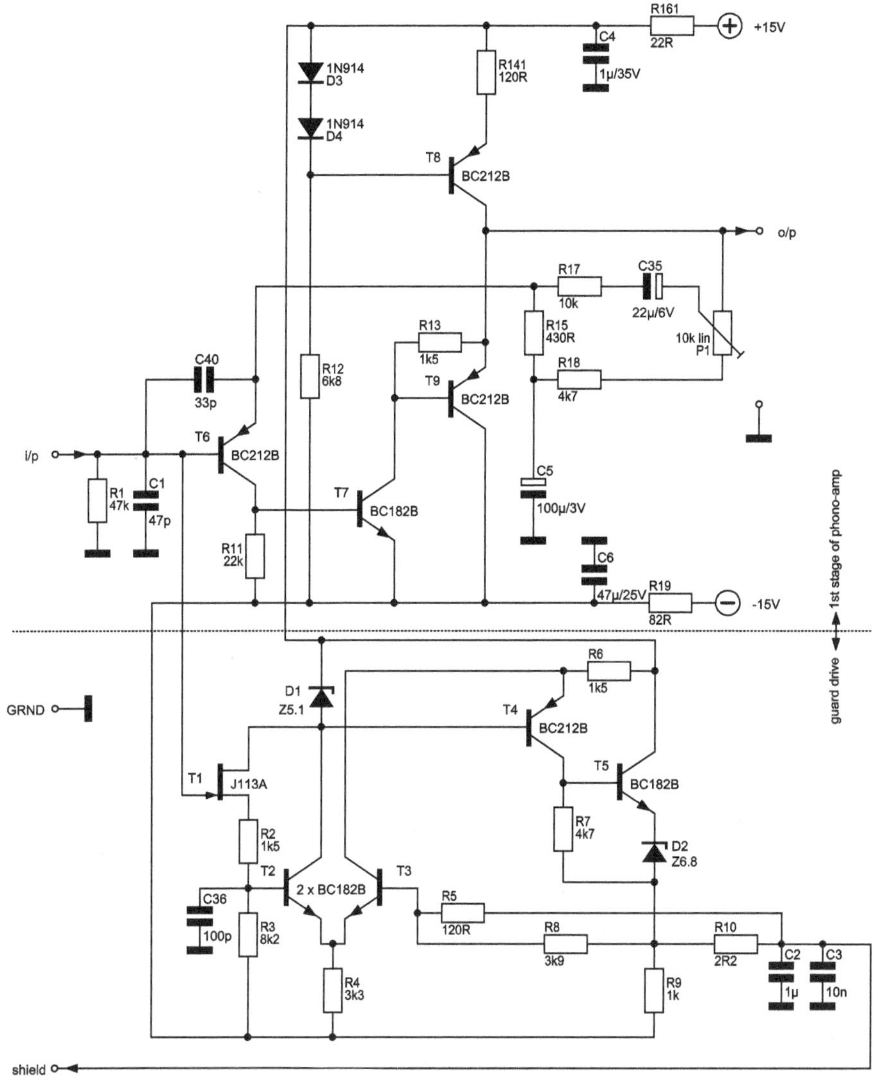


**Fig. 3.99** Neumann PUE 74 circuit of input (1st) stage and guard driver

etc.) via V15V, SME tonearm and PUE 74 phono-amp, thus, including all errors (rumble, resonances, etc.) of that chain. But, the flexibility of the whole cutting systems allows to compensate nearly all those amp chain errors. The final record's cut quality heavily depends on this calibration process. Therefore, those measurement and calibration records are THE essential tool to ensure perfect calibrated cutting lathes.

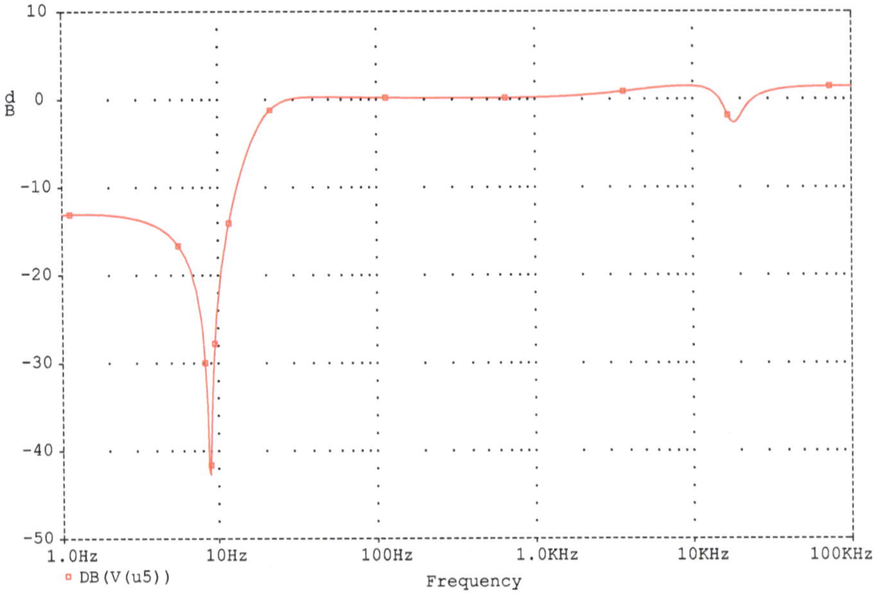


**Fig. 3.100** Frequency response of PUE 74 (all pots in middle position)

Curiously enough, there are not many left on the market and it is said that all respective production materials were destroyed – in consequence of the growing CD industry in the 80-ies of last century. In addition, those who know how to produce calibration records will "disappear" as well rather fast.

# Part II
# Best Practice

# Chapter 4
# Noise in MM Cartridges[1]

## Intro

The design of hum- and noiseless RIAA pre-amplifiers is really a kind of art. The existence of a great variety of circuit designs tells many stories about chases after high signal-to-noise ratios (SN), overload matters and fights for precision of the RIAA transfer. Solutions proposed in the past to use – as substitute – resistors or inductors[2] at the input of a RIAA amplifier for noise measurement purposes or to set on mathematical octave-band analysis[3] to get SN results close to the measured ones (with MM cartridge as input load) do not satisfy. In addition hum spoils many measurement attempts and the optimal loading capacitance of the MM cartridge is quite often not taken into account of calculations or measurements.

That's why I wanted to get the answer to the following question: "how can I calculate with relatively high precision the unweighted SN of a RIAA-equalized pre-amp, loaded with a particular MM cartridge at the input?" In this special case for me "relatively high precision" means a tolerable difference of max. 0.5 – 1.0 dB between results of the mathematical approach to find and measurement.

Several years ago, in another and much more complex mathematical struggle, I was confronted with a powerful mathematical software called MathCad[4] (MCD). I never forget it's easy-to-use and it's high speed to find solutions for such instruments of torture like differential and quadratic equations, integrals, magnitudes etc. Therefore, I thought it might be worth answering my question with the help of such a software.

But theory is only one side of the coin, the other side is craftsmanship: all mathematically-generated results have to be confronted with measurement results.

---

[1] Most parts of this chapter were published in EW 05-2005 under the title "Adventure: Noise – On how to mathematically outwit less nice aspects of an electronical and mechanical piece of art"

[2] "Low noise audio amplifiers", H. P. Walker, Wireless World 05-1972

[3] "AN-104", Jim Sherwin, National Semiconductor Application note, Linear Applications Data book 1986

[4] MathCad is a registered trademark of MathSoft Engineering & Education, Inc., USA, since 2006 part of Parametric Technology Corporation (PTC), MA., USA


DOI: 10.1007/978-3-540-76884-5, ©Springer 2008

Hence, before starting calculations with version 11 of MCD, the whole measurement setup must be developed, built and tested. Earlier versions of MCD will work equally but MCD11 has one giant advantage over all the others: it can be switched between the English and ones mother language version.

Well, to answer my question we need at least some electrical data of the cartridge and the amplifier themselves. A MM cartridge is a very sophisticated piece of electronics and mechanics. Rather high values for its resistance and inductance don't make it easy to develop the right input section of the appropriate amplifier nor to develop a good enough mathematical model.

### Comparison of Manufacturer's and Measured Data

Manufacturer's specifications about cartridges are mostly restricted to DC resistance ($R_1$), 1 kHz inductance ($L_1$), recommended load capacitance ($C_1$) and output voltage (U: in most cases given in $mV_{rms}$ at 1 kHz at 5 cm/s peak velocity). As one can see in Table 4.1 nearly all of these data have to be questioned since the reality, in many cases, is far away from the manufacturer's details (in this study all tested MM cartridges are made by Shure, the only ones I have in my tiny collection).

Concerning the data of the M44G that I've used, the big differences cannot be explained. Because of the many derivatives of this cartridge it might be possible that I've got wrong data for the right one. Nevertheless all measurements and calculations have been performed with the one I have on which is printed M44G. Another significant difference can also be found with the V15III cartridge[5].

Unfortunately this is not the end of the story: the resistance of $R_1$ seems not to have a fixed value at all. It is claimed that it grows proportional to growing frequencies[5] and, thus, taking an increasing part of the noise creation in the whole input network of a RIAA pre-amplifier which includes MM cartridge, $C_1$, input resistance $R_{in}$ (= 47 k) and input transistor of the phono-amp (Fig. 4.1). Fortunately,

**Table 4.1** Manufacturer data vs. measurements

| part | | Shure MM cartridges | | | | | | | |
|---|---|---|---|---|---|---|---|---|---|
| | | V15V MR | | V15IV | | V15III* | | M44G | |
| | | L | R | L | R | L | R | L | R |
| $R_1$ [Ω] | manufacturer | 815 | | 1380 | | 1350 | | 650 | |
| | measured | 791 | 793 | 1316 | 1347 | 1361 | 1359 | 640 | 641 |
| | delta abs. | 24 | 22 | 64 | 33 | 11 | 9 | 10 | 9 |
| | delta [%] | 2.9 | 2.7 | 4.6 | 2.4 | 0.8 | 0.7 | 1.5 | 1.4 |
| $L_1$ [mH] | manufacturer | 330 | | 500 | | 500 | | 650 | |
| | measured | 320.3 | 331.8 | 519.2 | 519.7 | 501.5 | 504.2 | 732.7 | 733.2 |
| | delta abs. | 9.7 | 1.8 | 19.2 | 19.7 | 1.5 | 4.2 | 82.7 | 83.2 |
| | delta [%] | 2.9 | 0.5 | 3.8 | 3.9 | 0.3 | 0.8 | 12.7 | 12.8 |

* values according to [5]: 1388.8 Ω/460 mH

[5] "Noise and moving-magnet cartridges", Marcel van Gevel, EW 10-2003

Fig. 4.1 Basic situation

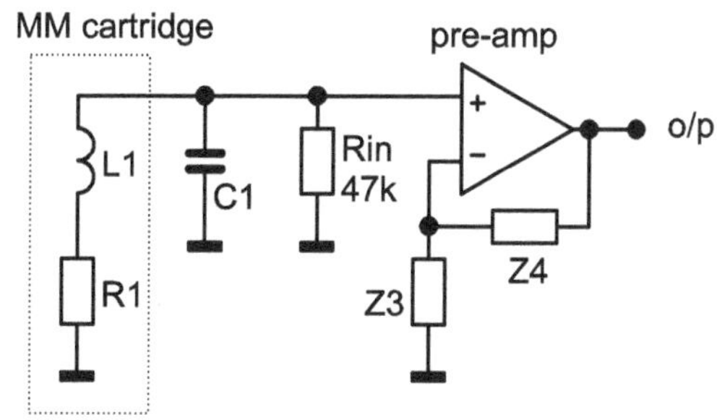


the influence of the input transistor can be made rather low. With a clever design according to the rules given in Chap. 1 for the most part it can be limited to it's noise contribution alone.

## Cartridge Impedance and Phase Measurement

If one of those very expensive network analysers is not available the frequency depended evolution of the magnitude of the impedance of the MM cartridge can be checked as of Fig. 4.2.

The frequency generator Gen1 feeds the MM cartridge via a high value resistor (2 M2), with that creating a current source. Gen1's output voltage $u(f)$ should be 0.5 $V_{rms}$ and it should be capable of handling the whole frequency range from 10 Hz to 20 kHz too. The cartridge is connected to a FET-input op-amp (e.g. OPA604) with low input capacitance (<10 pF) and very high input resistance (>10 M). The output of the op-amp must be connected to an appropriate measurement system, being capable to measure voltages and phase angles. Here, it's a CLIO 6.5[6,7]measuring system that runs very well on an old 500 MHz-Pentium computer under WIN 98. It also includes Gen1.

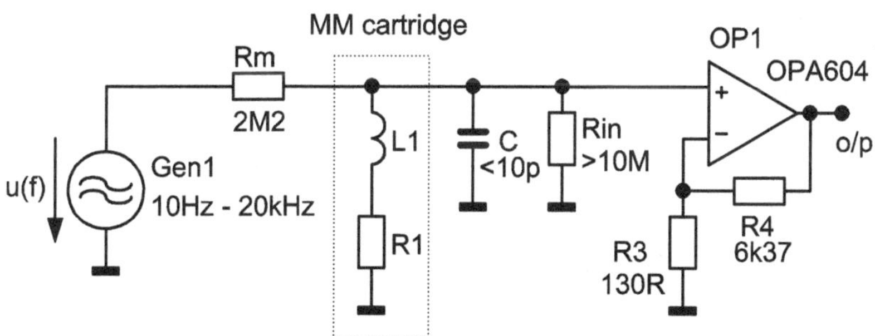


Fig. 4.2 Impedance measurement circuit

[6] "PC controlled audio measurement systems ", Elector Electronics 05-1995

[7] Clio 6.5 is a registered trade mark of AUDIOMATICA SRL, Italy

To create the trace of the magnitude of the MM cartridge the whole frequency range must be fed to the cartridge. The resulting voltage is proportional to the magnitude of $Z_1 = |R_1 + j\omega L_1|$ of the cartridge (Fig. 4.3), lower trace, left ordinate [dBV]. It can be transferred into [Ω] by calculation with the rule of three. Starting point for $Z_1$ is $1.01 \times R_1$ at 10 Hz which is an empirical value as a result of many performed measurements.

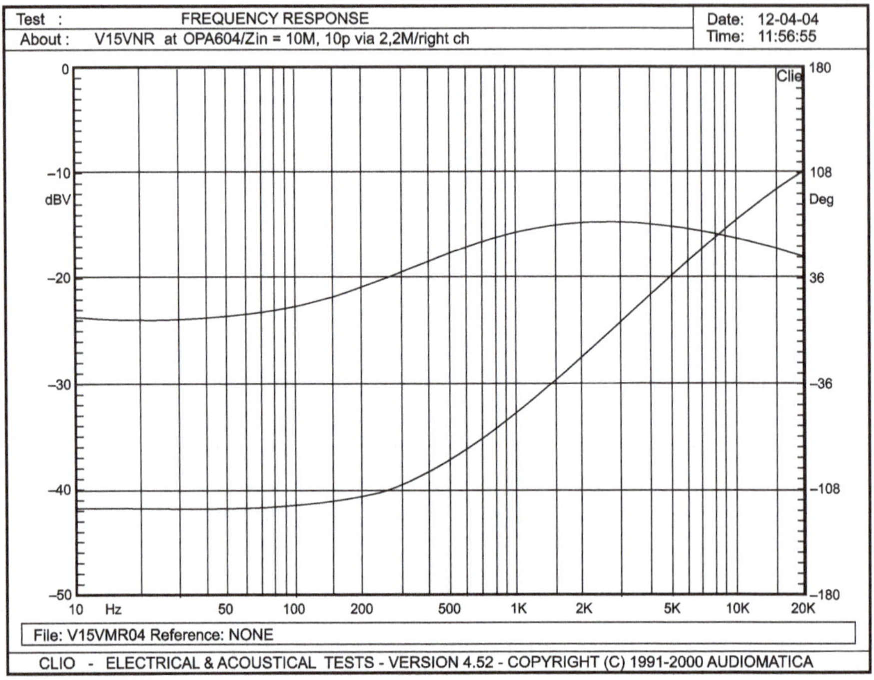


**Fig. 4.3** V15V MR – Impedance and phase

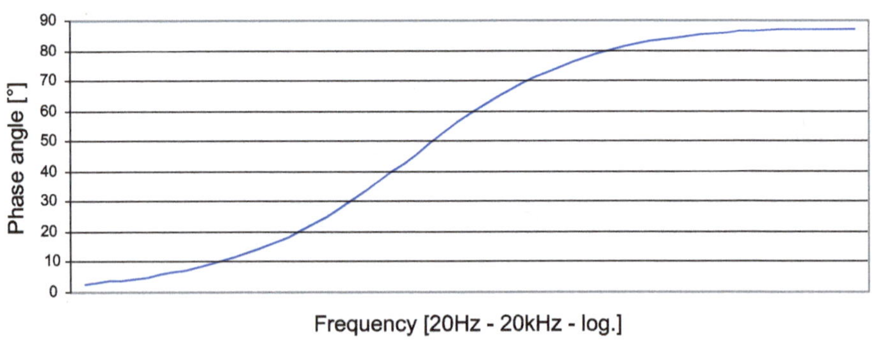


**Fig. 4.4** Phase of constant $R_1$ and $L_1$

### 3rd Octave Band Measurement

If $R_1$ and $L_1$ are a of constant value, then the phase angle $\phi$ (= angle between the magnitude of the impedance of the MM cartridge and it's real part $R_1$, Fig. 4.3 upper trace, right ordinate) should become values more and more close to 90° with frequencies above 10 kHz (Fig. 4.4, = pSpice simulation with constant values for the V15V resistance and inductance, transferred into EXCEL[8]). But this is not the case as one can see in Fig. 4.3 (>3 kHz). In contrast to [5] I guess that not only $R_1$ is frequency dependent, $L_1$ will be too (e.g. V15V – right channel: $L_1$(120 Hz) = 338.0 mH, $L_1$(1 kHz) = 331.8 mH). The manufacturers of MM cartridges don't give much usable informations about the substance of their creation. Therefore, any attempts will fail to dive deeper into the physical and chemical secrets by analysing skin effects and permeability à la[9]. But Fig. 4.5 shows several interesting looking

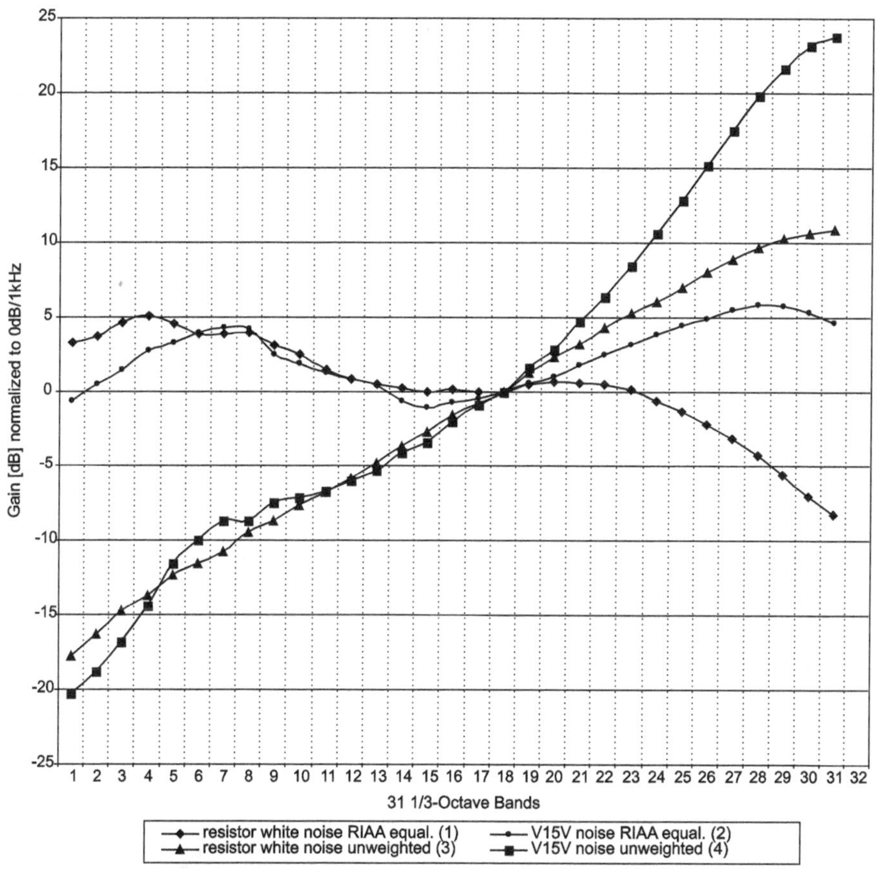


**Fig. 4.5** 3rd octave band measurements

[8] EXCEL is a registered trademark of Microsoft, Inc., USA

[9] "Simple formulae for skin effect", Leslie Green, EW 10-2003

traces for the V15V MR cartridge, which might give better hints for a mathematical model of a MM cartridge. Measured with CLIO's 3rd octave band analyzer (RTA) these 4 traces represent unweighted and RIAA-equalized resistor white noise and V15V noise. It can be seen that unweighted resistor white noise (3) and V15V noise (4) have the same slope <1 kHz. For frequencies >1 kHz the slope of the V15V noise becomes +6 dB and above, until it reaches the resonance frequency of the input network, whereas the slope of the resistor noise keeps its initial slope (+3 dB). The same difference applies to the RIAA-equalized situation [(1) and (2)]. That's why MM cartridge noise is always stronger than the noise of a resistor alone. Consequently, to get real good results close to reality when measuring SNs we'll never find a resistor that is able to replace a MM cartridge – other approaches have to be found.

## Cartridge Equivalent Model

In accordance with the above mentioned findings, it makes sense to start with the simplest mathematical model for a MM cartridge, which consists of a sequence of a constant value resistor $R_1$ and a constant value inductance $L_1$ (Fig. 4.6). At this point of survey Okham's razor forbids to spend too much time to discuss much more complex equivalent circuits and other theories[10]. As will be shown on the following pages the mathematical equivalent of the two component equivalent circuit will lead to satisfying results: for calculations of SNs all we need to know are the exact values of the noise-making components of the RIAA phono-amp and the MM cartridge itself.

## Complete Measurement Arrangement

To examine the results of the proposed mathematical model for a particular MM cartridge connected to a particular RIAA phono-amp, it must be possible for all readers to check these results. That's why details for all measurement tools are given. The measurements where carried out with different pieces of equipment, shown in the following comprehensive wiring diagram (Fig. 4.7).

The whole measurement arrangement consists of the following blocks:

1. A low-noise measurement pre-amplifier simulates the input stage of a RIAA phono-amp. It is not equalized at all. Its gain is set to +34 dB (input sensitivity: $5\,mV_{rms}/1\,kHz$).

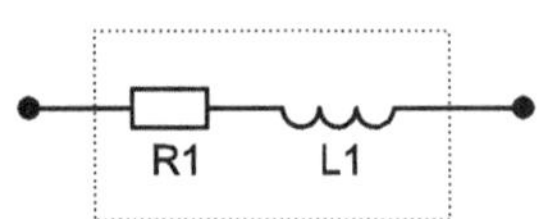


**Fig. 4.6** MM cartridge equivalent circuit

[10] "A puzzling model", Marcel van de Grevel, Letters EW 08-2005

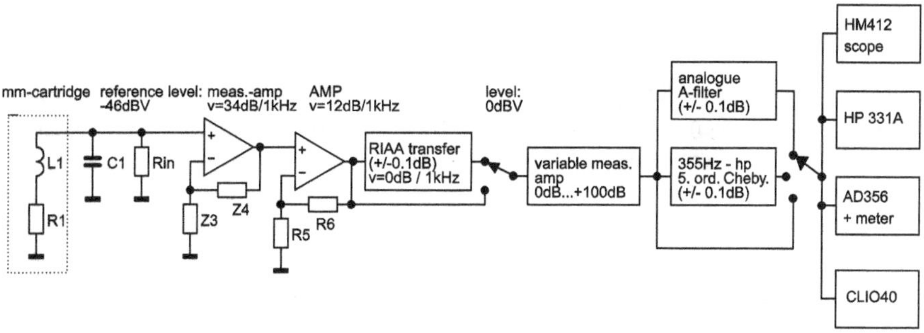


**Fig. 4.7** Measurement arrangement

2. A second amplifier stage (AMP) with a gain of +12 dB lifts up the output level of the 1st stage to a typical driving level (0 dBV) for power amps. To avoid overload it makes sense to design the output level of the first stage as low as possible[11] (say, at 1 kHz $125 - 300\,\mathrm{mV_{rms}}$).
3. To enable measurements with RIAA equalization, a low tolerance (±0.1 dB) RIAA equalizing stage with a gain of 0 dB at 1 kHz can be switched to the output of AMP.
4. To lift the very low-level noise signals an extremely low-noise variable gain (0 dB–100 dB) stage with 3 × LT1028 OPAs[12] follows.
5. A 3-position switch allows to select several weighting possibilities:
   a. a NAB-A-Filter (±0.1 dB),
   b. a 5th order ±0.1 dB Chebyshev high-pass filter ($f_c = 355\,\mathrm{Hz}$)[13] to enable measurements without hum interference[11]. However, the shielding efforts for the whole measurement arrangement should not be underestimated,
   c. no weighting.
6. Finally, 4 different measurement tools show results:
   a. CLIO6.5 is a 16-bit signal generation and measurement system for FFT, frequency response, RTA and much more. It also has a built-in low tolerance NAB-A-Filter,
   b. a RMS-voltmeter with AD536[14] followed by an analogue DC-meter,
   c. the voltmeter section of a HP 331A distortion analyzer,
   d. a Hameg HM 412 scope.
7. All resistors, inductances and capacitors where measured with a "ELC-131 D" L-C-R-meter (±0.5% tolerance) made by ESCORT.

---

[11] D/S, Chaps. 5 and 7

[12] "Designing low-noise audio amplifiers", Wilfried Adam, E&WW 06-1989

[13] "Rumble-/Subsonic-Filter", Elector Electronics 07/08-1990

[14] "Special Linear Reference Manual", Analog Devices 1992, p. 4–12, figure 10

## Mathematical Model with MathCad

We can easily calculate impedance networks with MathCad (the full calculation course will be given in the next chapter[15]). Another point makes work on a MCD worksheet very effective as well: elements once defined on the worksheet keep their value until the end of that definite worksheet, e.g. if the value of $R_1$ (some lines down this page) got changed to another value all following calculations on the worksheet will change accordingly.

The input impedance network $Z_{tot}(f)$ shown in Fig. 4.1 (MM cartridge, $C_1$, $R_{in}$) can be written as the sum of admittances, which is in MCD style:

$$Z_{tot}(f) := \left( \frac{1}{R_1 + 2j\pi f L_1} + 2j\pi f C_1 + \frac{1}{R_{in}} \right)^{-1} \tag{4.1}$$

To calculate the magnitude of $Z_{tot}(f)$ and it's phase angle all values of the components and the plot frequency range $f$ (e.g. 10 Hz steps from 10 Hz–20 kHz) have to be defined first, in this example case for the **Shure V15V MR** cartridge. The calculation results can be plotted in diagrams (Figs. 4.8, 4.9).

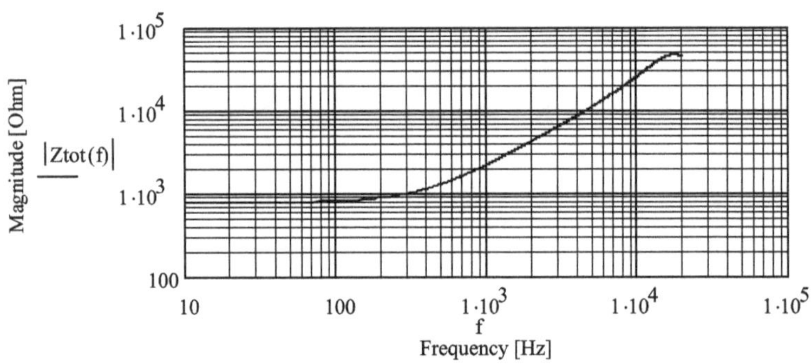


**Fig. 4.8** Impedance of input network

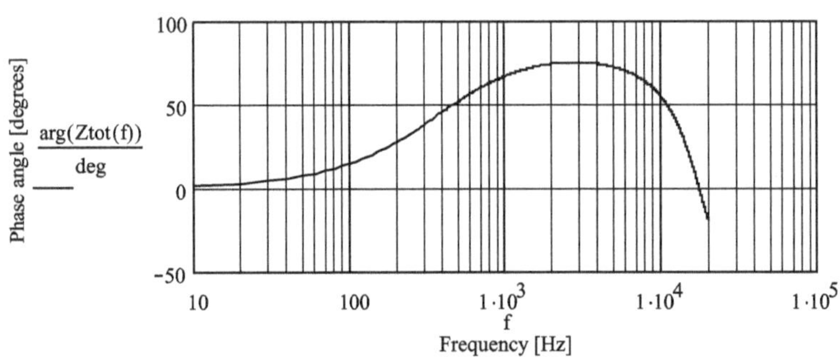


**Fig. 4.9** Phase of input network

[15] Worksheet I, Part II, Chap. 5

All values in the diagrams can be read out by applying the MCD-tool "$x-y$ trace". Values written in MCD style look as follows:

$$R_1 := 793\,\Omega$$
$$L_1 := 0.3318\,\mathrm{H}$$
$$C_1 := 250 \times 10^{-12}\,\mathrm{F}$$
$$R_{\mathrm{in}} := 47.5 \times 10^3\,\Omega$$
$$f := 10\,\mathrm{Hz}, 20\,\mathrm{Hz} \ldots 20{,}000\,\mathrm{Hz}$$

In MathCad, phase angles of complex figures are expressed as radians (rad) of the argument (arg). To get "degrees" the results in "rad" have to be divided by "deg" (Fig. 4.9). The definition of the frequency range "$f$" is essential for the creation of any plot. In this case it's a range from 10 Hz to 20 kHz in 10 Hz steps. $R_{\mathrm{in}} = 47\,\mathrm{k}5$ (and not 47 k) is chosen because of it's easy acquisition as low-noise metal resistor. In the following calculations from a noise point of view the difference of the two values is marginal.

The total noise voltage of $Z_{\mathrm{tot}}(f)$ consists of the two parts $e_{\mathrm{N1}}(f)$ & $e_{\mathrm{N2}}(f)$, which, as uncorrelated noise voltages, have to be summed up together with the other amplifier's uncorrelated noise voltages at the (+)-input of a noiseless amplifier (Fig. 4.10), according to the mathematical rules of the handling of noise voltages and currents[16]. In this case, $e_{\mathrm{N1}}(f)$ is the noise voltage of $R_1$ after it passed through the voltage divider formed by $Z_1(f)$ and $Z_2(f)$, $e_{\mathrm{N2}}(f)$ is the noise voltage of $R_{\mathrm{in}}$ after it passed through the voltage divider formed by $R_{\mathrm{in}}$ and $Z_1a(f)$[17].

## Noise Model of Measurement Amp Plus MM Cartridge

To continue in MCD style all physical constants and values have to be written down as well:

| | |
|---|---|
| $T := 300\,\mathrm{K}$ | = absolute (room) temperature in K |
| $k := 1.380651 \times 10^{-23}\,\mathrm{V\,A\,s\,K^{-1}}$ | = Boltzmann's constant |
| $B_{20\,\mathrm{k}} := 19{,}980\,\mathrm{Hz}$ | = frequency bandwidth |

Application of the Nyquist formula gives the noise voltages of the noise producing components $R_1$ and $R_{\mathrm{in}}$ within $B_{20\,\mathrm{k}}$:

$$e_{\mathrm{N.R1}} = \sqrt{4kTR_1B_{20\,\mathrm{k}}} = 5.124 \times 10^{-7}\,\mathrm{V} \tag{4.2}$$

$$e_{\mathrm{N.Rin}} = 3.965 \times 10^{-6}\,\mathrm{V} \tag{4.3}$$

16 T/S Chaps. 2 and 5

17 "AN-104", Jim Sherwin, National Semiconductor Application note, Linear Applications Data book 1986

The whole measurement set-up that includes MM cartridge under test, measurement amp and all noise sources is given in Fig. 4.10:

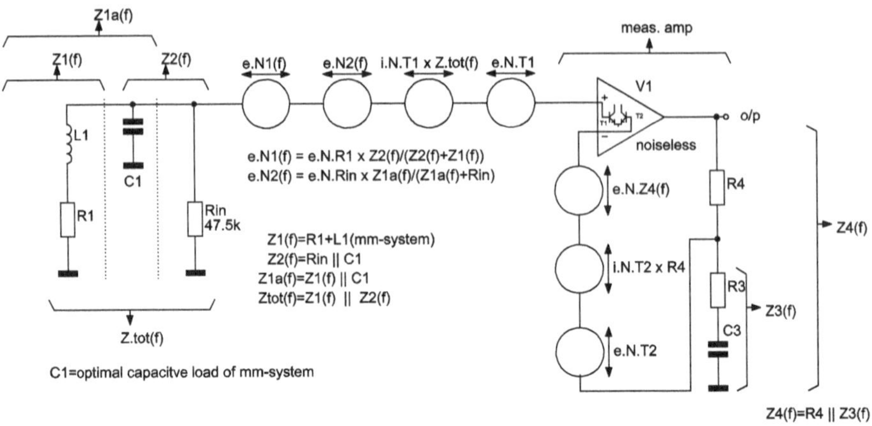


**Fig. 4.10** Measurement amplifier including all meaningful noise sources

The impedances that form the input voltage dividers in Fig. 4.10 are:

$$Z_1(f) := R_1 + 2j\pi f L_1 \tag{4.4}$$

$$Z_1a(f) := \left(\frac{1}{Z_1(f)} + 2j\pi f C_1\right)^{-1} \tag{4.5}$$

$$Z_2(f) := \left(\frac{1}{R_{\text{in}}} + 2j\pi f C_1\right)^{-1} \tag{4.6}$$

Consequently, the equations for $e_{\text{N1}}(f)$ and $e_{\text{N2}}(f)$ look like:

$$e_{\text{N1}}(f) := e_{\text{N.R1}} \left|\frac{Z_2(f)}{Z_1(f) + Z_2(f)}\right| \tag{4.7}$$

$$e_{\text{N2}}(f) := e_{\text{N.Rin}} \left|\frac{Z_1a(f)}{Z_1a(f) + R_{\text{in}}}\right| \tag{4.8}$$

Besides these two noise sources, there are several other equivalent and uncorrelated ones: equal noise voltages and currents $e_{\text{N.T1,2}}$, $i_{\text{N.T1,2}}$ of the long-tailed pair $T_1$, $T_2$ ("equal" if both transistors are carefully paired, $h_{\text{FE}}$ should be $>550$) and noise voltage sources from the feedback network itself or in conjunction with $i_{\text{N.T2}}$ as well as from the total input network in conjunction with $i_{\text{N.T1}}$ (details on the measurement amp will be shown Chap. 11). It is assumed that in the frequency band $B_{20\,\text{k}}$, the spectral noise densities are "white" in general and that there is no $1/f$-noise in $B_{20\,\text{k}}$. This assumption seems to be valid because the chosen transistors (2SC2546E)[18]

[18] Hitachi data sheet

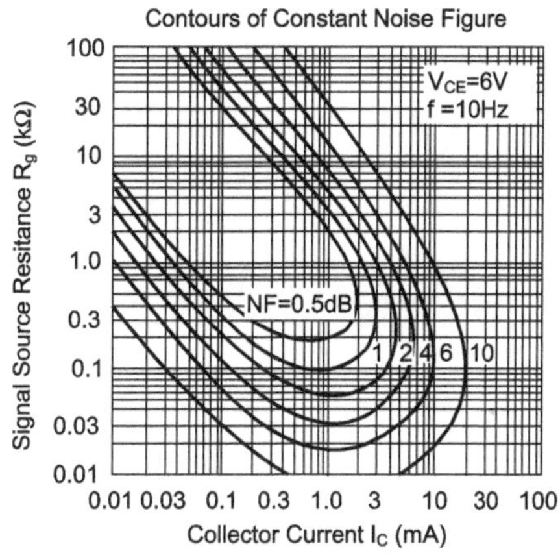


**Fig. 4.11** 2SC2546 contours of constant noise figure at 10 Hz

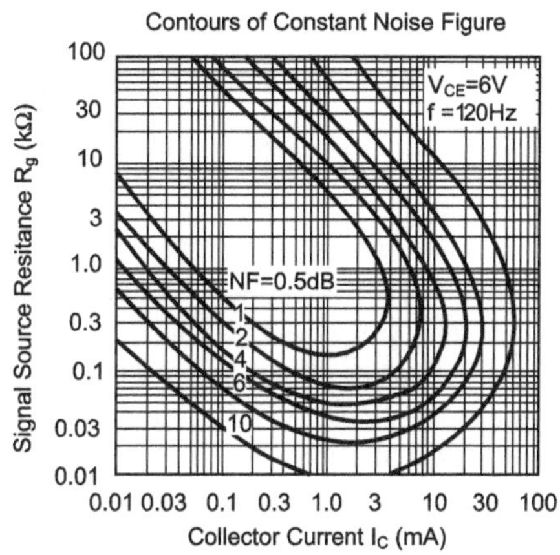


**Fig. 4.12** 2SC2546 contours of constant noise figure at 120 Hz

create noise figure traces (Figs. 4.11 … 4.13) which are very favourable for typical MM cartridge source resistances in the range of 700 R–40 k at $I_C = 100\,\mu A$.

But these findings do not give an answer to the question from where to get the input transistor noise voltage and noise current at a definite collector current. In the low-noise op-amp case you can find these figures or traces in the data sheets. For bipolar transistors a very good approximation can be calculated according to[19]. The values of $e_{N.T1}$, $e_{N.T2}$, $i_{N.T1}$ and $i_{N.T2}$ only depend on physical constants ($T$, $k$, $q =$ elementary charge $= 1.6022 \times 10^{-19}$ A s), $I_C$, $h_{FE}$ and base spreading resistor $r_{bb'}$ (all other internal transistor resistors can be neglected[20,21]).

---

[19] "AN 222", National Semiconductor Application Note, Linear Applications Data Book, 1986

[20] T/S Chap. 2

[21] "Noise in Transistor Circuits", P. J. Baxandall, WW 11& 12-1968

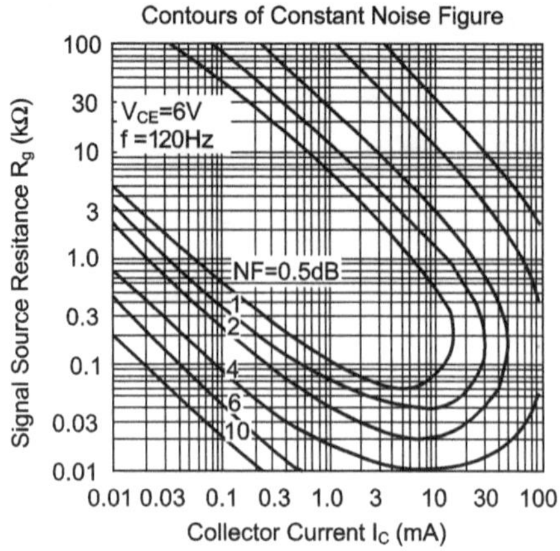


**Fig. 4.13** 2SC2546 contours of constant noise figure at 1 kHz

With Eqs. (3.69) and (3.70) we get as result:

$$e_{\mathrm{n.T1}} = 2.069 \times 10^{-7}\,\mathrm{V} \tag{4.9}$$

$$i_{\mathrm{N.T1}} = 3.267 \times 10^{-11}\,\mathrm{A} \tag{4.10}$$

To get $e_{\mathrm{N.T1}}$ ($= e_{\mathrm{n.T1}} + r_{\mathrm{bb'}}$-effect), further steps have to be taken to find the right value for $r_{\mathrm{bb'}}$ first. Application of the calculation rules given in[22] leads to a quadratic equation for $e_{\mathrm{N}}^2$ which can easily be solved with MCD. The 2SC2546 data sheet figures for the noise voltage $e_{\mathrm{N}}$ at the definite collector current $I_{\mathrm{K}}$ will be the basis for this calculation:

$$I_{\mathrm{K.2sc2546}} := 10^{-2}\,\mathrm{A}$$
$$e_{\mathrm{N.2sc2546}} := 0.5 \times 10^{-9}\,\mathrm{V/rtHz}$$

Noise voltage & current can be calculated with Eqs. (3.69) and (3.70) as well:

$$i_{\mathrm{n}}(I_{\mathrm{K}}) := \sqrt{\frac{2qI_{\mathrm{K}}}{h_{\mathrm{FE}}}B_{20\,\mathrm{k}}} \tag{4.11}$$

$$e_{\mathrm{n}}(I_{\mathrm{K}}) := kT\sqrt{\frac{2}{qI_{\mathrm{K}}}B_{20\,\mathrm{k}}} \tag{4.12}$$

## $\mathbf{r_{bb'}}$ Calculation

The quadratic equation's MCD solution for $r_{\mathrm{bb'}}$ looks as follows (including MCD-tool "solve, $x \rightarrow$")[23]:

---

[22] "AN 222"

[23] Worksheet II, see Chap. 5

$$\left[e_N^2 = e_N(I_K)^2 + (i_n(I_K)r_{bb'})^2 + 4kTr_{bb'}B_{20\,k}\right]$$
$$\text{solve}, r_{bb'} \rightarrow (-3115.98\,\Omega, +13.74\,\Omega) \tag{4.13}$$

**Noise Voltage and Current of Measurement Amp**

For further calculations only the positive solution for $r_{bb'} = 13\,\text{R74}$ makes sense. It lies in the range of the value shown in[24] (14 R0). A check of the calculation approach with LM394 creates a result close to the manufacturer's detail too: 40 R3 vs. 40 R0, thus, $e_{N.T1}(I_C = 100\,\mu\text{A})$ becomes:

$$e_{N.T1} := \sqrt{(e_{n.T1})^2 + (i_{N.T1}r_{bb'})^2 + 4kTr_{bb'}B_{20\,k}}$$
$$e_{N.T1} = 2.176 \times 10^{-7}\,\text{V} \tag{4.14}$$

The collection of important noise sources will be completed by inserting the influential factors of $Z_3(f)$ and $Z_4(f)$ into the whole calculation course:

$$Z_3(f) = R_3 + \frac{1}{2j\pi f C_3} \tag{4.15}$$

$$Z_4(f) = R_4 \,||\, |Z_3(f)| = \left(\frac{1}{R_4} + \frac{1}{|Z_3(f)|}\right)^{-1} \tag{4.16}$$

$i_{N.T2}$ flows through $R_4$ only, thus, $(i_{N.T2} \times R_4)^2$ is a noise voltage source which is independent of the noise gain of the amplifier[25]. To refer this term to the input it must be divided by the noise gain $G_N$

$$G_N = 1 + \frac{R_4}{|Z_3(f)|} \tag{4.17}$$

which leads to the following new noise source:

$$\frac{i_{N.T2}R_4}{G_N} = i_{N.T2}|Z_4(f)| = i_{N.T1}|Z_4(f)| \tag{4.18}$$

To get $Z_3(f)$ and $Z_4(f)$ you have to define the values of $R_3$, $C_3$, $R_4$ first:

$$R_3 := 130\,\Omega$$
$$C_3 := 122 \times 10^{-6}\,\text{F}$$
$$R_4 := 6.37 \times 10^3\,\Omega$$

---

[24] "Noise and moving-magnet cartridges", Marcel van Gevel, EW 10-2003

[25] "Noise in Operational Amplifier Circuits", Analogue Dialogue Vol. 3, March 1969, L. Smith and D. H. Sheingold

With Eqs. (4.15) and (4.16) the noise voltage of $Z_4(f)$ can be calculated:

$$e_{N.Z4}(f) := \sqrt{4kT\,|Z_4(f)|\,B_{20\,k}} \tag{4.19}$$

The sum (Eq. (4.20)) of all relevant noise voltages squared will lead to the input referred and frequency dependent noise voltage $e_{N.tot}(f)^2$. Its rms value is the basis of the signal-to-noise ratio SN with reference to an input voltage of e.g. $5\,mV_{rms}$ ($= -46\,dBV$). Consequently $SN_{ne}$ [dB] can be defined as SN of the unweighted and unequalized noise signal (ne = non equalized) $e_{N.tot}(f)$, which includes noise from the cartridge as well as from the measurement (or phono-) amp. $SN_{riaa}$ is the $SN$ of $e_{N.tot}(f)$ after equalization with the RIAA transfer, $SN_{ariaa} = SN_{riaa}$ + A-Filter weighting.

$$e_{N.tot}(f) := \sqrt{\begin{array}{l} 2(e_{N.T1})^2 + e_{N1}(f)^2 + e_{N2}(f)^2 \\ +(i_{N.T1}\,|Z_{tot}(f)|)^2 + (i_{N.T1}\,|Z_4(f)|)^2 + e_{N.Z4}(f)^2 \end{array}} \tag{4.20}$$

### *SN* Calculations

The rms form $e_N(f)$ of a noise voltage $e_{N.xy}(f)$ in a definite frequency bandwidth can be plotted as:

$$e_N(f) = \sqrt{\frac{1}{f_{high} - f_{low}} \int_{f_{low}}^{f_{high}} \left|e_{N.xy}(f)\right|^2 df} \tag{4.21}$$

Thus, $SN_{ne}$ [dB] referred to $5\,mV_{rms}/1\,kHz$ becomes:

$$SN_{ne} := 20\log\left(\frac{\sqrt{\frac{1}{B_{20\,k}} \int_{20\,Hz}^{20{,}000\,Hz} e_{N.tot}(f)^2\,df}}{5\,mV_{rms}}\right)$$

$$SN_{ne} = -65.1\,dB \tag{4.22}$$

$$\text{Measured: } SN_{ne} = -67.2\,dB$$

Before going further on at this point I have to go back to Fig. 4.10. There are two reasons for the inclusion of a hp pole (formed by $R_3$ and $C_3$) into the circuit:

1. heavy changes of DC voltages at the output can be minimized (caused by impedance changes at the input when measuring with different input loads)
2. this is an additional time constant, simulating the RIAA/IEC roll-off frequency at 20 Hz. I've chosen to shift this frequency to 10 Hz, because my V15V and V15IV driven RIAA phono-amps sound optimal with this configuration. Generally, this frequency doesn't give any heavy extra disturbance. It is kept at 10 Hz throughout the whole calculation and measurement process.

## RIAA Transfer Function and Respective SNs

Referenced to 0 dB at 1 kHz the magnitude of the RIAA transfer function is $R(f)$ (Eq. (2.6)). The plot in Fig. 4.14 allows to pick all values with the help of the respective MCD-tool: e.g. 20 Hz = +19.274 dB, 20 kHz = −19.62 dB.

$$R(f) := \left[ \frac{\sqrt{1+(2\pi f T_3)^2}}{\sqrt{1+(2\pi f T_1)^2}\sqrt{1+(2\pi f T_2)^2}} \right] R(10^3\,\text{Hz})^{-1} \tag{4.23}$$

$$R(10^3\,\text{Hz})^{-1} = 9.898 \tag{4.24}$$

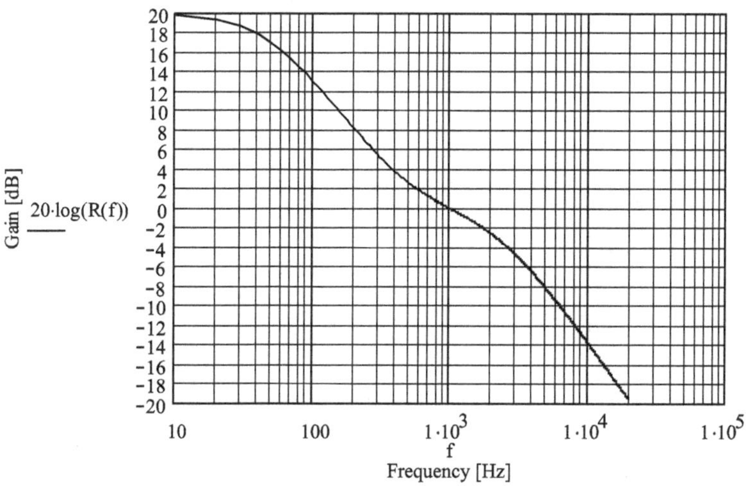


**Fig. 4.14** RIAA transfer

Thus, with the calculation rules for noise voltages passing through a given circuit block[26] the input voltage referred RIAA weighted SN becomes:

$$SN_{\text{riaa}} := 20\log\left( \frac{\sqrt{\frac{1}{B_{20\,\text{k}}} \int_{20\,\text{Hz}}^{20{,}000\,\text{Hz}} e_{\text{N.tot}}(f)^2 R(f)^2\,\mathrm{d}f}}{5\,\text{mV}_{\text{rms}}} \right)$$

$$SN_{\text{riaa}} = -78.4\,\text{dB} \tag{4.25}$$

Measured: $SN_{\text{riaa}} = -78.6\,\text{dB}$

[26] T/S Chap. 2

The precision RIAA transfer circuit of Fig. 4.7 was built according to the design rules set by Messrs van Dael and Kruithof[27] and will be explained in detail in Part III, Chap. 12.

## A-Filter Transfer Function and Respective SNs

RMS noise voltages passing through an A-filter according to NAB/ANSI standard (or IEC/CD 1672) with reference to a definite rms voltage level ($5\,\mathrm{mV_{rms}}$) produce the A-weighted signal-to-noise-ratio $SN_a$. Circuit, filter frequencies and transfer function $A(f)$ see Part III, Chap 12[28] and Fig. 4.15.

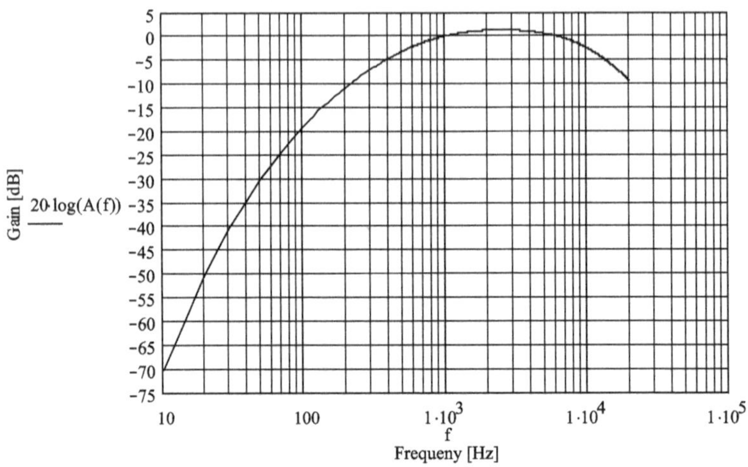


**Fig. 4.15** A-filter transfer

$SN_a$ for $e_{N.tot}(f)$ A-filter weighted becomes[29]:

$$SN_a := 20\log\left(\frac{\sqrt{\frac{1}{B_{20\,k}}\int_{20\,\mathrm{Hz}}^{20{,}000\,\mathrm{Hz}} e_{N.tot}(f)^2 A(f)^2\,\mathrm{d}f}}{5\,\mathrm{mV_{ms}}}\right)$$

$$SN_a = -70.1\,\mathrm{dB} \qquad (4.26)$$

$$\text{Measured: } SN_a = -70.9\,\mathrm{dB}$$

and $SN_{ariaa}$ for $e_{N.tot}(f)$ equalized with RIAA transfer plus A-filter weighting becomes:

[27] "RIAA-Isierung", J. W. van Dael, J. A. Kruithof, Elektor 11-1973 (written in German and Dutch only)

[28] "Precision A-weighting filter", Burkhard Vogel, EW 12-2004

[29] T/S Chap. 2

$$SN_{\text{ariaa}} := 20\log\left(\frac{\sqrt{\frac{1}{B}\int\limits_{20\,\text{Hz}}^{20{,}000\,\text{Hz}} e_{\text{N.tot}}(f)^2 R(f)^2 A(f)^2\,\mathrm{d}f}}{5\,\text{mV}_{\text{rms}}}\right)$$

$$SN_{\text{ariaa}} = -81.9\,\text{dB} \tag{4.27}$$

$$\text{Measured: } SN_{\text{ariaa}} = -81.4\,\text{dB}$$

It seems that RIAA equalization "smoothes" the mathematical SN results more towards the measured ones in comparison with the nonequalized cases.

## Measurement Amp Design

The measurement circuit consists of three different blocks. I guess their circuits are self-explanatory. They are all located on one small PCB that is fixed in a shielded Al-box with the dimensions of $170 \times 120 \times 60\,\text{mm}^3$. In Fig. 4.16 block 1 is the adaptation of a RIAA pre-amp circuit design described in National Semiconductor's Application Note An-222. Block 2 (Fig. 4.17) is the impedance measurement stage and block 3 (Fig. 4.18) is AMP according to Fig. 4.7.

To keep noise on the power supply lines as low as possible the respective circuit looks relatively extensive. VR1 & 2 stabilize the incoming $\pm 21$ V, which is fed in from a separate power supply unit through a 1 m shielded cable. Gyrators ($T_4$, $T_5$) form an extra power supply filter. The separate power supply unit consists of 1 toroidal transformer, 2 rectifiers and 2 high value Cs followed by 2 additional gyrators with high $h_{\text{FE}}$ Darlington transistors BD679 and BD680 (details see Chap. 6, Fig. 6.6).

The MM cartridge to measure keeps attached to its head shell, fixed by a SME connector to a very short piece of tonearm pipe on the top of a separate shielded Al-box (dimensions of $115 \times 65 \times 55\,\text{mm}^3$). The signal lines go out via BNC connectors. $u(f)$ fed through a BNC-L-connector into block 2 and cartridge box (Fig. 4.2) enables impedance measurements, while a very short BNC coupler connects block 1 with the cartridge box for *SN* measurements.

The circuit diagram of block 2 is shown in Fig. 4.17. For other measurement purposes, $S_3$ switches the input resistance from 10 M to 47 k5. The 1Hz-cut-off frequency of the high pass filter $C_3$, $C_4$ & $R_{15}$ is low enough to keep the amp free from gain errors in the 20 Hz … 20 kHz frequency band.

Block 3 (AMP) is a simple low-noise amplifier with it's gain setting components. $R_{18}$ simulates the resistor which might play a role in a two-stage RIAA pre-amp arrangement (75 µs low-pass filter: e.g. 750 R + 100 n). This stage's contribution to the overall noise can totally be neglected. A rule of thumb says: "if the input referred SN of an amplifier stage is more than 20 dB below the SN at the output of the stage in front of it, than this noise contribution can be neglected" (see also the "contribution allowed" issue in Chap. 3.2 "Noise in BJTs" with Tables 3.1 … 3.2 and Eq. (3.83)).

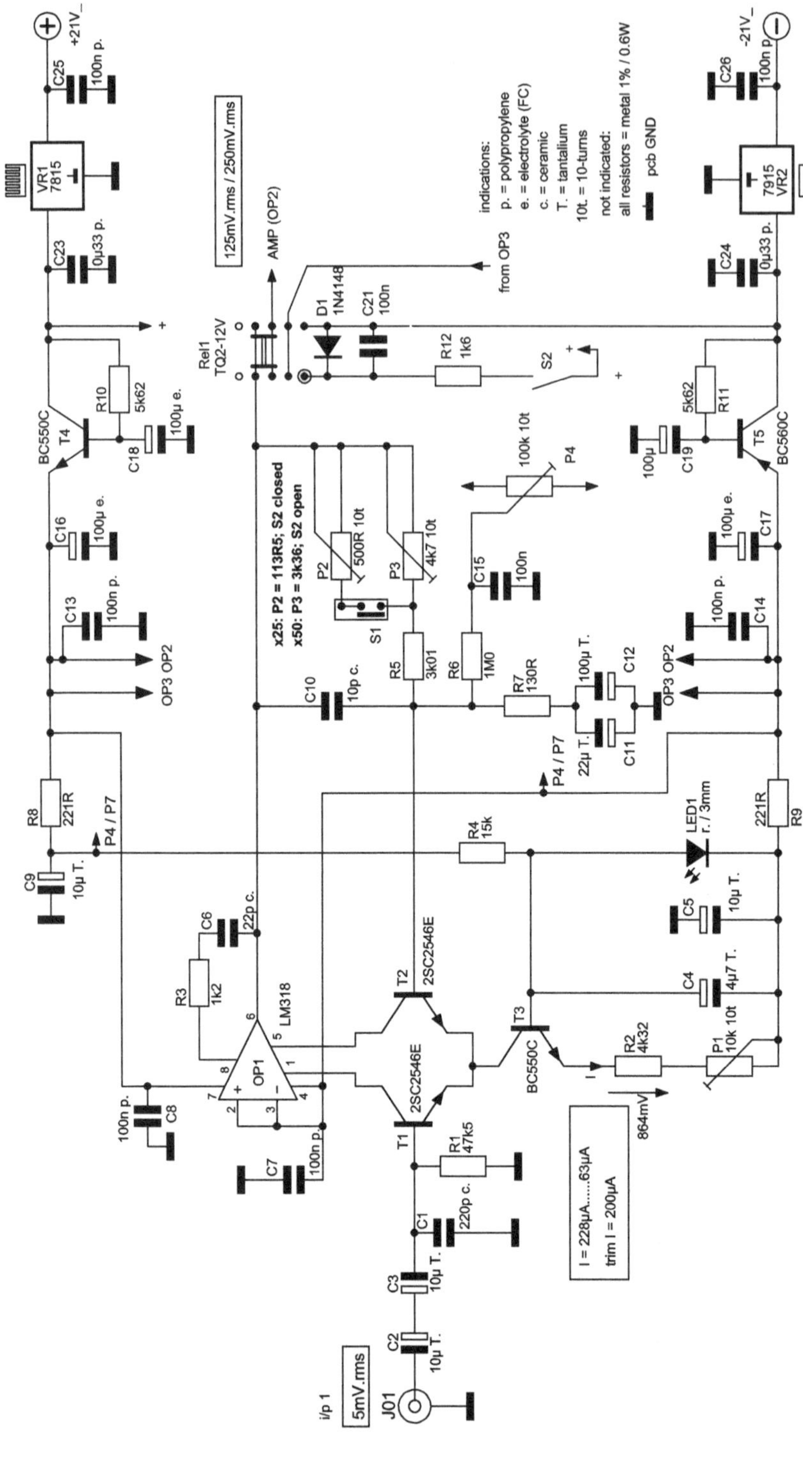


**Fig. 4.16** Circuit of measurement pre-amp and power supply for AMP and impedance measurement block

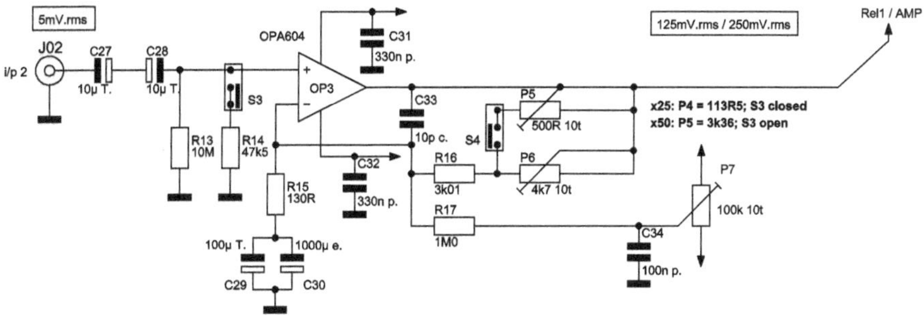


**Fig. 4.17** Impedance measurement circuit

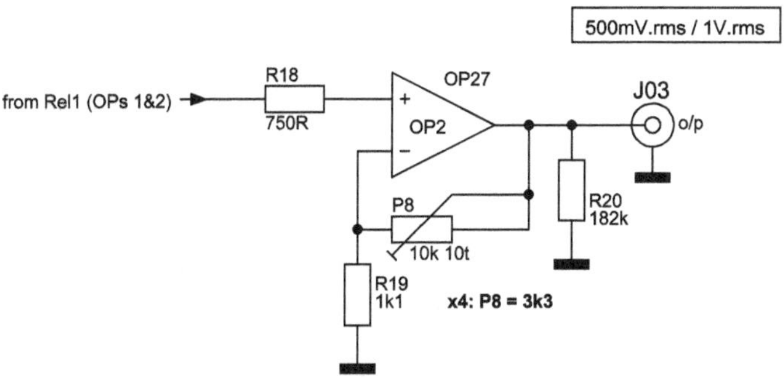


**Fig. 4.18** AMP circuit

## Results

To verify the proposed mathematical model many measurements had to be carried out, especially at late night, when the mains was cleaner and hum interference was less intensive. All calculation and measurement results are listed in Table 4.2. The most important lines are no. 13 (RIAA-equalized noise: $SN_{\mathrm{riaa}}$) and 16 (RIAA-equalized and A-weighted noise: $SN_{\mathrm{ariaa}}$). These deltas indicate that the claim at the beginning of this article becomes true that a max. 1.0 dB variance between mathematics and measurements could be possible. Another point is interesting as well: the measured results for the 1 k and 12 k resistors (lines 6, 9, 12,15) match perfectly with the calculated ones, which is a nice proof of the mathematical model for white noise. The results of the 0 R and 100 R resistors (lines 7, 10) indicate the problems shown in Figs. 4.11 … 4.13: very low source resistances and a low collector current (100 µA) don't match and will lead to additional noise, which is not reflected in the chosen mathematical approach.

For comparison reasons column "L" shows the calculated results of a so called "standard cartridge" which is used in test magazines to check SNs of RIAA amplifiers (e.g. "stereoplay"). It consists of a 1 k resistor series-connected with a 0.5 H inductance (which, of course, is not the same as a MM cartridge inductance of 0.5 H with it's resistance of 1 k! It's nearer to the truth than a resistor alone). But it might not be a good idea to compare test magazine results (with "standard" cartridge) with

**Table 4.2** Results

| 1/A | B | C | D | E | F | G | H | I | J | K | L |
|---|---|---|---|---|---|---|---|---|---|---|---|
| 2 | SN | Input load | MM cartridges | | | | Resistors | | | | MM |
| 3 | | | V15V MR | V15IV | V15III | M44G | 0R | 100 | 1k | 12k | Standard |
| 4 | [dB] | | Right channel only, $C_l = 250$ p | | | | $C_l = 220$ p | | | | 0.5 H + 1 k + 250 p |
| 5 | $SN_{ne}$ | calculated | −65.1 | −64.1 | −64.2 | −63.7 | −82.6 | −81.7 | −77.3 | −68.8 | −64.2 |
| 6 | | measured | −67.2 | −65.7 | −65.3 | −65.7 | −82.1 | −81.3 | −77.2 | −68.8 | |
| 7 | | delta | 2.1 | 1.6 | 1.1 | 2.0 | −0.5 | −0.4 | −0.1 | 0.0 | |
| 8 | $SN_a$ | calculated | −70.1 | −68.1 | −68.2 | −67.1 | −84.7 | −83.7 | −79.4 | −70.8 | −68.3 |
| 9 | | measured | −70.9 | −69.0 | −68.8 | −69.0 | −84.0 | −83.2 | −79.2 | −70.7 | |
| 10 | | delta | 0.8 | 0.9 | 0.6 | 1.9 | −0.7 | −0.5 | −0.2 | −0.1 | |
| 11 | $SN_{riaa}$ | calculated | −78.4 | −76.5 | −76.5 | −76.2 | −86.2 | −85.3 | −81.0 | −72.3 | −76.9 |
| 12 | | measured | −78.6 | −76.5 | −76.5 | −76.5 | −84.9 | −84.2 | −80.7 | −71.9 | |
| 13 | | delta | 0.2 | 0.0 | 0.0 | 0.3 | −1.3 | −1.1 | −0.3 | −0.4 | |
| 14 | $SN_{ariaa}$ | calculated | −81.9 | −79.3 | −79.4 | −78.0 | −90.6 | −89.6 | −85.3 | −76.6 | −79.7 |
| 15 | | measured | −81.4 | −79.2 | −79.2 | −79.0 | −89.9 | −89.0 | −85.1 | −76.5 | |
| 16 | | delta | −0.5 | −0.1 | −0.2 | 1.0 | −0.7 | −0.6 | −0.2 | −0.1 | |

self generated ones because there isn't enough information about $C_l$'s value in the measurement setup. This capacitor has a great influence on SN, which will be lined out a bit later.

The *SN*s shown in Table 4.2 are not the whole truth because each of the tested MM cartridges has it's definite sensitivity $U$, expressed – in most cases – in rms output voltage at 1 kHz at 5 cm/sec peak velocity. Taken this into account all *SN*s in Table 4.2 will be improved (as long as $U > 5\,\mathrm{mV_{rms}}/8\,\mathrm{cm/s}$: e.g. $U_{V15V} = 3.2\,\mathrm{mV_{rms}}/1\,\mathrm{kHz}/5\,\mathrm{cm/s}$; on a LP record the 0 dB level is at 8 cm/s peak velocity[30], therefore, with the rule of three $U_{V15V}$ becomes $5.12\,\mathrm{mV_{rms}}$ and thus, all V15V SNs improve by +0.21 dB. The M44G is much better: with it's output voltage of $U_{M44G} = 9.6\,\mathrm{mV_{rms}}/1\,\mathrm{kHz}/8\,\mathrm{cm/s}$ it improves all SNs by the factor of $20\log(9.6\,\mathrm{mV_{rms}}/1\,\mathrm{kHz}/5\,\mathrm{mV_{rms}}) = 5.67$ dB.

In line 4 of Table 4.2 there are different values of $C_l$. For MM cartridges it's 30 pF higher than for resistors because of the additional capacitance of the BNC connectors and cables inside the cartridge box. A test-wise increase to 250 pF for resistor measurements didn't change anything except for input loads >15 k.

### Influence of the Cartridge Loading Capacitor

A rather significant effect can be observed if we don't take $C_l$ into account. The $SN_{ne}$ of the V15V changes from −65.1 dB (250 p) to −68.5 dB (3 p), which is an

[30] "Reference, Trackability and Frequency Test Record", Deutsche Grammophon Gesellschaft no. 10 99 112

"invalid improvement" of +3.4 dB, $SN_{riaa}$'s improvement will be +1.2 dB. Similar invalid improvements will come up in the A-Filter case. But herein drowses a further potential SN reduction method for those who have access to a calibration and measurement record (e.g. DIN). With the calibration help of that record it would be possible to run a specific MM cartridge connected to a MM phono-amp without $C_1$. But the compensation of the cartridge's resonance has to be performed by an additional frequency response control amp stage (à la Neumann PUE 74) that follows the phono-amp stages. The disadvantage of that method is the fact that the frequency response calibration process has to be performed each time after a cartridge change. Depending on the cartridge the improvements would end up in app. 1 – 2 dB additional *SN*.

## Influence of the Gain Setting Resistor and of Temperature

$R_7$ of Fig. 4.16 has an influence on the *SN*s too. Provided that $C_{11}$, $C_{12}$ and $R_5$, $P_3$ or $P_4$ have been changed adequately a change from 130 R to 10 R improves the $SN_{riaa}$ of a V15V cartridge by a factor of 0.2 dB whereas a change to 499 R worsens it by a factor of 0.7 dB. In the RIAA+A-filter case the respective figures are 0.2 dB improvement/0.5 dB worsening.

Cooling of the phono-amp (e.g. down to $-18\,^{\circ}\mathrm{C} = 255.2\,\mathrm{K}$) leads to an SN improvement of only 0.5 dB for $SN_{riaa}$ and $SN_{ariaa}$ because the cartridge can't be cooled down the same way.

## Summary

With a software like MathCad and the formulae given in this study to calculate unweighted or weighted signal-to-noise-ratios of MM cartridges connected to a RIAA-transfer forming pre-amplifier, we only need 7 basic parameters to get very good calculation results that are close to reality:

- the cartridge's DC resistance
- the cartridge's inductance
- the cartridge's output voltage
- the optimal load capacitance of the cartridge
- the phono-amp's input referred noise voltage
- the phono-amp's input referred noise current
- the gain setting components of the feedback network of the phono-amp.

*SN* measurements with resistors at the input never reflect the MM cartridge's noise reality and those carried out with values $<10\,\mathrm{k}$ will lead to *SN*s that are too optimistic.

Doubling of the input transistors or minimizing the resistors in the feedback network (e.g. in Fig. 4.7: $R_7 \sim 1\,\mathrm{R}$) do not produce that big difference in noise reduction, at all.

As MCD Worksheets the whole mathematical course is given on the following pages!

A detailed schematic for a lowest-noise MM phono-amp is shown in the MC phono-amp chapter (Chap. 6). It serves as a MM and – with transformer input – as a MC phono-amp. The respective MM phono-amp noise data look the same like the ones in Table 4.2.

# Chapter 5
# Noise in MM Cartridges – Mathematical Calculation Course

The following pages show two MCD worksheets with all detailed calculation steps of the previous chapter:

- **Worksheet I** contains the calculations with units included. This is necessary to get an exact overview for the right results. Exception will be the calculation of the base spreading resistor $r_{bb'}$. In worksheet I $r_{bb'}$ is given as a defined value only.
- The calculation of $r_{bb'}$ with units will lead to an over-sized result expression. Therefore I've prepared a second worksheet (**Worksheet II**) with the $r_{bb'}$ calculation without units.
- To get the results of Table 4.2 the values of the components $R_1$ and $L_1$ are the only ones that have to be changed according to Table 4.1 and Table 4.2 line 3 columns H . . . K.
- Any other component change effect will be explained on the worksheets.

**Note 1:** MCD 11 has no built-in unit "rtHz" or "$\sqrt{\text{Hz}}$". To get $\sqrt{1\,\text{Hz}}$ based noise voltage and noise current densities the rms noise voltage and current in a specific frequency range $B > 1$ Hz must be multiplied by $\sqrt{1\,\text{Hz}}$ and divided by the root of that specific frequency range $\sqrt{B}$!

**Note 2:** MCD 11 offers no "dB unit". This is available from MCD 13 on!


DOI: 10.1007/978-3-540-76884-5, 

## ➤ MCD Worksheet I: Noise in MM Cartridges - Page 1

**Step 1 Definition of physical constants and frequency ranges:**

$T := 300\cdot K$ $\quad k := 1.38065\cdot 10^{-23}\cdot V\cdot A\cdot s\cdot K^{-1}$ $\quad q := 1.6021765\cdot 10^{-19}\cdot A\cdot s$

$B_{20k} := 19980\cdot Hz$ $\quad B_1 := 1Hz$ $\quad f := 10Hz, 20Hz .. 20000Hz$

**Step 2 Definition of components of input impedance, input impedance and nominal input voltage [equ. (4.1)]:**

$R1 := 793\Omega$ $\quad L1 := 0.3318H$ $\quad C1 := 250\cdot 10^{-12}F$ $\quad Rin := 47.5\cdot 10^{3}\Omega$

$$Z_{tot}(f) := \left(\frac{1}{R1 + 2j\cdot\pi\cdot f\cdot L1} + 2j\cdot\pi\cdot f\cdot C1 + \frac{1}{Rin}\right)^{-1} \qquad e_{in.nom} := 5\cdot 10^{-3}V$$

**Step 3 graphs of input impedance:**

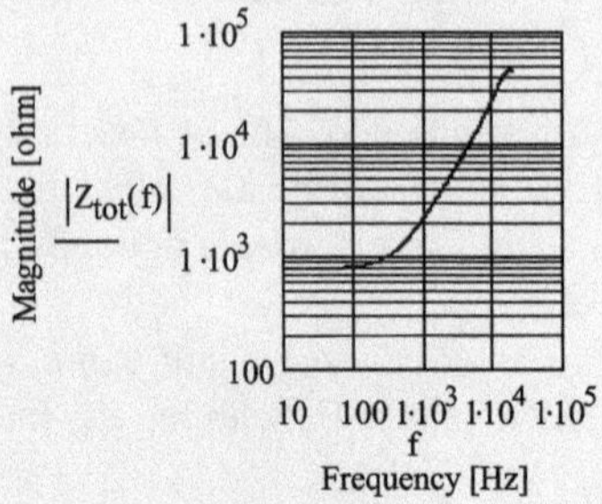


Figure 5.1 = Figure 4.8

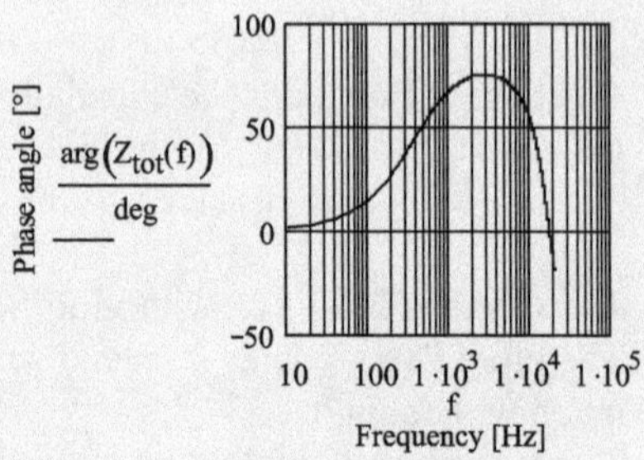


Figure 5.2 = Figure 4.9

**Step 4 calculation of noise voltage of input impedance components (Nyquist formula) [equ. (4.2 ... 4.3)]:**

$$e_{N.R1} := \sqrt{4\cdot k\cdot 300K\cdot R1\cdot B_{20k}} \qquad e_{N.R1} = 5.124\times 10^{-7}V \qquad \frac{e_{N.R1}\cdot\sqrt{B_1}}{\sqrt{B_{20k}}} = 3.625\times 10^{-9}V$$

$$e_{N.Rin} := \sqrt{4\cdot k\cdot T\cdot Rin\cdot B_{20k}} \qquad e_{N.Rin} = 3.965\times 10^{-6}V \qquad \frac{e_{N.Rin}\cdot\sqrt{B_1}}{\sqrt{B_{20k}}} = 2.805\times 10^{-8}V$$

**Step 5 calculation of input voltage devider and graph [equations (2.4 ... 2.8)]:**

$$Z1(f) := R1 + 2j\cdot\pi\cdot f\cdot L1 \qquad Z1a(f) := \left(\frac{1}{Z1(f)} + 2j\cdot\pi\cdot f\cdot C1\right)^{-1} \qquad Z2(f) := \left(\frac{1}{Rin} + 2j\cdot\pi\cdot f\cdot C1\right)^{-1}$$

$$e_{N1}(f) := \left|e_{N.R1}\cdot\frac{Z2(f)}{Z1(f) + Z2(f)}\right|$$

$$e_{N2}(f) := \left|e_{N.Rin}\cdot\frac{Z1a(f)}{Z1a(f) + Rin}\right|$$

➢ **MCD Worksheet I:** Noise in MM Cartridges - Page 2

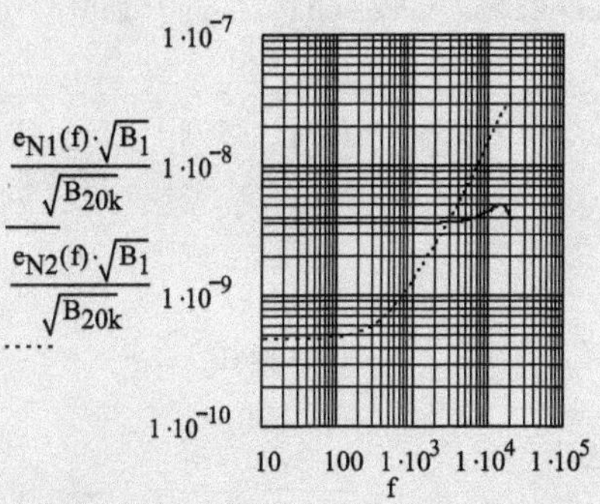


Figure 5.3 Noise voltage densitiy of input voltage dividers

**Step 6 calculation of noise current of input transistor [equ. (4.10)]:**

$I_C := 100\cdot10^{-6}A$ $\qquad h_{FE} := 600$

$i_{N.T1} := \sqrt{\frac{2\cdot q\cdot I_C}{h_{FE}}\cdot B_{20k}}$ $\qquad i_{N.T1} = 3.267\times10^{-11}A$ $\qquad \frac{i_{N.T1}\cdot\sqrt{B_1}}{\sqrt{B_{20k}}} = 2.311\times10^{-13}A$

**Step 7 calculation of noise voltage of input transistor - including the steps to calculate $r_{bb'}$ [equ. (4.9, 4.11 ... 4.14)]:**

$e_{n.T1} := k\cdot T\cdot\sqrt{\frac{2}{q\cdot I_C}\cdot B_{20k}}$ $\qquad e_{n.T1} = 2.069\times10^{-7}V$ $\qquad \frac{e_{n.T1}\cdot\sqrt{B_1}}{\sqrt{B_{20k}}} = 1.463\times10^{-9}V$

2SC2546 values from data sheet:

$I_K := 10^{-2}A$ $\qquad i_n(I_K) := \sqrt{\frac{2\cdot q\cdot I_K}{h_{FE}}\cdot B_1}$ $\qquad i_n(I_K) = 2.311\times10^{-12}A$

$e_N := 0.5\cdot10^{-9}V$ $\qquad e_n(I_K) := k\cdot T\cdot\sqrt{\frac{2}{q\cdot I_K}\cdot B_1}$ $\qquad e_n(I_K) = 1.463\times10^{-10}V$

Because of the enormous size of the result representation of the $r_{bb'}$ calculation with units the calculation for $r_{bb'}$ will take place on a second MCD worksheet without units.

$r_{bb} := 13.74\Omega$ $\qquad e_{N.T1} := \sqrt{e_{n.T1}^2 + (i_{N.T1}\cdot r_{bb})^2 + 4\cdot k\cdot T\cdot r_{bb}\cdot B_{20k}}$

$e_{N.T1} = 2.176\times10^{-7}V$ $\qquad \frac{e_{N.T1}\cdot\sqrt{B_1}}{\sqrt{B_{20k}}} = 1.539\times10^{-9}V$

**step 8 calculation of noise effect of feedback network [equ. (4.15 ... 4.19)]:**

$R3 := 130\Omega$ $\qquad C3 := 122\cdot10^{-6}F$ $\qquad R4 := 6.37\cdot10^3\Omega$

$Z3(f) := R3 + \frac{1}{2j\cdot\pi\cdot f\cdot C3}$ $\qquad Z4(f) := \left(\frac{1}{R4} + \frac{1}{|Z3(f)|}\right)^{-1}$ $\qquad e_{N.Z4}(f) := \sqrt{4\cdot k\cdot T\cdot|Z4(f)|\cdot B_{20k}}$

➢ **MCD Worksheet I:** Noise in MM Cartridges - Page 3

**step 9 graph and calculation of total input referred noise voltage [equ. (4.20)]:**

$$e_{N.tot}(f) := \sqrt{2\cdot e_{N.T1}^2 + e_{N1}(f)^2 + e_{N2}(f)^2 + \left(i_{N.T1}\cdot\left|Z_{tot}(f)\right|\right)^2 + \left(i_{N.T1}\cdot\left|Z4(f)\right|\right)^2 + e_{N.Z4}(f)^2}$$

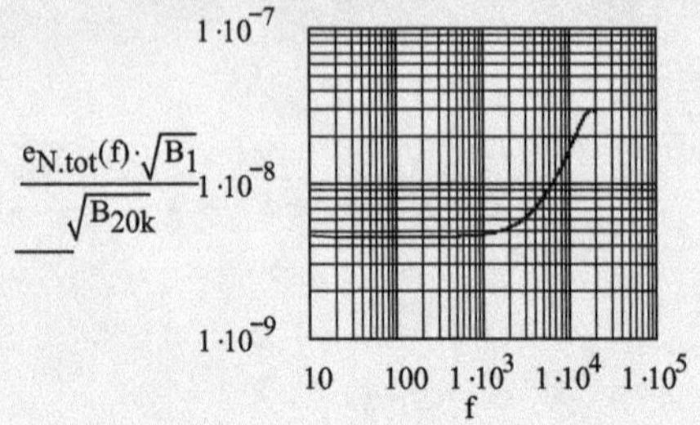

$$h := 1000\text{Hz}$$

$$\frac{e_{N.tot}(h)\cdot\sqrt{B_1}}{\sqrt{B_{20k}}} = 4.643\times10^{-9}\,\text{V}$$

Figure 5.4 Total input noise voltage density

**Step 10graph and calculation of total noise voltage of V15V and input network:**

$$e_{N.V15V}(f) := \sqrt{e_{N1}(f)^2 + e_{N2}(f)^2}$$

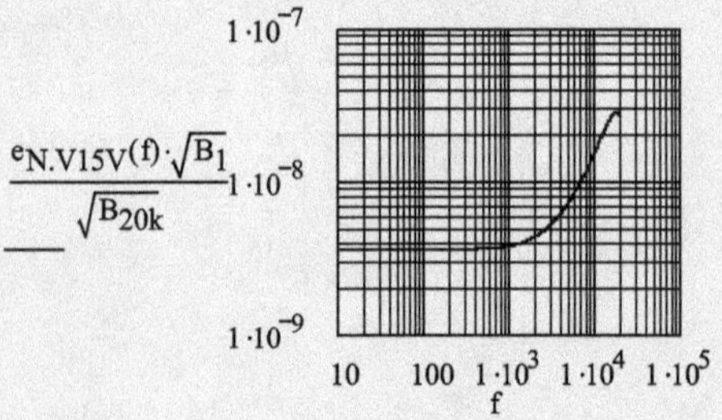

$$\frac{e_{N.V15V}(h)\cdot\sqrt{B_1}}{\sqrt{B_{20k}}} = 3.802\times10^{-9}\,\text{V}$$

Figure 5.5 Noise voltage density of MM cartridge V15V

**Step 11 calculation of SNs [dB] - [equ. (4.21 ... 4.27)]:**

$$e_N(f) = \sqrt{\frac{1}{f_{high} - f_{low}}\cdot\int_{f_{low}}^{f_{high}}\left(\left|e_{Nxy}(f)\right|\right)^2 df}$$

**Step 11a non-equalized:**

$$SN_{ne} := 20\cdot\log\left[\frac{\sqrt{\frac{1}{B_{20k}}\left(\int_{20\text{Hz}}^{20000\text{Hz}} e_{N.tot}(f)^2\,df\right)}}{e_{in.nom}}\right] \qquad SN_{ne} = -65.075$$

measured: -67.2dB

➢ **MCD Worksheet I:** Noise in MM Cartridges - Page 4

**Step 11b** **RIAA-equalized and graph:**

$$R_{1000} := \left[\frac{\sqrt{1+\left(2\cdot\pi\cdot10^{3}\text{Hz}\cdot318\cdot10^{-6}\text{s}\right)^{2}}}{\sqrt{1+\left(2\cdot\pi\cdot10^{3}\text{Hz}\cdot3180\cdot10^{-6}\text{s}\right)^{2}}\cdot\sqrt{1+\left(2\cdot\pi\cdot10^{3}\text{Hz}\cdot75\cdot10^{-6}\text{s}\right)^{2}}}\right]^{-1} \qquad R_{1000} = 9.898$$

$$R(f) := \left[\frac{\sqrt{1+\left(2\cdot\pi\cdot f\cdot318\cdot10^{-6}\text{s}\right)^{2}}}{\sqrt{1+\left(2\cdot\pi\cdot f\cdot3180\cdot10^{-6}\text{s}\right)^{2}}\cdot\sqrt{1+\left(2\cdot\pi\cdot f\cdot75\cdot10^{-6}\text{s}\right)^{2}}}\right]\cdot R_{1000}$$

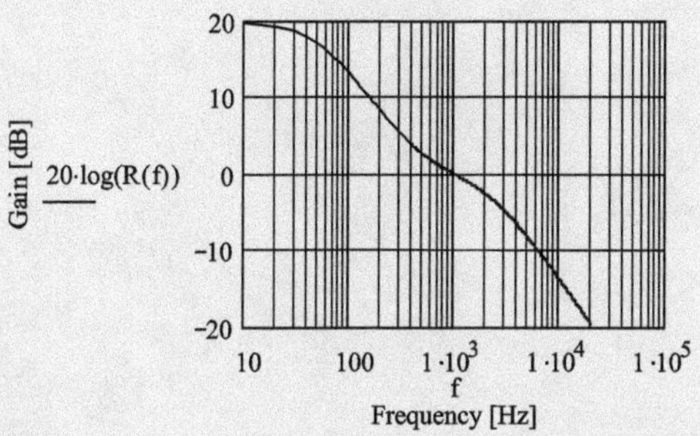


Figure 5.6 = Figure 4.14

$$SN_{riaa} := 20\cdot\log\left(\frac{\sqrt{\frac{1}{B_{20k}}\cdot\int_{20\text{Hz}}^{20000\text{Hz}} e_{N.tot}(f)^{2}\cdot R(f)^{2}\,df}}{e_{in.nom}}\right) \qquad SN_{riaa} = -78.344$$

measured: -78,6dB

**Step 11c** **non-equalized but A-weighted and graph:**

definition of A-filter frequencies and transfer function:

$f_1 := 20.6\text{Hz}$ $f_2 := f_1$ $f_3 := 107.7\text{Hz}$ $f_4 := 737.9\text{Hz}$ $f_5 := 12200\text{Hz}$ $f_6 := f_5$ $f_G := 1000\text{Hz}$

$$v_{1000} := \left[\sqrt{1+\left(\frac{f_1}{f_G}\right)^{2}}\right]^{2}\cdot\sqrt{1+\left(\frac{f_3}{f_G}\right)^{2}}\cdot\sqrt{1+\left(\frac{f_4}{f_G}\right)^{2}}\cdot\left[\sqrt{1+\left(\frac{f_G}{f_5}\right)^{2}}\right]^{2}$$

$v_{1000} = 1.259$

$$A(f) := v_{1000}\left[\frac{1}{\sqrt{1+\left(\frac{f_1}{f}\right)^{2}}}\right]^{2}\cdot\frac{1}{\sqrt{1+\left(\frac{f_3}{f}\right)^{2}}}\cdot\frac{1}{\sqrt{1+\left(\frac{f_4}{f}\right)^{2}}}\cdot\left[\frac{1}{\sqrt{1+\left(\frac{f}{f_5}\right)^{2}}}\right]^{2}$$

➢ **MCD Worksheet I:** Noise in MM Cartridges - **Page 5**

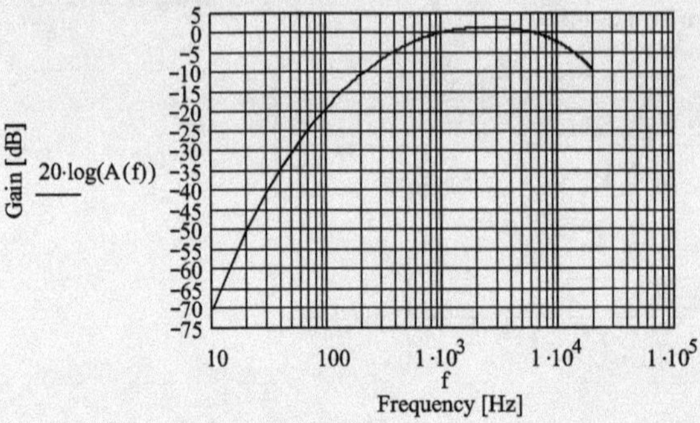


Figure 5.7 = Figure 4.15

$$SN_a := 20 \cdot \log\left(\frac{\sqrt{\frac{1}{B_{20k}} \cdot \int_{20Hz}^{20000Hz} e_{N.tot}(f)^2 \cdot A(f)^2 \, df}}{e_{in.nom}}\right) \qquad SN_a = -70.055$$

measured: -70,9dB

**Step 11d RIAA-equalized and A-weighted:**

$$SN_{ariaa} := 20 \cdot \log\left(\frac{\sqrt{\frac{1}{B_{20k}} \cdot \int_{20Hz}^{20000Hz} e_{N.tot}(f)^2 \cdot A(f)^2 \cdot R(f)^2 \, df}}{e_{in.nom}}\right) \qquad SN_{ariaa} = -81.844$$

measured: -81,4dB

**Step 12 Influence of C1 on SNs:**

A changed C1 value from 250pF to 3pF in the whole calculation course will lead different SN results:

"improvements":

$SN_{ne.3pF} := -68.457$ $\qquad |SN_{ne.3pF} - SN_{ne}| = 3.382$

$SN_{riaa.3pF} := -79.604$ $\qquad |SN_{riaa.3pF} - SN_{riaa}| = 1.26$

$SN_{a.3pF} := -72.750$ $\qquad |SN_{a.3pF} - SN_{a}| = 2.695$

$SN_{ariaa.3pF} := -82.908$ $\qquad |SN_{ariaa.3pF} - SN_{ariaa}| = 1.064$

## ➤ **MCD Worksheet I:** Noise in MM Cartridges - Page 6

**Step 13 Influence of R7 on SNs:**

We have to repeat Steps 8 ... 11 for the two R3 values 10R and 499R:

$R3a := 10\Omega \quad F_{cor.a} := \frac{R3}{R3a} \quad F_{cor.a} = 13 \qquad R3b := 499\Omega \quad F_{cor.b} := \frac{R3}{R3b} \quad F_{cor.b} = 0.261$

$C3a := F_{cor.a} \cdot 122 \cdot 10^{-6} F \qquad C3b := F_{cor.b} \cdot 122 \cdot 10^{-6} F$

$R4a := \frac{6.37 \cdot 10^{3} \Omega}{F_{cor.a}} \qquad R4b := \frac{13 \cdot 6.37 \cdot 10^{3} \Omega}{F_{cor.b}}$

$Z3a(f) := R3a + \frac{1}{2j \cdot \pi \cdot f \cdot C3a} \qquad Z3b(f) := R3b + \frac{1}{2j \cdot \pi \cdot f \cdot C3b}$

$Z4a(f) := \left(\frac{1}{R4a} + \frac{1}{|Z3a(f)|}\right)^{-1} \qquad Z4b(f) := \left(\frac{1}{R4b} + \frac{1}{|Z3b(f)|}\right)^{-1}$

$e_{N.Z4a}(f) := \sqrt{4 \cdot k \cdot T \cdot |Z4a(f)| \cdot B_{20k}} \qquad e_{N.Z4b}(f) := \sqrt{4 \cdot k \cdot T \cdot |Z4b(f)| \cdot B_{20k}}$

$$e_{N.tot.a}(f) := \sqrt{2 \cdot e_{N.T1}^{2} + e_{N1}(f)^{2} + e_{N2}(f)^{2} + i_{N.T1}^{2} \cdot \left[\left(|Z_{tot}(f)|\right)^{2} + \left(|Z4a(f)|\right)^{2}\right] + e_{N.Z4a}(f)^{2}}$$

$$e_{N.tot.b}(f) := \sqrt{2 \cdot e_{N.T1}^{2} + e_{N1}(f)^{2} + e_{N2}(f)^{2} + i_{N.T1}^{2} \cdot \left[\left(|Z_{tot}(f)|\right)^{2} + \left(|Z4b(f)|\right)^{2}\right] + e_{N.Z4b}(f)^{2}}$$

$$SN_{ne.a} := 20 \cdot \log\left[\frac{\sqrt{\frac{1}{B_{20k}}\left(\int_{20Hz}^{20000Hz} e_{N.tot.a}(f)^{2}\, df\right)}}{e_{in.nom}}\right] \qquad SN_{ne.a} = -65.097$$

$SN_{ne} - SN_{ne.a} = 0.022$

$$SN_{ne.b} := 20 \cdot \log\left[\frac{\sqrt{\frac{1}{B_{20k}}\left(\int_{20Hz}^{20000Hz} e_{N.tot.b}(f)^{2}\, df\right)}}{e_{in.nom}}\right] \qquad SN_{ne.b} = -65.007$$

$SN_{ne} - SN_{ne.b} = -0.068$

$$SN_{a.a} := 20 \cdot \log\left(\frac{\sqrt{\frac{1}{B_{20k}} \cdot \int_{20Hz}^{20000Hz} e_{N.tot.a}(f)^{2} \cdot A(f)^{2}\, df}}{e_{in.nom}}\right) \qquad SN_{a.a} = -70.098$$

$SN_{a} - SN_{a.a} = 0.043$

$$SN_{a.b} := 20 \cdot \log\left(\frac{\sqrt{\frac{1}{B_{20k}} \cdot \int_{20Hz}^{20000Hz} e_{N.tot.b}(f)^{2} \cdot A(f)^{2}\, df}}{e_{in.nom}}\right) \qquad SN_{a.b} = -69.922$$

$SN_{a} - SN_{a.b} = -0.133$

➢ **MCD Worksheet I:** Noise in MM Cartridges - Page 7

$$SN_{riaa.a} := 20\cdot\log\left(\frac{\sqrt{\frac{1}{B_{20k}}\cdot\int_{20Hz}^{20000Hz} e_{N.tot.a}(f)^2\cdot R(f)^2\,df}}{e_{in.nom}}\right)$$

$SN_{riaa.a} = -78.553$

$SN_{riaa} - SN_{riaa.a} = 0.209$

$$SN_{riaa.b} := 20\cdot\log\left(\frac{\sqrt{\frac{1}{B_{20k}}\cdot\int_{20Hz}^{20000Hz} e_{N.tot.b}(f)^2\cdot R(f)^2\,df}}{e_{in.nom}}\right)$$

$SN_{riaa.b} = -77.744$

$SN_{riaa} - SN_{riaa.b} = -0.6$

$$SN_{ariaa.a} := 20\cdot\log\left(\frac{\sqrt{\frac{1}{B_{20k}}\cdot\int_{20Hz}^{20000Hz} e_{N.tot.a}(f)^2\cdot A(f)^2\cdot R(f)^2\,df}}{e_{in.nom}}\right)$$

$SN_{ariaa.a} = -82.014$

$SN_{ariaa} - SN_{ariaa.a} = 0.17$

$$SN_{ariaa.b} := 20\cdot\log\left(\frac{\sqrt{\frac{1}{B_{20k}}\cdot\int_{20Hz}^{20000Hz} e_{N.tot.b}(f)^2\cdot A(f)^2\cdot R(f)^2\,df}}{e_{in.nom}}\right)$$

$SN_{ariaa.b} = -81.348$

$SN_{ariaa} - SN_{ariaa.b} = -0.496$

**Step 14 Influence of temperature T on SNs:**

A changed T value from 300K to 255.2K in the whole calculation course will lead to different SN results. The calculation course goes like this:

keep T = 300K for the noise voltage of R1 (the cartridge can't be cooled down the same way like the amp), for all other noise voltages change T to 255.2K.

"improvements":

$SN_{ne.255} := -65.656$ $\left|SN_{ne.255} - SN_{ne}\right| = 0.581$

$SN_{riaa.255} := -78.827$ $\left|SN_{riaa.255} - SN_{riaa}\right| = 0.483$

$SN_{a.255} := -70.622$ $\left|SN_{a.255} - SN_{a}\right| = 0.567$

$SN_{ariaa.255} := -82.343$ $\left|SN_{ariaa.255} - SN_{ariaa}\right| =$

## ➤ MCD Worksheet II: $r_{bb'}$ calculation - Page 1

**Step 1 Definition of physical constants and frequency ranges (without units):**

$T := 300$ $\quad k := 1.38065 \cdot 10^{-23}$ $\quad q := 1.6021765 \cdot 10^{-19}$

$B_{20k} := 19980$ $\quad B_1 := 1$ $\quad f := 10, 20 .. 20000$

**Step 6 calculation of noise current of input transistor (without units):**

$I_C := 100 \cdot 10^{-6}$ $\quad h_{FE} := 600$

$$i_{N.T1} := \sqrt{\frac{2 \cdot q \cdot I_C}{h_{FE}} \cdot B_{20k}} \qquad i_{N.T1} = 3.267 \times 10^{-11} \qquad \frac{i_{N.T1}}{\sqrt{B_{20k}}} = 2.311 \times 10^{-13}$$

**Step 7 calculation of noise voltage of input transistor and $r_{bb'}$ (without units) [equ. (4.11 ... 4.14)]:**

$$e_{n.T1} := k \cdot T \cdot \sqrt{\frac{2}{q \cdot I_C} \cdot B_{20k}} \qquad e_{n.T1} = 2.069 \times 10^{-7} \qquad \frac{e_{n.T1} \cdot B_1}{\sqrt{B_{20k}}} = 1.463 \times 10^{-9}$$

$$I_K := 10^{-2} \qquad i_n(I_K) := \sqrt{\frac{2 \cdot q \cdot I_K}{h_{FE}} \cdot B_1} \qquad i_n(I_K) = 2.311 \times 10^{-12}$$

$$e_N := 0.5 \cdot 10^{-9} \qquad e_n(I_K) := k \cdot T \cdot \sqrt{\frac{2}{q \cdot I_K} \cdot B_1} \qquad e_n(I_K) = 1.463 \times 10^{-10}$$

$$e_N{}^2 = e_n(I_K)^2 + \left(i_n(I_K)^2 \cdot r_{bb}{}^2\right) + 4 \cdot k \cdot T \cdot r_{bb} \cdot B_1{}^2 \text{ solve}, r_{bb} \rightarrow \begin{pmatrix} -3115.9785701523207747 \\ 13.736089376950520619 \end{pmatrix}$$

$$r_{bb} = \begin{pmatrix} -3.116 \times 10^3 \\ 1.374 \times 10^1 \end{pmatrix}$$

$$r_{bb} := 13.74$$

$$e_{N.T1} := \sqrt{e_{n.T1}{}^2 + \left(i_{N.T1} \cdot r_{bb}\right)^2 + 4 \cdot k \cdot T \cdot r_{bb} \cdot B_{20k}}$$

$$e_{N.T1} = 2.176 \times 10^{-7} \qquad \frac{e_{N.T1}}{\sqrt{B_{20k}}} = 1.539 \times 10^{-9}$$

# Chapter 6
# Noise in MC Phono-Amps[1]

## Intro

Despite the fast growing digitisation of the electronic world the vinyl and analogue aficionado is still alive. Moreover, it's a world-wide growing market with increasing demand for excellent phono amplifiers for moving coil (MC) cartridges. Unfortunately, many so-called high-end (and high-price) products show lousy signal-to-noise (*SN*) figures, far away from what is theoretically achievable. Table 3.12 gives an overview.

But there are solutions for lowest-noise phono-amplifiers with a specific MC-cartridge as their input load. There are two alternatives to connect a MC cartridge with an output voltage lower than $1\,mV_{rms}$ to an amplifier chain (see Figs. 1.4 ... 1.5). The two-stage solution consists of a step-up transformer or a solid-state/valve pre-pre-amp (stage 1) attached to a moving magnet (MM) RIAA phono-amp (= stage 2) with an input sensitivity of app. $5\,mV_{rms}$. As such, it creates enough output voltage to drive a pre-amp.

The other option is the full solid-state or valve solution in one stage, including the two different stages. Although the valve driven pre-pre-amp or one-stage solution might be possible to design, inherent obstacles make it impossible to meet the high *SN* ratios that can be achieved with other approaches. For low output voltage producing MC cartridges it's extremely hard to find the right lowest-noise input section. An additional noise-driving factor is the cartridge's source resistance, while its inductance doesn't play an influential role.

As a typical representative of phono-amp challenging MC cartridges, I've chosen a cartridge with output voltage of only $0.35\,mV_{rms}$ at 1 kHz and 5 cm/s velocity, with source resistance = 43 R0 and coil inductance = 56 µH: the Denon DL-103 (no. 2798). It's not too expensive and it sounds excellent.

[1] Most parts of this chapter were published in EW 10-2006 under the title: "The sound of silence: transformer or solid-state? In this article, the author analyses different approaches to solutions for lowest-noise RIAA pre-amplifiers for use with a specific MC-cartridge"

B. Vogel, *The Sound of Silence*
DOI: 10.1007/978-3-540-76884-5, ©Springer 2008

## Benchmarks and Related MC Phono-Amp Problems

Before I started developing a suitable amplifier, I tried to get a benchmark on *SN*s of excellent MC pre-pre-amps and phono-amps. I've checked several test magazines about test reports on MC pre-pre-amps (two-stage approach) and MC phono-amps (one-stage-approach). This survey ended up in very mixed messages: it is nearly impossible to compare *SN* measurement results from one magazine with those of another. In one case they've measured with the input shorted, in another with an input load of 20 R, in other with 25 R and so forth.

The next obstacle is measurements with A-weighting filters: with or without IEC roll-off at 20 Hz. Standard A-weighting filters are allowed to have enormous tolerances. They are developed for sound measurement purposes only and not for amp noise measurements. So, if you have two results from different measurement set-ups you can't compare them exactly.

Viewed from an *SN* standpoint alone – for me the most important one, the best phono-amp I could find was the Linn Linto. A-weighted measurement results for $SN_{\text{riaa.Linto.25}}$ were claimed to be $-81.0\,\text{dBA}/0.5\,\text{mV}_{\text{rms}}$ with 25 R input load in a frequency band of $B_{20\,\text{k}}$[2]. When I wrote the Sound-of-Silence-article in 2005 I had no access to specific technical informations about the Linto phono-amp, especially the input impedance. Therefore I assumed an input impedance of 1 k, thus producing wrong results in Table 1 of the article! Today (YE 2007), I know the Linto's input impedance is 150 R || 4n7[3]. This whole matter forces me now to re-calculate all Linto related equations and table figures. Related results are given in Table 6.1, columns G, H, I.

At this point of the survey it must be pointed out that, in reality, it makes no sense to calculate the Linto's noise performance for a 43 R input load because it is not designed for input loads >15 R (= 1/10 of input impedance). Even a measurement approach with a 25 R input load is questionable because this input load lies outside the recommended input load range. Nevertheless, from an insight point of view I'll go through the calculation course in Chap. 7. Further down this lines a special factor (= adapted input impedance) will help to calculate the Linto noise performance with two changed input impedances.

With reference to an input voltage of $0.5\,\text{mV}_{\text{rms}}$ and a frequency band $B_{20\,\text{k}}$ RIAA-equalized and A-weighted $SN_{\text{ariaa.25}}$ for a 25 R resistor is[4]

$$SN_{\text{ariaa.25}} = -82.737\,\text{dBA} \tag{6.1}$$

Thus the SN-delta $\text{DN}_{\text{ariaa.Linto.25}}$ for the RIAA equalized and A-weighted Linn phono-amp becomes

$$\begin{aligned} DN_{\text{ariaa.Linto.25}} &= SN_{\text{ariaa.Linto.25}} - SN_{\text{ariaa.25}} \\ &= -81.0\,\text{dB} - (-82.737\,\text{dB}) \\ &= 1.737\,\text{dB} \end{aligned} \tag{6.2}$$

[2] "stereoplay" 04-1998, measured with an Audio Precision measurement instrument

[3] Linn Linto data sheet

[4] Worksheet I, Chap. 7

Note: DN should not be mistaken for the noise figure $NF$ of an amplifier. $NF$ relates to the noise factor $F$, which as a power ratio, includes both noise voltages and noise currents [Eqs. (3.46)ff]. But in the case of an unclear noise current situation DN sufficiently shows differences in SN-performance of amplifiers.

The noise voltage $e_{\mathrm{N.R}}$ of a resistor $R$ can be calculated with the Nyquist formula

$$e_{\mathrm{N.R}} = \sqrt{4kTRB} \tag{6.3}$$

With reference to a rms voltage $U$ and a frequency band $B = (f_{\mathrm{high}} - f_{\mathrm{low}})$, the non-equalised and non-weighted $SN_{\mathrm{ne}}$ [dB] of $R$ with

$$X_0(f)^2 = 1 \tag{6.4}$$

becomes:

$$SN_{\mathrm{ne.R}} = 20\log\left(\frac{\sqrt{\frac{1}{B}\left[\int\limits_{f_{\mathrm{low}}}^{f_{\mathrm{high}}} e_{\mathrm{N.R}}^2 X_{\mathrm{n}}(f)^2\,\mathrm{d}f\right]}}{U}\right) \tag{6.5}$$

### Mathematical Rules to Calculate any Kind of *SN* by Simple Means

Inclusion of the factors

$$X_1(f)^2 = R(f)^2 \tag{6.6}$$

or

$$X_2(f)^2 = R(f)^2 \times A(f)^2 \tag{6.7}$$

into Eq. (6.5) enables to calculate any RIAA-equalized ($SN_{\mathrm{riaa.R}}$, $SN_{\mathrm{sriaa.R}}$) and A-weighted ($SN_{\mathrm{ariaa.R}}$) $SN$s of $R$ as well.

- $R(f)$ = RIAA transfer [Eq. (2.6)]
- $A(f)$ = transfer of A-weighting filter[5]
- $k$ = Boltzmann's constant
- $T$ = the absolute room temperature (300 K = 27 °C).

If the Linn Linto is really the record holder the phono-amp I tried to develop, it should beat these figures or at least it should hit the same bottom line. But how can I find out the right approach to meet this goal for an input load of 43 R?

In the past, quite seldomly, electronic magazines specifically picked up the MC phono-amp noise problem. In the middle of the eighties of last century Mr Douglas

[5] Details see Chap. 12 and Worksheet II in Chap. 7

Self did a remarkable job on this issue[6]. That's why his design of a MC pre-pre-amp became mathematically compared with the (more modern?) design of the Linn and my own MC phono-amps (Figs. 6.3 and 6.4).

To convert the measurement figures of the two amps into those with an input load of $R_{0,43} = 43\,\text{R}$, we have to "calculate back" the A-weighted RIAA-equalized $SN$ figure of the Linn for a $R_{0,25} = 25\,\text{R}$ input load (−81.0 dBA) by subtracting the RIAA & A factor = −7.935 dB. The calculation rules for these factors go as the following:

If $SN_{\text{ariaa.AMP}}$[dBA] of a specific amplifier AMP is given, hence, we can calculate certain $SN$s [dB] with the following formulae and SN-factors [dB]. The equations are valid for white noise and purely resistive input loads only. We always have to calculate back to $SN_{\text{ne.AMP}}$ [dB] first (see Eq. (6.8)), followed by an addition of the chosen equalization or weighting factor[7]!

$$SN_{\text{ariaa.AMP}} - SN_{\text{ar}} = SN_{\text{ne.AMP}} \tag{6.8}$$
$$SN_{\text{a.AMP}} = SN_{\text{ne.AMP}} + SN_{\text{a}} \tag{6.9}$$
$$SN_{\text{riaa.AMP}} = SN_{\text{ne.AMP}} + SN_{\text{r}} \tag{6.10}$$
$$SN_{\text{sriaa.AMP}} = SN_{\text{ne.AMP}} + SN_{\text{sr}} \tag{6.11}$$

$SN$s of the transfer functions $R(f)$, $R(f)+A(f)$, $A(f)$ in $B_{20\,\text{k}}$ become:

$$SN_{\text{r}} = -3.646\,\text{dB} \tag{6.12}$$
$$SN_{\text{ar}} = -7.935\,\text{dB} \tag{6.13}$$
$$SN_{\text{a}} = -2.046\,\text{dB} \tag{6.14}$$

$SN$ of $S(f)$ (S-filter) in the frequency range 355 Hz … 20 kHz becomes:

$$SN_{\text{sr}} = -7.853\,\text{dB} \tag{6.15}$$

The following calculations will lead to wrong results:

$$SN_{\text{riaa.AMP}} \neq SN_{\text{ariaa.AMP}} - SN_{\text{a}} \tag{6.16}$$
$$SN_{\text{a.AMP}} \neq SN_{\text{ariaa.AMP}} - SN_{\text{r}} \tag{6.17}$$

Consequently, the non-weighted and non-equalized $SN_{\text{ne.Linto.25}}$ becomes:

$$\begin{aligned} SN_{\text{ne.Linto.25}} &= SN_{\text{ariaa.Linto.25}} - SN_{\text{ar}} \\ &= -81.0\,\text{dB} - (-7.935\,\text{dB}) \\ &= -73.065\,\text{dB} \end{aligned} \tag{6.18}$$

That is the basis for the calculation of the input referred noise voltage $e_{\text{N.Linto.25}}$, including the noise voltage of the source resistance of 25 R and the noise voltage of

[6] "Design of moving-coil head amplifiers", D/S, EW&WW 12-1987

[7] Worksheet II, Chap. 7

the phono-amp. Subtraction of the noise voltage of the 25 R resistor leads to $e_{N.Linto}$, the input referred noise voltage of the Linto phono-amp. Referenced to $\sqrt{1\,\text{Hz}}$ and with an assumption for the input referred noise current $i_{N.Linto}$, we can calculate the input referred noise voltage $e_{N.Linto.43}$ of the Linto with an input load of 43 R and with an adapted input impedance of

$$R_{in,Linto.2} = \frac{R_{in,Linto.1}}{R_{0,25}} R_{0,43} = 258\,\Omega \tag{6.19}$$

The results of this process are[8]:

$$SN_{ariaa.Linto.43.2} = -79.425\,\text{dBA} \tag{6.20}$$
$$SN_{ariaa.43} = -80.381\,\text{dBA} \tag{6.21}$$

Thus, $DN_{ariaa.Linto.43.2}$ becomes:

$$\begin{aligned} DN_{ariaa.Linto.43.2} &= SN_{ariaa.Linto.43.2} - SN_{ariaa.43} \\ &= 0.956\,\text{dB} \end{aligned} \tag{6.22}$$

The input noise current doesn't play a vital role in the calculations. A typical value for a solid-state multi input transistor solution with DC current gain >400 is approximately $4\,\text{pA}/\sqrt{\text{Hz}}$. A change from 2 to $6\,\text{pA}/\sqrt{\text{Hz}}$ shows no significant effect on all *SN*s (in any case $\leq 0.1$ dB).

I've also gone through a 3rd calculation process to check the Linto's noise performance with a hypothetical input impedance equal to the one of my solid-state MC phono-amp:

$$R_{in,Linto.3} = 475\,\Omega \tag{6.23}$$

The results of this process are[9]:

$$SN_{ariaa.Linto.43.3} = -79.207\,\text{dBA} \tag{6.24}$$
$$SN_{ariaa.43} = -80.381\,\text{dBA} \tag{6.25}$$

Thus, $DN_{ariaa.Linto.43.3}$ becomes:

$$\begin{aligned} DN_{ariaa.Linto.43.3} &= SN_{ariaa.Linto.43.3} - SN_{ariaa.43} \\ &= 1.174\,\text{dB} \end{aligned} \tag{6.26}$$

### S-Filter

It's much more difficult to calculate Mr Self's design. I assume *SN* was measured with input shorted ($SN = -139.5$ dBu in the frequency range of 400 Hz – 30 kHz). Otherwise, the *SN* would have been very much lower (= less good). This result

[8] Worksheet III, Chap. 7

[9] dto.

of the non-equalized pre-pre-amp is measured with an hp-filter which cuts away most of the lower frequencies ($f_c = 400\,\text{Hz}$). It is the best way to overcome the measurement difficulties that, in most cases, will occur when measuring solid-state devices being driven by high collector currents – typically for low source resistance designs. Unfortunately, Mr Self's filter is not a standard at all – but it should be one. I call this *SN* measurement method S(Self)-weighting and I will use it throughout the whole exercise in comparing its results with the A-weighting case (in $B_{20\,\text{k}}$ no big differences could be observed). As a standard application this kind of method would make things very much easier, especially comparisons. My own S-filter is a 5th order +0/−0.1 dB Chebishev hp-filter with a cut-off frequency of 355 Hz (Fig. 6.1).

The figures of the conversion of Mr Self's (DOSE) data into the ones I wanted to get for comparisons look as follows. They are referenced to an input voltage of $0.5\,\text{mV}_{\text{rms}}$ at 1 kHz, input load = 43 R, input resistor = 510 R (the reason for that will be explained a bit later), frequency band $B_{20\,\text{k}}$, MM RIAA phono-amp with 5534 op-amp[10]:

$$SN_{\text{riaa.dose.43}} = -73.515\,\text{dB} \tag{6.27}$$
$$SN_{\text{ariaa.dose.43}} = -77.805\,\text{dBA} \tag{6.28}$$
$$SN_{\text{sriaa.dose.43}} = -77.723\,\text{dB} \tag{6.29}$$
$$DN_{\text{riaa.dose.43}} = 2.577\,\text{dB} \tag{6.30}$$
$$DN_{\text{ariaa.dose.43}} = 2.576\,\text{dB} \tag{6.31}$$
$$DN_{\text{sriaa.dose.43}} = 2.576\,\text{dB} \tag{6.32}$$

It's funny but real: by accident my S-filter nearly creates the same *SN* figures like my precision A-weighting filter[11]. To show the tiny differences I express calculated

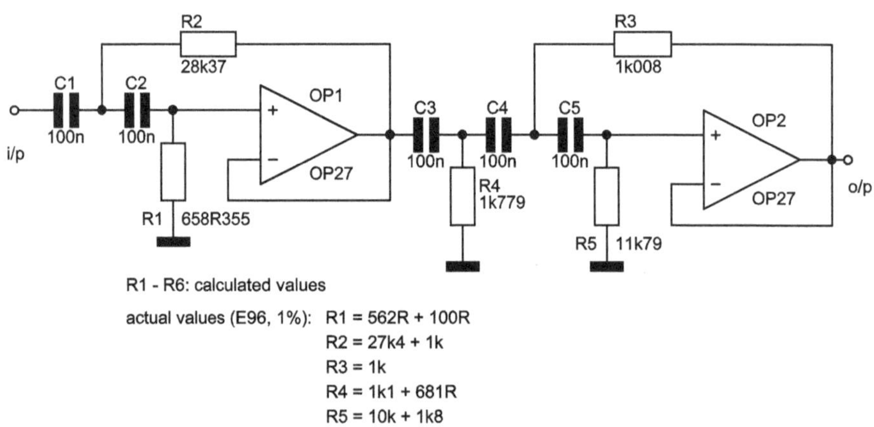


**Fig. 6.1** S-filter with hp cut-off frequency 355 Hz

[10] Worksheet IV, Chap. 7

[11] EW 12-2004

results with 3 digits after the decimal point and round them to 1 digit in tables. In my handicraft box I found five 100 n capacitors with tolerance $<1\%$ and the resistors shown in Fig. 6.1. This is the only reason why the S-filter design ended up with $f_c = 355\,\text{Hz}$.

For the Self 3-stage-design (MC pre-pre-amp plus MM phono-amp plus IEC high-pass filter), the *SN* data are shown in Table 6.1, columns D and E[12]. The calculations were carried out including Mr Self's additional IEC-filter input impedance only[13]. Because of its voltage divider situation at the input, this filter arrangement worsens the *SN* figures by additional 0.4 dB. With that, the total negative impact on $SN_{\text{ariaa}}$ of the Self amp chain (ppa + MM p-a + IEC-hp) becomes 1.3 dB/1.6 dB (MM phono-amp op-amps 5534/5532) by comparing it with the non-impact situation of column F, Table 6.1.

## Calculation Results of a Variety of MC Phono-Amp Design Solutions

Hence, if I would take the Self design approach as the input section of a one-stage MC phono-amp (Table 6.1, column F), the *SN* figures of columns D, E could be improved by 1.3 dB or 1.6 dB. I also calculated the Self pre-pre-amp connected to my own (BUVO) MM phono-amp[14] (Fig. 6.3 without transformer and $R_x$, $C_x$). The resulting *SN* numbers are shown in column C of Table 6.1. They look a bit better than those of Mr Self's two-stage design because my MM phono-amp has lower input noise voltage and input noise current than the 5534 or 5532 op-amps ($\sqrt{2} \times 1.54\,\text{nV/rtHz}$ and $0.23\,\text{pA/rtHz}$). But these results are not as good as if I would develop a new MC phono-amp with an adapted Self input section. That's what I finally did – but not before I tried to find out the *SN* situation of a transformer driven MC phono-amp.

## The Transformer Solution

Without big mechanical work a step-up transformer could be fixed inside the MM phono-amp case. A very good help on the selection of the right step-up transformer is offered on the Jensen Transformers web-site[15]. With the given equations it is easy to understand how transformers work and perform. I checked the products of several transformer suppliers: Lundahl (Sweden), Pikatron and experience electronics (Germany), Sowter (UK) and Jensen (USA).

Source resistance $R_0$ plus primary coil resistance $R_P$ plus secondary coil resistance $R_S$ parallel with the input resistance $R_{\text{in}}$ of the MM phono-amp form the new total input load of the MM phono-amp. The impedance of the $R_x$ and $C_x$ sequence

---

[12] Worksheet IV, Chap. 7

[13] "Self on Audio", D/S, 2000

[14] Worksheet V, Chap. 7

[15] www.jensen-transformers.com

**Table 6.1** Calculated *SN*s in [dB] of various MC phono-amp solutions with input load = 43 R (except Linn Linto, column G), input reference voltage 0.5 $mV_{rms}$, frequency band $B_{20\,k}$

| 1/A | B | C | D | E | F | G | H | I | J | K | L |
|---|---|---|---|---|---|---|---|---|---|---|---|
| 2 | Type of | Douglas Self Design (3 × 2N4403) | | | | | Linn | | Musical | Resistor | Resistor |
| 3 | phono-amp | + BUVO-MM | + DOSE-MM | + DOSE-MM | $h_{FE} = 200$ | | Linto | | Fidelity | $R_{0,25}$ | $R_{0,43}$ |
| 4 | | 2SC2546 | 5534 | 5532 | $I_{C3} = 3mA$ | 25R* | 43R** | 43R*** | XLP | 25R | 43R |
| 5 | *SN* + type of equalization | Calculated | | | | | | | | | |
| 6 | $SN_{riaa}$ | −74.4 | −73.5 | −73.2 | −74.8 | −76.7 | −75.1 | −74.9 | −69.4 | −78.4 | −76.1 |
| 7 | $SN_{ariaa}$ | −78.7 | −77.8 | −77.5 | −79.1 | −81.0 | −79.4 | −79.2 | −73.7 | −82.7 | −80.4 |
| 8 | $SN_{sriaa}$ | −78.6 | −77.7 | −77.4 | −79.0 | −80.9 | −79.3 | −79.1 | −73.6 | −82.7 | −80.3 |
| 9 | Deltas | Amp calculated minus $R_0$ | | | | | | | | | |
| 10 | $DN_{riaa}$ | 1.7 | 2.6 | 2.9 | 1.3 | 1.7 | 1.0 | 1.2 | 6.7 | | |
| 11 | $DN_{ariaa}$ | 1.7 | 2.6 | 2.9 | 1.3 | 1.7 | 1.0 | 1.2 | 6.7 | | |
| 12 | $DN_{sriaa}$ | 1.7 | 2.6 | 2.9 | 1.3 | 1.7 | 1.0 | 1.2 | 6.7 | | |

* input imp. 150R
** input imp. 258R
*** input imp. 475R

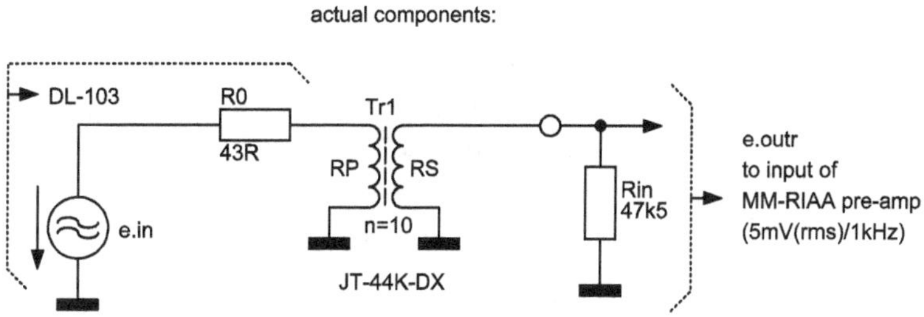


**Fig. 6.2a** Transformer circuit

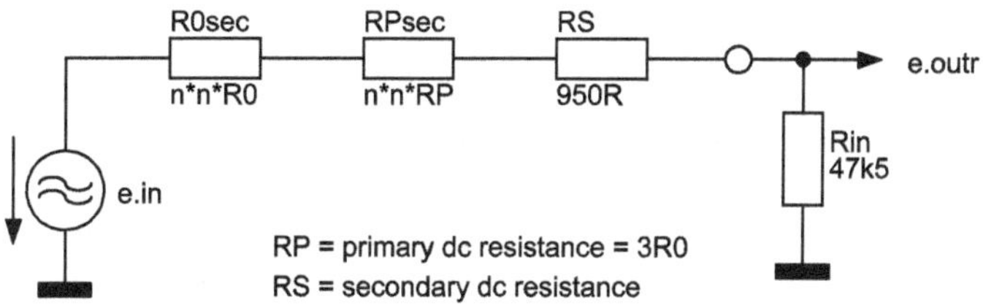


**Fig. 6.2b** Equivalent transformer circuit

can be ignored in this calculation (Fig. 6.3a). It only manages the resonance and keeps the frequency response flat at the upper end of the audio band. Unfortunately, $R_0$ and $R_P$ have to be multiplied with the square of the coil turns $n^2$ ($n = 10$, $n^2 = 100$, see Figs. 6.2a, b). Thus, to create lowest noise at the input of the MM phono-amp we have to try to find a transformer with $R_P$ and $R_S$ as low as possible.

## Equivalent Circuit

Here, I won't discuss other selection parameters like distortion, frequency response, overload etc. The respective (paper) figures of the transformers I've checked were all good enough. They don't play the same role like the noise making ones.

$R_P = 3\,R$ and $R_S = 950\,R$ for the Jensen transformer are by far the best figures, followed by the Pikatron: $R_P = 9\,R4$ and $R_S = 380\,R$. The two other types have very much higher values, Sowter only offers a step-up transformer for high source resistances with $n = 12.5$.

## Trafo Plus BUVO MM Phono-Amp

With the 47 k5 input resistor of my MM phono-amp the 43 R input loaded Jensen transformer produces a total input load of 4k969[16] (Pikatron: 5 k045). The respective calculated and measured *SN*s are given in Table 6.2, columns H, I. These figures are not as good as those of the Self pre-pre-amp plus the BUVO MM phono-amp

[16] Jensen Transformers Application Note AS040

(Table 6.1, column C). But this transformer/MM phono-amp chain configuration has an advantage over all the full-blown solid-state designs: its low frequency noise [mainly shot- and $1/f$- (flicker-) noise] is very much lower than that of the solid-state case (Table 6.2, columns C-F, lines 6, 10). That's why there is a big difference between measured (RTA and FFT with 50 times averaging) and calculated values of $SN_{\mathrm{riaa}}$ for the two MC phono-amps I've built up: 2.7 dB and 3.3 dB. In the transformer case it's only 0.5 dB.

The complete transformer driven MC phono-amp circuit is shown in Fig. 6.3a. For those who try to avoid any capacitor inside the amplification chain in Fig. 6.3b a C3 replacing circuit is shown as well. This circuit keeps the output of OP2 free of any DC voltage. I could not hear – nor noise-measure – a difference between the two approaches. The relatively simple power-supply-unit (inside the phono-amp case) is given in Fig. 6.3c. It is fed with $\pm 21$ V via J02 in Fig. 6.7 or via a separate PSU à la Fig. 6.6.

The transformer with its high-frequency resonance managing $R_{\mathrm{x}} - C_{\mathrm{x}}$ network at the secondary winding is coupled directly to the input of the phono-amp and $R_1$. The input stage is built-up with the same transistor types like those of Fig. 4.16. They are also driven by the same collector current (100 µA – set by $P_1$), thus, producing the same amount of input noise voltage and noise current. Used as a MM phono-amp it creates the *SN* figures given in Table 4.2. The basic circuit design of this very low-distortion amp was developed by E. F. Taylor in the seventies of last century[17].

In contrast to the Taylor design (with its simple two-transistor current mirror) the long-tailed input pair's collector currents are carefully balanced by a precision Wilson current mirror. Besides a very high input resistance the advantage of that kind of current mirror is the fact that it produces the lowest amount of balancing errors[18] between the two currents that feed the long-tailed pair, thus, keeping the noise producing mechanisms of the two transistors absolutely equal – provided that the current gains were carefully paired. This first high-gain stage is followed by an op-amp with its (+)-input fixed at app. +5.2 V, thus, via the op-amp (-)-input forcing the collector voltage of T1 as well to the same amount. This keeps the power consumption and temperature of this transistor rather low. To fix $T_2$'s temperature and power consumption the same way a test-wise insertion of a resistor between collectors of $T_2$ and $T_4$ didn't improve – nor worsened – the noise effects.

The RIAA transfer is built-up in two steps: bass-boost with OP1 and $T_1$, $T_2$, high-frequency cut by OP2 and the R–C network in front of it[19]. To keep the overload problems as low as possible[20] the total gain of the 1st stage is fixed at a rather low +28 dB. To keep the noise on the power supply lines as low as possible they are filtered with gyrators ($T_{7,8}$). Their audio-band filter-effect is app. $-50$ dB[21].

---

[17] "Distortion in low-noise amplifiers", E. F. Taylor, WW 09-1977

[18] T/S Chap. 4

[19] A detailed analysis of RIAA transfer producing networks is given in Chaps. 8 ... 9

[20] "Phono preamp", D/S, Letter EW 05-2001

[21] "AN222", National Semiconductor Application Note, Linear Applications Databook, 1986

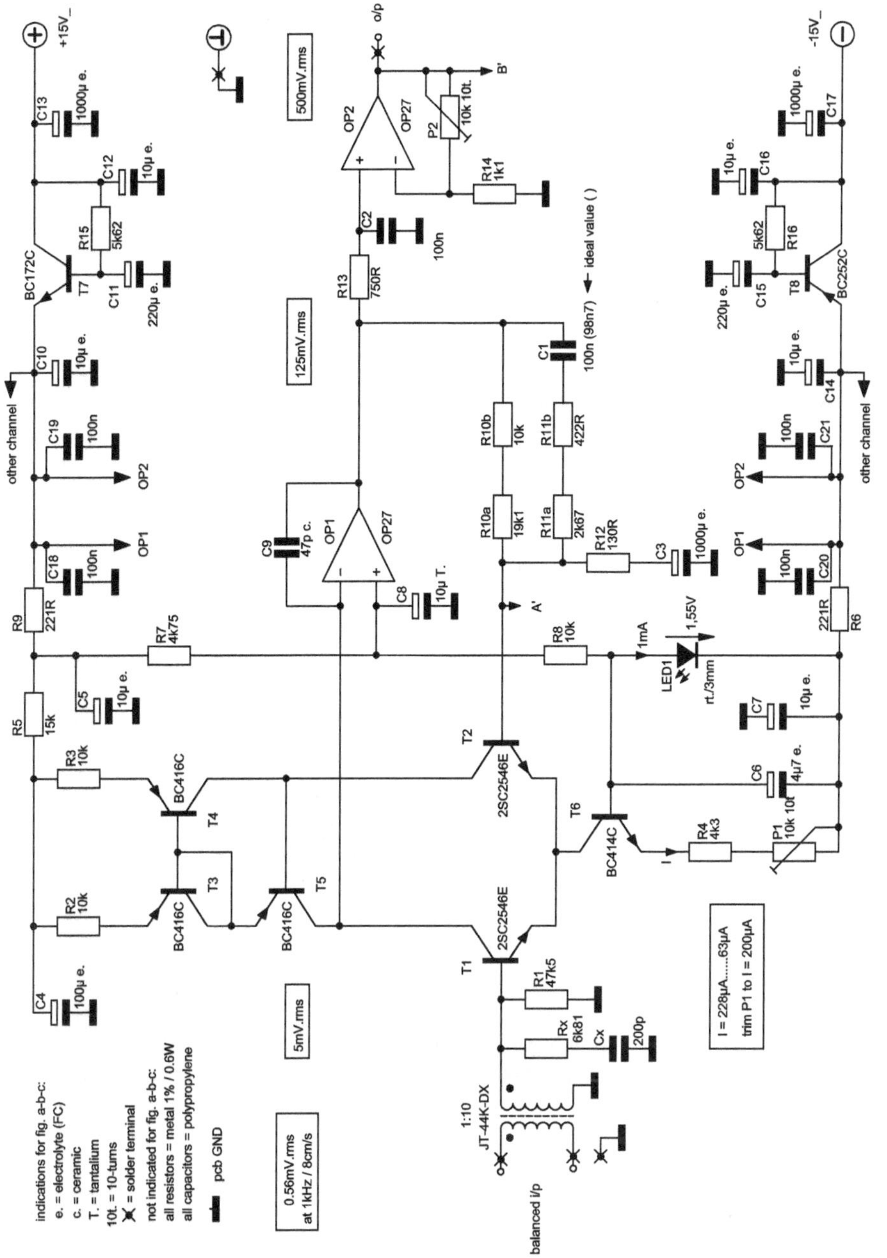


**Fig. 6.3a** MC phono-amp formed by a step-up transformer plus BUVO MM phono-amp

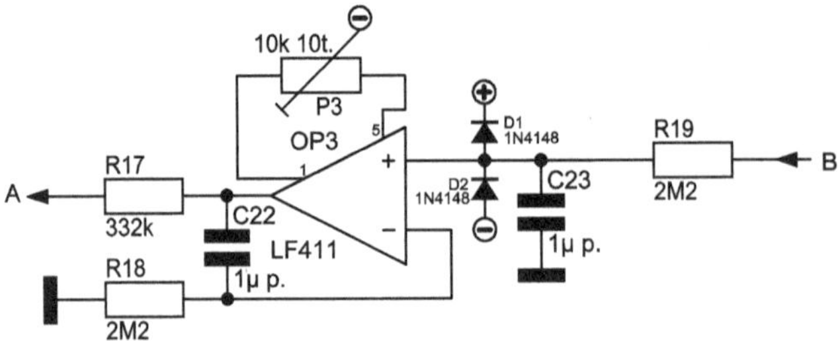


1. bridge C3
2. insert extra circuitry around OP3
3. connect points A and A' + B and B'

**Fig. 6.3b** Alternative for C3 of Fig. 6.2a

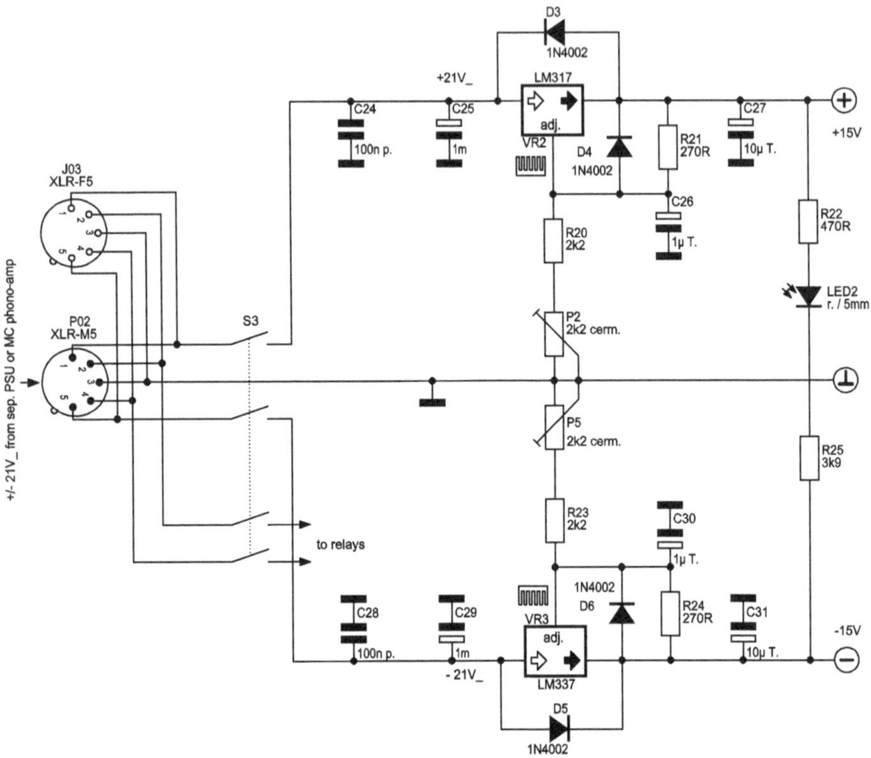


**Fig. 6.3c** In-case power-supply-unit for Fig. 6.3a, b circuits

## Power Supply

The transformer set-up also triggered the choice of the input resistor of the solid state phono-amp: 510 R, thus making its input resistance approximately 475 R. $R_{in}$ is transformed down by division with $n^2$. If the connecting cable between transformer output and MM phono-amp input is too long (>20 cm), then there will be a good chance to catch mains interferences because of the high output impedance of the transformer (>5 k) and the higher input resistance of the MM phono-amp. That's why I strongly recommend to directly put the transformer into the MM phono-amp case. All shielded cables inside the case are RG176.

### Wiring Between Turntable and Phono-Amp

Another "dangerous" point is the wiring between turntable and transformer input. The best thing to do is to take the symmetrical cable solution. I've installed Twin-BNC plugs and jacks[22,23] as well as XLR[24] types as connectors, and the cable is Mogami n. 2549. Deviation from the exact RIAA-transfer is less than $\pm 0.1$ dB. The whole arrangement sounds excellent. However, I still wanted to beat the Linto's noise figures with a better solid-state solution.

### The Solid-State Approach

The solid-state design is shown in Fig. 6.4. The measurement and calculation results are given in Table 6.2, columns C–F. A front-end with 4 parallel-working transistors ($T_{1-4}$) feed OP1. It was easier to select two double-transistors than three or four single ones (by accident I've installed two LM394 with a heavy $h_{FE}$-difference of 250: all SN figures deteriorated by only 2 dB). The LM394 was chosen because it's base spreading resistance $r_{bb'}$ (= 40 R) equals that of Mr Self's transistors (2N4403). The MAT02/SSM2210 is claimed to be the follower of the LM394, with better low frequency noise data. It really does a bit better with its $r_{bb'}$ of 30 R.

### Selection of the Phono-Amp Input BJTs

Normally, low-$r_{bb'}$ BJTs in lowest-noise designs are superior to those with higher $r_{bb'}$. This only is true for very low source resistances. But for a source resistance of 43 R a device with $r_{bb'}$ of only 4 R0 and $h_{FE} = 40$ (BFW16A) is not able to beat the *SN*s of BJTs with higher $r_{bb'}$ and much higher $h_{FE}$ (see calculated results in Table 6.2, column G). It's a hard thing to find a lowest-$r_{bb'}$ transistor with $h_{FE} \gg 40$. The latter is essential for low input noise currents. Basically, the (base) input noise current of any BJT can be calculated with Eqs. (3.70 ... 3.73).

[22] Emerson/Vitelec, UK

[23] L-Com, Inc., MA. USA

[24] Neutrik GmbH, Lichtenstein

**Table 6.2** Calculation and measurement results in dB (rounded to 1 digit after the dec. point) for various input devices with reference to an input load of 43 R, an input voltage of 0.5 $mV_{rms}/1$ kHz and $B_{20\,k}$

| 1/A | B | C | D | E | F | G | H | I | J |
|---|---|---|---|---|---|---|---|---|---|
| 2 | Type of phono-amp | 2 LM394 | | SSM2210/MAT02 | | 1 BFW16A | Jensen transform. | | Resistor |
| 3 | | $h_{FE} = 530$, $r_{bb} = 40R$ | | $h_{FE} = 680$, $r_{bb} = 30R$ | | $r_{bb} = 4R$** | + BUVO-MM | | $R_0$ |
| 4 | | $I^*_{C4} = 12.05mA$ | | $I_{C4} = 6.7mA$ | | $I_{C1} = 25mA$ | 2 × 2SC2546 | | 43R |
| 5 | $SN$ + type of equalization | Calc. | Meas. | Calc. | Meas. | Calc. | Calc. | Meas. | Calc. |
| 6 | $SN_{riaa}$ | −75.0 | −71.7 | −75.2 | −72.5 | −73.9 | −74.1 | −73.6 | −76.1 |
| 7 | $SN_{ariaa}$ | −79.3 | −78.9 | −79.5 | −79.2 | −78.2 | −78.4 | −78.4 | −80.4 |
| 8 | $SN_{sriaa}$ | −79.2 | −78.8 | −79.4 | −79.2 | −78.1 | −78.3 | −78.4 | −80.4 |
| 9 | Deltas | | | \|Amp calculated minus amp measured\| | | | | | |
| 10 | $SN_{riaa}$ | 3.3 | | 2.7 | | | 0.5 | | |
| 11 | $SN_{ariaa}$ | 0.4 | | 0.3 | | | 0.0 | | |
| 12 | $SN_{sriaa}$ | 0.4 | | 0.2 | | | 0.1 | | |
| 13 | Deltas | | | \|$R_0$ calculated minus amp calculated\| | | | | | |
| 14 | $DN_{riaa}$ | 1.0 | | 0.9 | | 2.2 | 2.0 | | |
| 15 | $DN_{ariaa}$ | 1.1 | | 0.9 | | 2.2 | 2.0 | | |
| 16 | $DN_{sriaa}$ | 1.1 | | 1.0 | | 2.3 | 2.1 | | |
| 17 | Deltas | | | \|$R_0$ calculated minus amp measured\| | | | | | |
| 18 | $DN_{riaa}$ | 4.4 | | 3.6 | | | 2.5 | | |
| 19 | $DN_{ariaa}$ | 1.5 | | 1.2 | | | 2.0 | | |
| 20 | $DN_{sriaa}$ | 1.6 | | 1.2 | | | 2.0 | | |

$I^*_{cn}$ = sum of $n$ collector currents; ** $h_{FE} = 40$

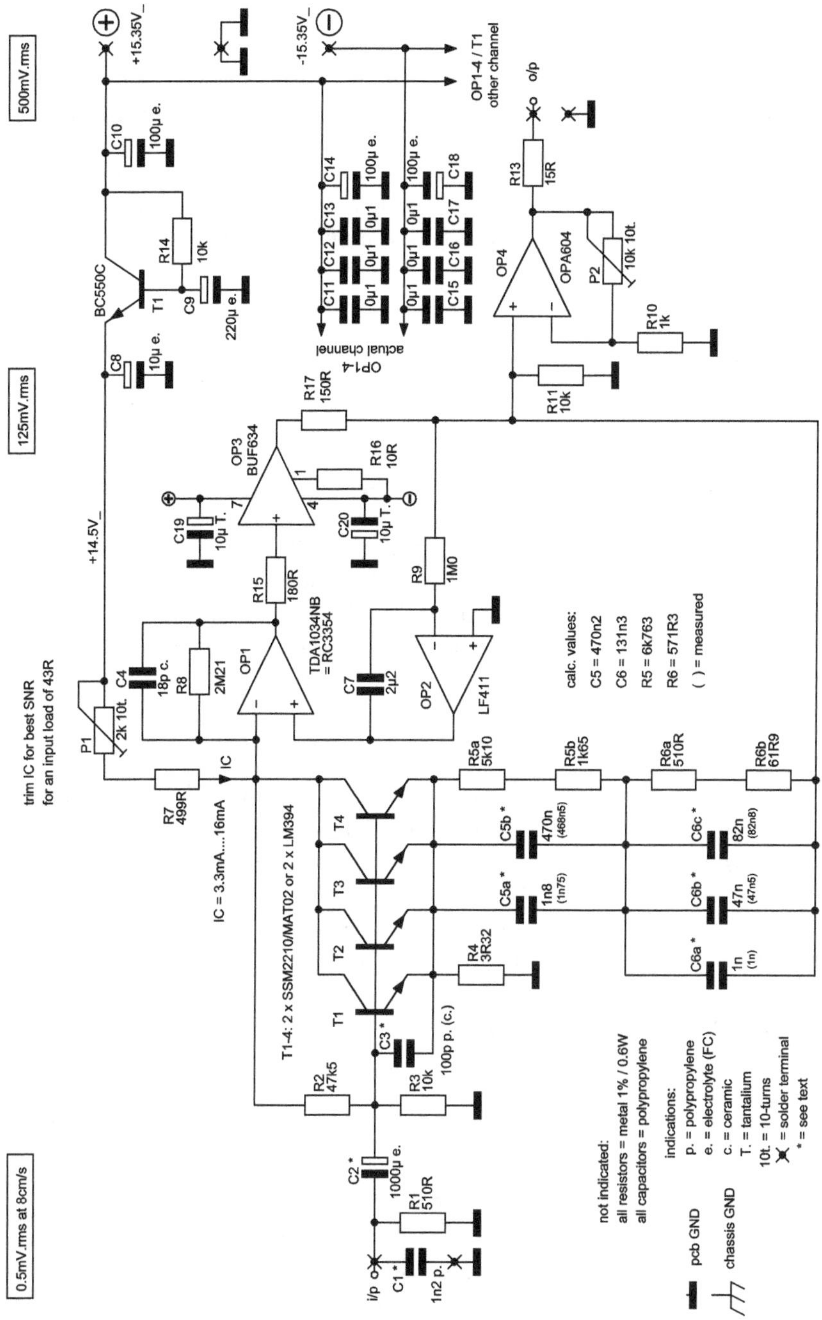


**Fig. 6.4** Solid-state NPN-BJT MC RIAA phono-amp

### Calculation and Measurement Results of Various I/P Devices

For comparison reasons, in column I of Table 6.1 you'll find the calculated figures of a typical representative of many MC-RIAA pre-amps with $SN_{ariaa}$ between $-72.0$ dBA and $-78.0$ dBA with 43 R as input load: Musical Fidelity XLP. With a 25 R input load the measured $SN_{ariaa} = -74.1\,\mathrm{dBA}/0.5\,\mathrm{mV_{rms}}/B_{20\,k}$ (see Table 3.12). Its 2SC550 input transistors are high frequency power devices from Toshiba. I guess they have very low $r_{bb'}$. But in this case it doesn't help much to extremely reduce the noise level.

### Circuit of the BUVO Solid-State MC Phono-Amp

In Fig. 6.4 OP3 (BUF634) serves as current booster to enable the connection of a very low-impedance RIAA feedback network (77 R at 20 kHz) between OP3's output and the emitters of $T_{1-4}$ and $R_4$. To avoid ringing, some precautions have to be implemented around OP3: $R_{17}$ at the output of the buffer, $R_{15}$ at its input. The bandwidth is set to nearly maximum by $R_{16}$. $C_{18}$ and $C_{19}$ have to be placed as near as possible to OP3. To minimise the noise voltage level of the amp, the optimal collector current can be trimmed with $P_1$. This is checked by inserting a S-filter (Fig. 6.1) between the output of the phono-amp and an appropriate measuring tool.

### Test Circuit for the I/P Capacitance

The high value of the input capacitor $C_2$ is essential to keep its low-frequency impedance as low as possible. For any reader who has problems with electrolyte capacitors inside the amplification chain I strongly recommend to read the respective letter debate in E&WW in 1988, after Mr Self published his pre-pre-amp article in 12-1987. I only did hear tiny differences between very expensive, cheaper and bridged capacitors that I tested with the circuit of Fig. 6.5. Finally, I've chosen one

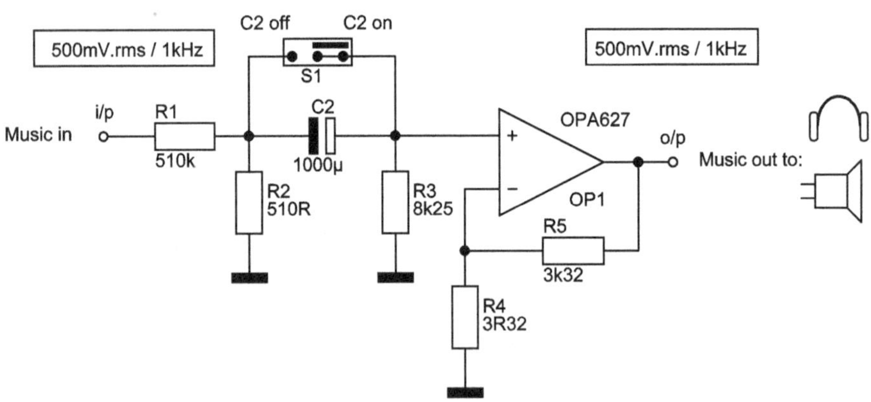


**Fig. 6.5** Test circuit for input capacitor $C_2$ of Fig. 6.4

from Panasonic (FC/25 V). I didn't hear any difference between the on and off state of $S_1$. To exactly hit the RIAA transfer, the value of $C_7$ (2 µ2) should not be changed. With that, the deviation of the phono-amp's RIAA transfer is max. ±0.1 dB.

## Power Supply

Concentration on noise reduction in amplifiers makes no sense, as long as the power supply is not noise-optimised as well. A short description will help to understand how it works. Attached to a separate Al-case a Talema toroidal transformer (Fig. 6.6) is followed by a rectifier, high value capacitors and two gyrators. They generate the main positive and negative supply voltages that are stabilized by LM 317/337 devices (±21 V). A (not necessarily) shielded 1.5 m cable connects power supply and phono-amps.

To get extreme low noise on the voltage lines inside the phono-amp case, a further supply voltage noise reduction treatment takes place with a µA723 in lowest noise mode, followed by a 2N3055 (Fig. 6.7). This arrangement produces (short-circuit protected) ±400 mA max. DC current for the final output voltages of ±15.35 V DC (positive voltage set by $P_3$, negative voltage set by $P_4$). Negative voltage is drawn off from the final positive output voltage line via an OP27-inverter and a PNP-Darlington transistor configuration with 2N2905A and 2N3055.

## Sound

Both types of MC phono-amps sound – for my ears – very well. The SN difference between the transformer and solid-state solution is apparent (especially at the lower end of the frequency band), but marginal in the frequency range <100 Hz (see Figs. 6.8 … 6.9). With the loudness pot set in a 2 pm position noise can't be heard in a listening distance of 30 cm from the loudspeakers. It really sounds silent.

Finally, it will be a question of expenses and listening tests to decide the alternative. With a DL-103 as input device, any further attempts to reduce noise make no sense because of fast growing physical limitations. To get capacitor-free direct coupling, a long-tailed pair of input transistors of the solid-state solution (e.g. four MAT02/SSM2210) would downgrade all *SN*s by approximately 1 dB. Further *SN* improvements can only be achieved with MC cartridges with lower source resistance and/or higher output voltage [e.g. Ortofon "Rondo" (its price is five times higher than that of the DL-103, $R_0 = 6\,\mathrm{R0}$ and $U = 0.9\,\mathrm{mV_{rms}}/1\,\mathrm{kHz}/8\,\mathrm{cm/s}$]. With reference to a nominal phono-amp input voltage of $0.9\,\mathrm{mV_{rms}}/1\,\mathrm{kHz}$ this type of MC cartridge would improve all *SN*s of Table 6.2, columns E and F by $20\log(U/0.5\,\mathrm{mV_{rms}}) = 5.11\,\mathrm{dB}$. For transformer coupling a different transformer is needed, being able to work with 6 R0 input load.

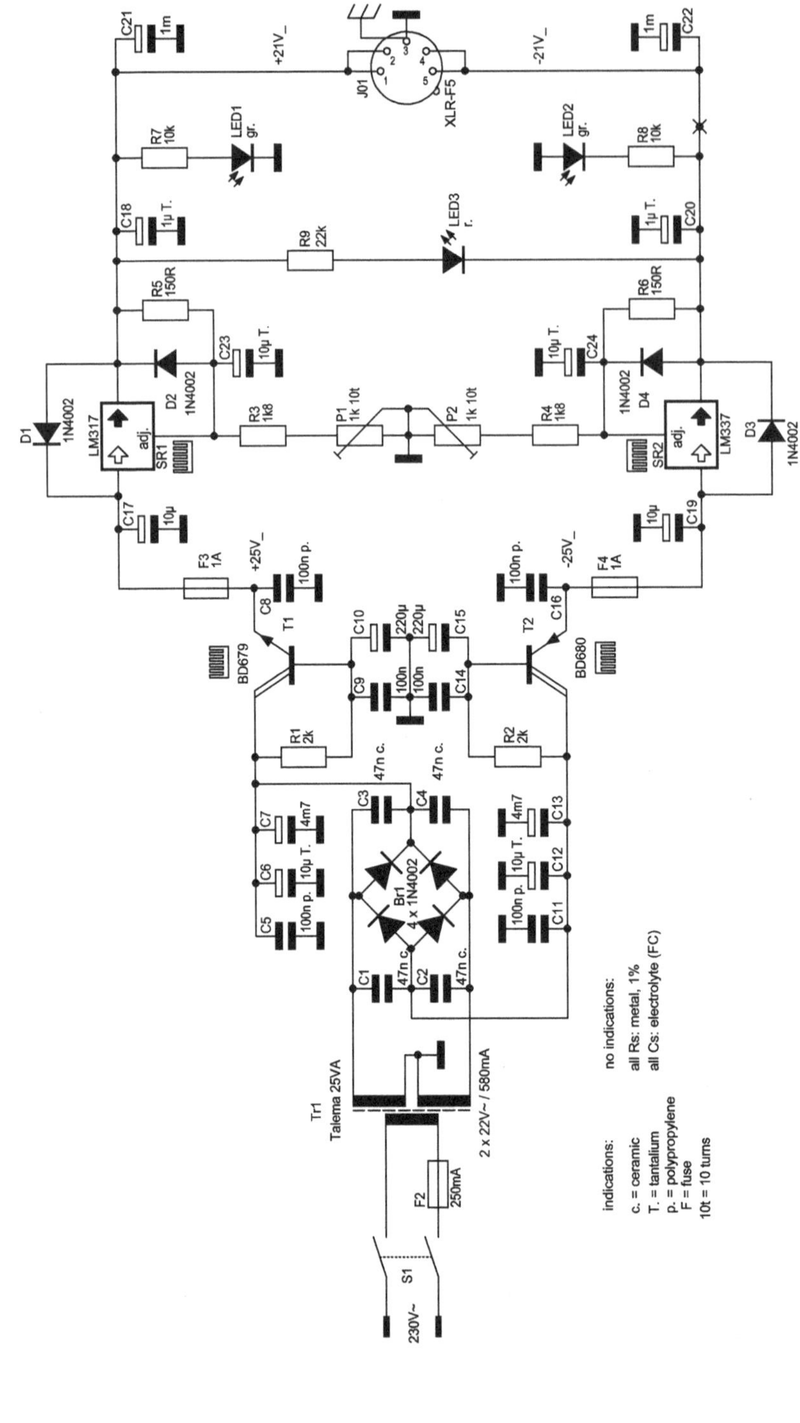


**Fig. 6.6** Separate power-supply-unit

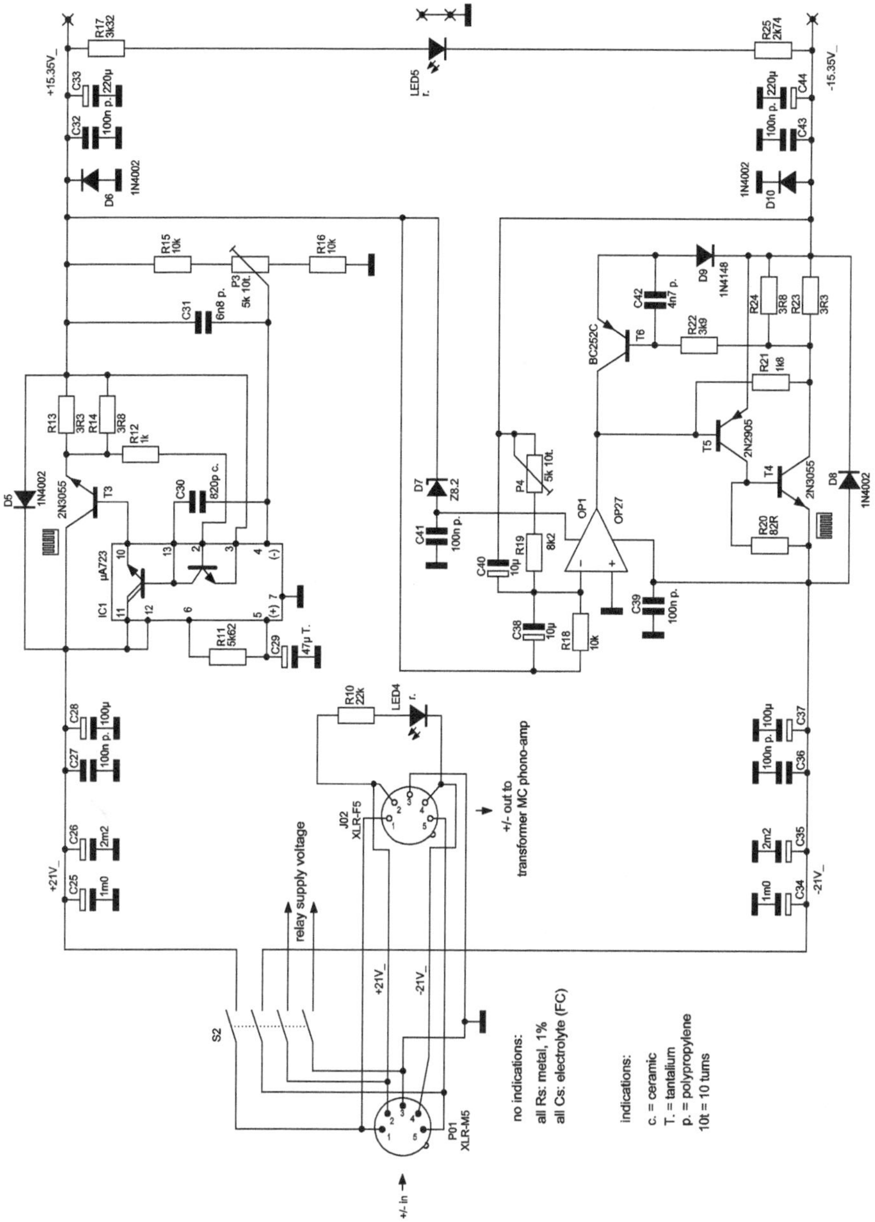


**Fig. 6.7** Power-supply-unit inside the solid-state MC phono-amp case

### Frequency Response Defining Components

The sound of an amp will also be affected by the right or wrong choice of the frequency response defining capacitors. Following M Cyril Bateman's[25,26] advises on low-distortion and low-intermodulation capacitors the following decisions were made:

Figure 6.3: Together with the transformer $C_x$ was delivered by Jensen Transformers and it's a polystyrene type of C, which is in line with Bateman's advises.
$C_1$ and $C_2$ are WIMA MKP4/10 or Epcos B32652 400 V/630 V polypropylene Cs. The aim of this RIAA network was to find a solution with only one type of capacitor for all 3 time constants. Bateman's statement: very low distortion types.
$C_3$ is a 25 V Panasonic FC (see respective comments on Fig. 6.5)

Figure 6.4: $C_1$ should be a 1n2 polystyrene type of C soldered at the input solder terminals of the pcb. I've gone through several design evolutions on that issue, starting with ceramic Cs at the input terminals like in Figure 4 of the original EW article and ending up with the polystyrene solution. With the chosen final pcb set-up it also guarantees proper amp stability with low-resistance input loads.
$C_2$: same as C3 above.
Depending on the pcb layout $C_3$ is not a must at all. In my final pcb lay-out I could skip it. If needed a ceramic solution would not be the best choice. A polystyrene C would do better. But I could not hear any difference between the two alternatives.
$C_{5,6}$: First choice were the Epcos B32652 types of C. But they are very big and therefore in danger to catch hum interferences. I've also tried the very much smaller Siemens 7.5 mm/5% MKT types, allowing to drastically reduce the size of the pcb. No difference in sound could be detected. But from a distortion and temperature drift point of view they are not comparable with the MKP types from WIMA or Epcos.

Calculation rules for the several kinds of RIAA networks will be given in Chap. 8.

### Additional Measurement Results

- Measured with a precision Anti-RIAA[27] circuit very low phase angles could be observed [28]:

| | | |
|---|---|---|
| at 20 Hz: | Solid-state: | $\leq +7.0°$ |
| | Transformer: | $\leq +5.5°$ |

[25] "Understanding Capacitors" Parts 1 ... 7, EW 12-1997 ... EW 08-1998

[26] "Capacitor Sounds" Parts 1 ... 5, EW 05 ... 11 – 2002

[27] "Precision Anti-RIAA", EW 05-2007 and Chap. 12

[28] Measurement results are heavily depending on measurement set-up: see Chap. 10

| | | |
|---|---|---|
| at 20 kHz: | Solid state: | $\leq -3.0°$ |
| | Transformer: | $\leq -6.0°$ |

- In addition, the bottom line set by the Linto could be hit

  1. mathematically – with a 43 R input load and an virtual change of the Linto input impedance from 150 R to 475 R:

     | | |
     |---|---|
     | Linn: | –79.425 dBA |
     | SSM2210: | –79.483 dBA |

  2. measurement-wise – with a 25 R input load:

     | | | |
     |---|---|---|
     | Linn: | – 81.0 dBA | as of [29] |
     | SSM2210: | – 80.6 dBA | still with $I_C = 6.7$ mA, the optimal collector current for 43 R input load – see Table 6.2 |

  3. a change of the input resistor of the BUVO MC phono-amp from 510 R to 154 R (to get an input impedance of app. 150 R like the Linto) created the following measured *SN* results (no change of $I_C$!):

     | | | | | |
     |---|---|---|---|---|
     | input load | 43 R : | SSM2210: | – 79.9 dBA | (a) |
     | input load | 25 R : | SSM2210: | – 81.3 dBA | (b) |

     (a) indicates that the Linto must have low-$h_{FE}$ input devices, driven with a rather high collector current, thus, making input noise current as well rather high. Hence, this is – besides the rather low input impedance – a 2nd reason that the Linto is not the right amp for MC cartridges with $R_0 > 15$ R.
     (b) indicates that with a different design approach (capacitor-coupled vs. DC-coupled) very good lowest-noise *SN* results can be achieved – provided that the design, the capacitors and the mechanical set-up were carefully chosen.

### Graphs of I/P Noise Voltage Densities

- The noise performance of the two phono-amp approaches will be demonstrated with the following measurement results in graph format and the following explanatory notes on Figs. 6.8 ... 6.9:

  – Figures 6.8 ... 6.9 have the *y*-ordinate in dBmV format! This comes from the measurement amplifier with a gain-set of +60 dB. It is connected be-

[29] "stereoplay" 04-1998

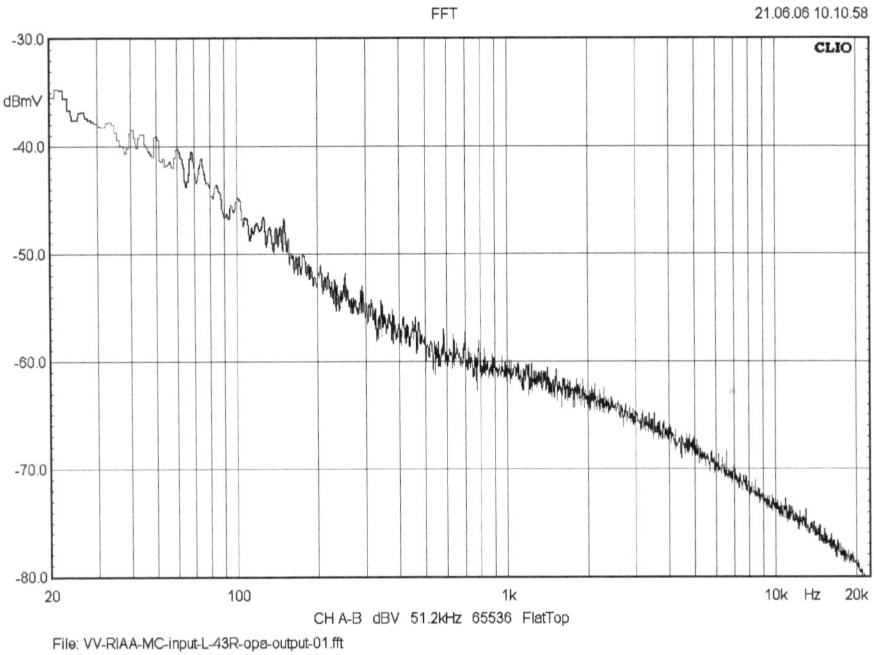


**Fig. 6.8** Input noise voltage density of MC solid-state phono-amp with SSM2210 input transistors

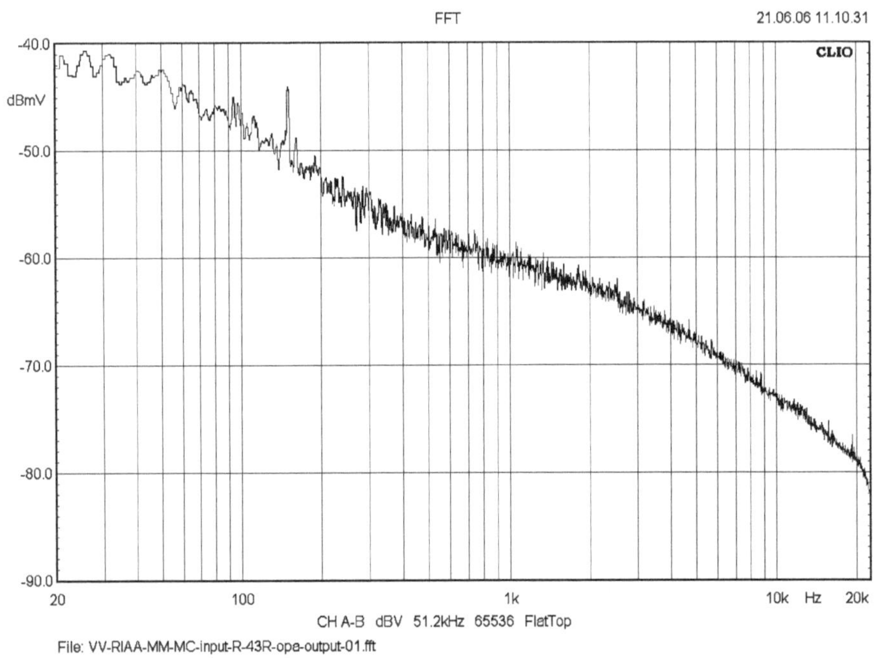


**Fig. 6.9** Input noise voltage density of transformer driven input of BUVO MM phono-amp

tween phono-amp output and input of the Clio 6.5 measurement instrument[30].

- In both figures the spikes at 150 Hz (−104 dBV, −107 dBV) are completely covered by the total noise of that phono-amp. They are PC-generated only and they will appear on all graphs. No other hum interferences could be detected.

[30] Details see Part III

# Chapter 7
# Noise in MC Phono-Amps – Mathematical Calculation Course

The following pages show all relevant MCD worksheets to enable the reader to follow the calculation results given in the previous chapter:

- Worksheet I: Various SNs of resistors
- Worksheet II: SNs of various transfer functions (RIAA, A-weighting, S-filter)
- Worksheet III: Change of input load of the Linn Linto
- Worksheet IV: SN evaluation of the DOSE MC pre-pre-amp
- Worksheet V: SN evaluation of the BUVO MM phono-amp + DOSE MC ppa
- Worksheet VI: SN evaluation of the BUVO MC trafo-coupled phono-amp
- Worksheet VII: SN evaluation of the BUVO MC solid-state phono-amp
- Worksheet VIII: SN evaluation of a BFW16A MC solid-state phono-amp

**Note 1:** MCD 11 has no built-in unit "rtHz" or "$\sqrt{\mathrm{Hz}}$". To get $\sqrt{1\,\mathrm{Hz}}$ based noise voltage and noise current densities the rms noise voltage and current in a specific frequency range $B > 1\,\mathrm{Hz}$ must be multiplied by $\sqrt{1\,\mathrm{Hz}}$ and divided by the root of that specific frequency range $\sqrt{B}$!

**Note 2:** MCD 11 offers no "dB unit". This is available from MCD 13 on!

B. Vogel, *The Sound of Silence*
DOI: 10.1007/978-3-540-76884-5, ©Springer 2008

➤ **MCD Worksheet I :** Various SNs of resistors : Page 1

**Step 1 Definition of physical constants and frequency ranges:**

$T := 300 \cdot K$ $\quad k := 1.38065 \cdot 10^{-23} \cdot V \cdot A \cdot s \cdot K^{-1}$ $\quad q := 1.6021765 \cdot 10^{-19} \cdot A \cdot s$

$B_{20k} := 19980 \cdot Hz$ $\quad B_1 := 1Hz$ $\quad h := 1000Hz$

**Step 2 Calculation of SNs [dB] of a resistor of 25R [equations [6.1 ... 6.7)]:**

$R := 25\Omega$ $\quad e_{N.25} := \sqrt{4 \cdot k \cdot T \cdot R \cdot B_{20k}}$ $\quad e_{N.25} = 9.097 \times 10^{-8} V$

$e_{in.nom} := 0.5 \cdot 10^{-3} V$ $\quad e_{N.25} \cdot \frac{B_1}{B_{20k}} = 4.553 \times 10^{-12} V$

$T1 := 318 \cdot 10^{-6} s$ $\quad T2 := 3180 \cdot 10^{-6} \cdot s$ $\quad T3 := 75 \cdot 10^{-6} s$

$$R(f) := \left[\frac{\sqrt{\left[1 + (2 \cdot \pi \cdot f \cdot T1)^2\right]}}{\sqrt{\left[1 + (2 \cdot \pi \cdot f \cdot T2)^2\right]} \cdot \sqrt{\left[1 + (2 \cdot \pi \cdot f \cdot T3)^2\right]}}\right] \cdot \left[\frac{\sqrt{\left[1 + (2 \cdot \pi \cdot h \cdot T1)^2\right]}}{\sqrt{\left[1 + (2 \cdot \pi \cdot h \cdot T2)^2\right]} \cdot \sqrt{\left[1 + (2 \cdot \pi \cdot h \cdot T3)^2\right]}}\right]^{-1}$$

$$A(f) := \frac{\left[1 + \left(\frac{20.6Hz}{h}\right)^2\right] \cdot \sqrt{1 + \left(\frac{107.7Hz}{h}\right)^2} \cdot \sqrt{1 + \left(\frac{737.9Hz}{h}\right)^2} \cdot \left[1 + \left(\frac{h}{12200Hz}\right)^2\right]}{\left[1 + \left(\frac{20.6Hz}{f}\right)^2\right] \cdot \sqrt{1 + \left(\frac{107.7Hz}{f}\right)^2} \cdot \sqrt{1 + \left(\frac{737.9Hz}{f}\right)^2} \cdot \left[1 + \left(\frac{f}{12200Hz}\right)^2\right]}$$

$$SN_{ariaa.25} := 20 \cdot \log\left(\frac{\sqrt{\frac{1}{B_{20k}} \cdot \int_{20Hz}^{20000Hz} e_{N.25}{}^2 \cdot A(f)^2 \cdot R(f)^2\, df}}{e_{in.nom}}\right) \qquad SN_{ariaa.25} = -82.736$$

➢ **MCD Worksheet II:** SNs of various transfer functions (RIAA, A-weighting, S-filter) - Page 1

**Step 1 Definition of frequency ranges:**

$B_{20k} := 19980\text{Hz}$ $\quad h := 1000\text{Hz}$ $\quad B_{355} := 20000\text{Hz} - 355\text{Hz}$ $\quad B_{355} = 19645\,\text{Hz}$

**Step 2 Calculation of $SN_r$ [equation [6.12)] :**

$T1 := 318\cdot 10^{-6}\text{s}$ $\quad T2 := 3180\cdot 10^{-6}\cdot\text{s}$ $\quad T3 := 75\cdot 10^{-6}\text{s}$

$$R(f) := \left[\frac{\sqrt{\left[1+(2\cdot\pi\cdot f\cdot T1)^2\right]}}{\sqrt{\left[1+(2\cdot\pi\cdot f\cdot T2)^2\right]}\cdot\sqrt{\left[1+(2\cdot\pi\cdot f\cdot T3)^2\right]}}\right]\cdot\left[\frac{\sqrt{\left[1+(2\cdot\pi\cdot h\cdot T1)^2\right]}}{\sqrt{\left[1+(2\cdot\pi\cdot h\cdot T2)^2\right]}\cdot\sqrt{\left[1+(2\cdot\pi\cdot h\cdot T3)^2\right]}}\right]^{-1}$$

$$SN_r := 20\cdot\log\left[\sqrt{\frac{1}{B_{20k}}\cdot\int_{20\text{Hz}}^{20000\text{Hz}}(R(f))^2\,df}\right] \qquad SN_r = -3.646$$

**Step 3 Calculation of $SN_{ar}$ [equation (6.13)]:**

$$A(f) := \frac{\left[1+\left(\frac{20.6\text{Hz}}{h}\right)^2\right]\cdot\sqrt{1+\left(\frac{107.7\text{Hz}}{h}\right)^2}\cdot\sqrt{1+\left(\frac{737.9\text{Hz}}{h}\right)^2}\cdot\left[1+\left(\frac{h}{12200\text{Hz}}\right)^2\right]}{\left[1+\left(\frac{20.6\text{Hz}}{f}\right)^2\right]\cdot\sqrt{1+\left(\frac{107.7\text{Hz}}{f}\right)^2}\cdot\sqrt{1+\left(\frac{737.9\text{Hz}}{f}\right)^2}\cdot\left[1+\left(\frac{f}{12200\text{Hz}}\right)^2\right]}$$

$$SN_{ar} := 20\cdot\log\left(\sqrt{\frac{1}{B_{20k}}\cdot\int_{20\text{Hz}}^{20000\text{Hz}}A(f)^2\cdot R(f)^2\,df}\right) \qquad SN_{ar} = -7.935$$

**Step 4 Calculation of $SN_a$ [equation (6.14)]**

$$SN_a := 20\cdot\log\left(\sqrt{\frac{1}{B_{20k}}\cdot\int_{20\text{Hz}}^{20000\text{Hz}}A(f)^2\,df}\right) \qquad SN_a = -2.046$$

**Step 5 Calculation of $SN_{sr}$ [equation (6.15)]:**

$$SN_{sr} := 20\cdot\log\left(\sqrt{\frac{1}{B_{355}}\cdot\int_{355\text{Hz}}^{20000\text{Hz}}R(f)^2\,df}\right) \qquad SN_{sr} = -7.853$$

➢ **MCD Worksheet III:** Change of input load of the Linn Linto - Page 1

**Step 1 Definition of physical constants, frequency ranges and standard SNs [dB] for various transfers:**

$T := 300 \cdot K$ $\quad k := 1.38065 \cdot 10^{-23} \cdot V \cdot A \cdot s \cdot K^{-1}$ $\quad q := 1.6021765 \cdot 10^{-19} \cdot A \cdot s$

$B_{20k} := 19980 \cdot Hz$ $\quad B_1 := 1Hz$ $\quad e_{in.nom} := 0.5 \cdot 10^{-3} V$

$SN_r := -3.64566$ $\quad SN_{ar} := -7.93532$ $\quad SN_a := -2.04623$ $\quad SN_{sr} := -7.853$

**Step 2 Evaluation of Linto's input noise voltage (assuming that the Linto produces no flicker (1/f) noise) [equation (6.18)]:**

$SN_{ariaa.Linto.25} := -81.0$ $\quad R0_{25} := 25\Omega$ $\quad Rin_{Linto.1} := 150\Omega$

$$Rin_{res1.Linto} := \left(\frac{1}{Rin_{Linto.1}} + \frac{1}{R0_{25}}\right)^{-1} \qquad Rin_{res1.Linto} = 21.429\Omega$$

$$SN_{ne.Linto.25} := SN_{ariaa.Linto.25} - SN_{ar} \qquad SN_{ne.Linto.25} = -73.065$$

$$e_{N.Linto.25} := \frac{e_{in.nom}}{10^{\frac{|SN_{ne.Linto.25}|}{20}}} \qquad e_{N.Linto.25} = 1.111 \times 10^{-7} V$$

$$\frac{e_{N.Linto.25} \cdot \sqrt{B_1}}{\sqrt{B_{20k}}} = 7.86 \times 10^{-10} V$$

$$e_{N.Rin.res1} := \sqrt{4 \cdot k \cdot T \cdot B_{20k} \cdot Rin_{res1.Linto}} \qquad e_{N.Rin.res1} = 8.422 \times 10^{-8} V$$

$$\frac{e_{N.Rin.res1} \cdot \sqrt{B_1}}{\sqrt{B_{20k}}} = 5.958 \times 10^{-10} V$$

**Assumtion:** $i_{N.Linto} := \frac{4 \cdot 10^{-12} A}{\sqrt{B_1}} \cdot \sqrt{B_{20k}}$ $\qquad i_{N.Linto} = 5.654 \times 10^{-10} A$

**To check the variations in SNs of the Linto as a dependency of $i_{N.Linto}$ change the value for the assumption only! The results (2pA ... 6pA) will always be <0.1dB!**

$$e_{N.Linto} := \sqrt{e_{N.Linto.25}^2 - e_{N.Rin.res1}^2 - i_{N.Linto}^2 \cdot Rin_{res1.Linto}^2}$$

$$e_{N.Linto} = 7.144 \times 10^{-8} V \qquad \frac{e_{N.Linto} \cdot \sqrt{B_1}}{\sqrt{B_{20k}}} = 5.054 \times 10^{-10} V$$

**Step 3 Evaluation of Linto's SNs and DNs with 25R input load:**

$$e_{N.25} := \sqrt{4 \cdot k \cdot T \cdot B_{20k} \cdot R0_{25}} \qquad e_{N.25} = 9.097 \times 10^{-8} V \qquad \frac{e_{N.25} \cdot \sqrt{B_1}}{\sqrt{B_{20k}}} = 6.436 \times 10^{-10} V$$

$$SN_{ne.25} := 20 \cdot \log\left(\frac{e_{N.25}}{e_{in.nom}}\right) \qquad SN_{ne.25} = -74.801$$

➤ **MCD Worksheet III:** Change of input load of the Linn Linto - **Page 2**

$SN_{riaa.25} := SN_{ne.25} + SN_r \quad SN_{riaa.25} = -78.447$

$SN_{riaa.Linto.25} := SN_{ne.Linto.25} + SN_r \quad SN_{riaa.Linto.25} = -76.71$

$SN_{ariaa.25} := SN_{ne.25} + SN_{ar} \quad SN_{ariaa.25} = -82.737$

$SN_{ariaa.Linto.25} := SN_{ne.Linto.25} + SN_{ar} \quad SN_{ariaa.Linto.25} = -81$

$SN_{sriaa.25} := SN_{ne.25} + SN_{sr} \quad SN_{sriaa.25} = -82.654$

$SN_{sriaa.Linto.25} := SN_{ne.Linto.25} + SN_{sr} \quad SN_{sriaa.Linto.25} = -80.918$

$DN_{riaa.Linto.25} := SN_{riaa.Linto.25} - SN_{riaa.25} \qquad DN_{riaa.Linto.25} = 1.737$

$DN_{ariaa.Linto.25} := SN_{ariaa.Linto.25} - SN_{ariaa.25} \qquad DN_{ariaa.Linto.25} = 1.737$

$DN_{sriaa.Linto.25} := SN_{sriaa.Linto.25} - SN_{sriaa.25} \qquad DN_{sriaa.Linto.25} = 1.737$

**Step 4 Insertion of new input load $R0_{43}$ = 43R into the calculation course [equ. (6.19 ... 6.21) ] and a changed Linto input impedance of 258R:**

$$R0_{43} := 43\Omega \qquad Rin_{Linto.2} := \frac{Rin_{Linto.1}}{R0_{25}} \cdot R0_{43} \qquad Rin_{Linto.2} = 258\,\Omega$$

$$Rin_{res2.Linto} := \left(\frac{1}{Rin_{Linto.2}} + \frac{1}{R0_{43}}\right)^{-1} \qquad Rin_{res2.Linto} = 36.857\,\Omega$$

$$e_{N.Rin.res2} := \sqrt{4 \cdot k \cdot T \cdot B_{20k} \cdot Rin_{res2.Linto}} \qquad e_{N.Rin.res2} = 1.105 \times 10^{-7}\,V$$

$$\frac{e_{N.Rin.res2} \cdot \sqrt{B_1}}{\sqrt{B_{20k}}} = 7.814 \times 10^{-10}\,V$$

$$e_{N.Linto.43.2} := \sqrt{e_{N.Linto}^2 + \left(i_{N.Linto} \cdot Rin_{res2.Linto}\right)^2 + e_{N.Rin.res2}^2}$$

$$e_{N.Linto.43.2} = 1.332 \times 10^{-7}\,V$$

$$\frac{i_{N.Linto} \cdot Rin_{res2.Linto} \cdot \sqrt{B_1}}{\sqrt{B_{20k}}} = 1.474 \times 10^{-10}\,V \qquad \frac{e_{N.Linto.43.2} \cdot \sqrt{B_1}}{\sqrt{B_{20k}}} = 9.423 \times 10^{-10}\,V$$

$$SN_{ne.Linto.43.2} := 20 \cdot \log\left(\frac{e_{N.Linto.43.2}}{e_{in.nom}}\right) \qquad SN_{ne.Linto.43.2} = -71.49$$

$$SN_{riaa.Linto.43.2} := SN_{ne.Linto.43.2} + SN_r \qquad SN_{riaa.Linto.43.2} = -75.136$$

➤ **MCD Worksheet III:** Change of input load of the Linn Linto - Page 3

$SN_{ariaa.Linto.43.2} := SN_{ne.Linto.43.2} + SN_{ar}$ $\quad SN_{ariaa.Linto.43.2} = -79.425$

$SN_{sriaa.Linto.43.2} := SN_{ne.Linto.43.2} + SN_{sr}$ $\quad SN_{sriaa.Linto.43.2} = -79.343$

$e_{N.43} := \sqrt{4 \cdot k \cdot T \cdot B_{20k} \cdot R0_{43}}$ $\quad e_{N.43} = 1.193 \times 10^{-7} V$

$$\frac{e_{N.43} \cdot \sqrt{B_1}}{\sqrt{B_{20k}}} = 8.44 \times 10^{-10} V$$

$SN_{ne.43} := 20 \cdot \log\left(\frac{e_{N.43}}{e_{in.nom}}\right)$ $\quad SN_{ne.43} = -72.446$

$SN_{riaa.43} := SN_{ne.43} + SN_r$ $\quad SN_{riaa.43} = -76.092$

$SN_{ariaa.43} := SN_{ne.43} + SN_{ar}$ $\quad SN_{ariaa.43} = -80.381$

$SN_{sriaa.43} := SN_{ne.43} + SN_{sr}$ $\quad SN_{sriaa.43} = -80.299$

$DN_{riaa.Linto.43.2} := SN_{riaa.Linto.43.2} - SN_{riaa.43}$ $\quad DN_{riaa.Linto.43.2} = 0.956$

$DN_{ariaa.Linto.43.2} := SN_{ariaa.Linto.43.2} - SN_{ariaa.43}$ $\quad DN_{ariaa.Linto.43.2} = 0.956$

$DN_{sriaa.Linto.43.2} := SN_{sriaa.Linto.43.2} - SN_{sriaa.43}$ $\quad DN_{sriaa.Linto.43.2} = 0.956$

**Step 5 Insertion of new input load $R0_{43}$ = 43R into the calculation course [equ. (6.19 ... 6.21)] and a changed Linto input impedance of 475R:**

$R0_{43} := 43\Omega$ $\quad Rin_{Linto.3} := 475\Omega$

$Rin_{res3.Linto} := \left(\frac{1}{Rin_{Linto.3}} + \frac{1}{R0_{43}}\right)^{-1}$ $\quad Rin_{res3.Linto} = 39.431\,\Omega$

$e_{N.Rin.res3} := \sqrt{4 \cdot k \cdot T \cdot B_{20k} \cdot Rin_{res3.Linto}}$ $\quad e_{N.Rin.res3} = 1.142 \times 10^{-7} V$

$$\frac{e_{N.Rin.res3} \cdot \sqrt{B_1}}{\sqrt{B_{20k}}} = 8.083 \times 10^{-10} V$$

$$e_{N.Linto.43.3} := \sqrt{e_{N.Linto}^2 + \left(i_{N.Linto} \cdot Rin_{res3.Linto}\right)^2 + e_{N.Rin.res3}^2}$$

$e_{N.Linto.43.3} = 1.366 \times 10^{-7} V$

$\frac{i_{N.Linto} \cdot Rin_{res3.Linto} \cdot \sqrt{B_1}}{\sqrt{B_{20k}}} = 1.577 \times 10^{-10} V$ $\quad \frac{e_{N.Linto.43.3} \cdot \sqrt{B_1}}{\sqrt{B_{20k}}} = 9.662 \times 10^{-10} V$

➢ **MCD Worksheet III:** Change of input load of the Linn Linto - Page 4

$SN_{ne.Linto.43.3} := 20 \cdot \log\left(\frac{e_{N.Linto.43.3}}{e_{in.nom}}\right)$ $SN_{ne.Linto.43.3} = -71.272$

$SN_{riaa.Linto.43.3} := SN_{ne.Linto.43.3} + SN_r$ $SN_{riaa.Linto.43.3} = -74.917$

$SN_{ariaa.Linto.43.3} := SN_{ne.Linto.43.3} + SN_{ar}$ $SN_{ariaa.Linto.43.3} = -79.207$

$SN_{sriaa.Linto.43.3} := SN_{ne.Linto.43.3} + SN_{sr}$ $SN_{sriaa.Linto.43.3} = -79.125$

$e_{N.43} := \sqrt{4 \cdot k \cdot T \cdot B_{20k} \cdot R0_{43}}$ $e_{N.43} = 1.193 \times 10^{-7} V$

$\frac{e_{N.43} \cdot \sqrt{B_1}}{\sqrt{B_{20k}}} = 8.44 \times 10^{-10} V$

$DN_{riaa.Linto.43.3} := SN_{riaa.Linto.43.3} - SN_{riaa.43}$ $DN_{riaa.Linto.43.3} = 1.174$

$DN_{ariaa.Linto.43.3} := SN_{ariaa.Linto.43.3} - SN_{ariaa.43}$ $DN_{ariaa.Linto.43.3} = 1.174$

$DN_{sriaa.Linto.43.3} := SN_{sriaa.Linto.43.3} - SN_{sriaa.43}$ $DN_{sriaa.Linto.43.3} = 1.174$

➢ **MCD Worksheet IV:** SN evaluation of the DOSE MC pre-pre-amp - Page 1

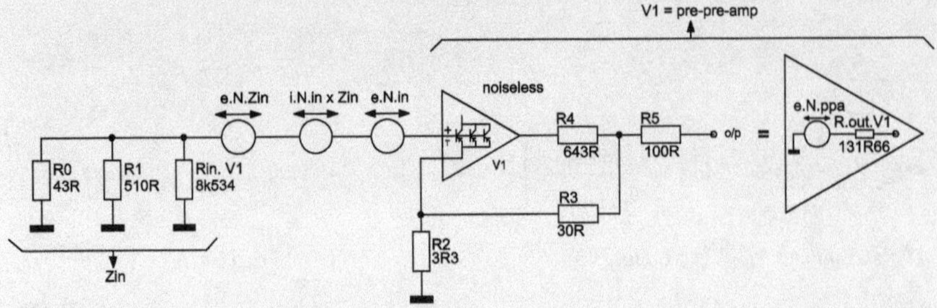


Figure 7.1 Equivalant circuit of the DOSE design (with all meaningful noise sources)

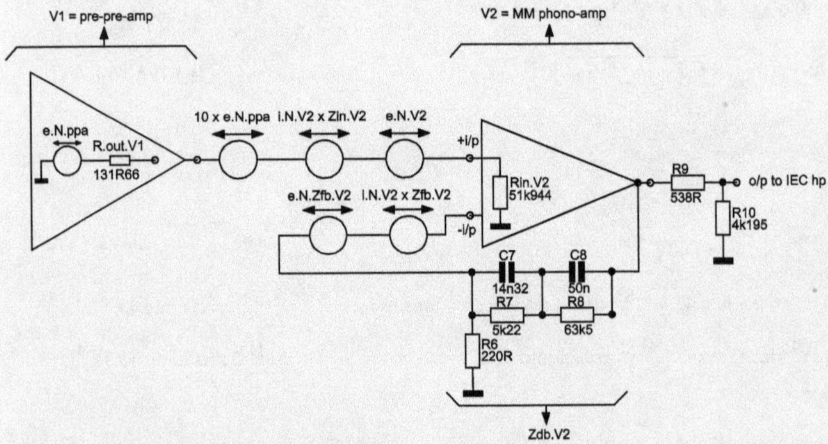


Figure 7.2 Equivalant circuit of the DOSE design connected to a MM phono-amp (with all meaningful noise sources)

**Step 1 Definition of all meaningful constants, components, etc.:**

Given: D. Self Design with $SN_{ne}$= -139.5dB / 775m$V_{rms}$ in $B_{30k}$ = (30000Hz-400Hz)

$T := 300 \cdot K$ $\quad k := 1.38065 \cdot 10^{-23} \cdot V \cdot A \cdot s \cdot K^{-1}$ $\quad q := 1.6021765 \cdot 10^{-19} \cdot A \cdot s$

$B_{20k} := 19980 \cdot Hz$ $\quad B_{30k} := 29600Hz$ $\quad B_1 := 1Hz$ $\quad B_{19k} := 19400Hz$

$SN_r := -3.64566$ $\quad SN_{ar} := -7.93532$ $\quad SN_a := -2.04623$ $\quad SN_{sr} := -7.853$

$R0 := 43\Omega$ $\quad R1 := 510\Omega$ $\quad R2 := 3.3\Omega$ $\quad R3 := 30\Omega$

$Rin_{V1} := 8.534 \cdot 10^3 \Omega$ $\quad R4 := 643\Omega$ $\quad R5 := 100\Omega$ $\quad I_C := 3 \cdot 10^{-3} A$

$e_{in.nom} := 0.5 \cdot 10^{-3} V$ $\quad RA := 10 \cdot 10^3 \Omega$ $\quad RB := 56 \cdot 10^3 \Omega$

**Step 2 Assumtions:** according to Tables 3.3 & 6.1 2N4403 with $h_{FE} := 200$

input shorted to get the measurement result of -139.5 dBu
instead of R1 = 100R I've chosen the same input impedance like the BUVO MC phono-amp with R1 = 510R
Gain of pre-pre-amp (ppa): $G_{ppa} := 10$

➢ **MCD Worksheet IV:** SN evaluation of the DOSE MC pre-pre-amp - Page 2

**Step 3 Evaluation of input noise voltage and pure ppa SNs with reference to $B_{20k}$, $0.5mV_{rms}$ /1kHz and including any 1/f effect:**

$$SN_{ne.dose.30k.dBu.0} := -139.5$$

$$e_{N.in.dose.1} := \frac{\left(\dfrac{775\cdot10^{-3}V}{10^{\frac{|SN_{ne.dose.30k.dBu.0}|}{20}}}\right)}{\sqrt{B_{30k}}}\cdot\sqrt{B_1} \qquad e_{N.in.dose.1} = 4.772\times10^{-10}V$$

$$e_{N.in.dose} := e_{N.in.dose.1}\cdot\sqrt{\frac{B_{20k}}{B_1}} \qquad e_{N.in.dose} = 6.745\times10^{-8}V$$

$$SN_{ne.dose.30k.0} := 20\cdot\log\left(\frac{e_{N.in.dose.1}\cdot\sqrt{\dfrac{B_{30k}}{B_1}}}{e_{in.nom}}\right) \qquad SN_{ne.dose.30k.0} = -75.693$$

$$SN_{ne.dose.0} := 20\cdot\log\left(\frac{e_{N.in.dose.1}\cdot\sqrt{\dfrac{B_{20k}}{B_1}}}{e_{in.nom}}\right) \qquad SN_{ne.dose.0} = -77.4$$

with equations (3.94) and (3.5) and $f_{ci}$ = 2000Hz:

$$f_{ci} := 2000Hz$$

$$i_{n.in.dose} := \sqrt{\frac{2\cdot q\cdot I_C}{h_{FE}}\cdot B_{20k} + \frac{4\cdot k\cdot T}{\left(\dfrac{1}{RA}+\dfrac{1}{RB}\right)^{-1}}\cdot B_{20k}} \qquad i_{n.in.dose} = 3.675\times10^{-10}A$$

$$\frac{i_{n.in.dose}\cdot\sqrt{B_1}}{\sqrt{B_{20k}}} = 2.6\times10^{-12}A$$

$$i_{N.in.dose} := \frac{i_{n.in.dose}}{\sqrt{B_{20k}}}\cdot\sqrt{f_{ci}\cdot\ln\left(\frac{20000}{20}\right)+B_{20k}} \qquad i_{N.in.dose} = 4.779\times10^{-10}A$$

$$\frac{i_{N.in.dose}\cdot\sqrt{B_1}}{\sqrt{B_{20k}}} = 3.381\times10^{-12}A$$

$$Zin_{V1} := \left(\frac{1}{R0}+\frac{1}{R1}+\frac{1}{Rin_{V1}}\right)^{-1} \qquad Zin_{V1} = 39.473\Omega \qquad e_{N.Zin.V1} := \sqrt{4\cdot k\cdot T\cdot B_{20k}\cdot Zin_{V1}}$$

$$e_{N.R2} := \sqrt{4\cdot k\cdot T\cdot B_{20k}\cdot R2}$$

$$e_{N.in.tot} := \sqrt{e_{N.Zin.V1}^2 + \left[i_{N.in.dose}\cdot(Zin_{V1}+R2)\right]^2 + e_{N.in.dose}^2 + e_{N.R2}^2}$$

➢ **MCD Worksheet IV:** SN evaluation of the DOSE MC pre-pre-amp - Page 3

$e_{N.in.tot} = 1.383 \times 10^{-7} V$ $\qquad \frac{e_{N.in.tot} \cdot \sqrt{B_1}}{\sqrt{B_{20k}}} = 9.784 \times 10^{-10} V$

$SN_{ne.dose.43.ppa} := 20 \cdot \log\left(\frac{e_{N.in.tot}}{e_{in.nom}}\right)$ $\qquad SN_{ne.dose.43.ppa} = -71.163$

$SN_{riaa.dose.43.ppa} := SN_{ne.dose.43.ppa} + SN_r$ $\qquad SN_{riaa.dose.43.ppa} = -74.809$

$SN_{ariaa.dose.43.ppa} := SN_{ne.dose.43.ppa} + SN_{ar}$ $\qquad SN_{ariaa.dose.43.ppa} = -79.099$

$SN_{sriaa.dose.43.ppa} := SN_{ne.dose.43.ppa} + SN_{sr}$ $\qquad SN_{sriaa.dose.43.ppa} = -79.016$

**Step 3a** **Cross-check to evalute the input noise voltage of the DOSE ppa by applying the respective formulae of Chapter 3, Section 3.2:**

with equation (3.93) and $r_{bb} := \frac{40}{3}\Omega$:

$e_{N.in.ppa} := \sqrt{\frac{2 \cdot k^2 \cdot T^2}{q \cdot I_C} \cdot B_{20k} + 4 \cdot k \cdot T \cdot (r_{bb} + R2) \cdot B_{20k}}$ $\qquad e_{N.in.ppa} = 8.326 \times 10^{-8} V$

$e_{N.in.ppa.43} := \sqrt{e_{N.Zin.V1}^2 + \left[i_{N.in.dose} \cdot (Zin_{V1} + R2)\right]^2 + e_{N.in.ppa}^2 + e_{N.R2}^2}$

$SN_{ne.ppa.43} := 20 \cdot \log\left(\frac{e_{N.in.ppa.43}}{e_{in.nom}}\right)$ $\qquad SN_{ne.ppa.43} = -70.653$

$SN_{ne.ppa.43} - SN_{ne.dose.43.ppa} = 0.51$

**Step 4** **Evaluation of output impedance Rout and its effect on the input noise voltage of a MM phono-amp with input devices 5534 / 5532 (by ignoring the rather small 1/f effects of the op-amps):**

$e_{N.ppa} := G_{ppa} \cdot e_{N.in.tot}$ $\qquad e_{N.ppa} = 1.383 \times 10^{-6} V$ $\qquad \frac{e_{N.ppa} \cdot \sqrt{B_1}}{\sqrt{B_{20k}}} = 9.784 \times 10^{-9} V$

$Rout_{V1} := R5 + \frac{(R2 + R3) \cdot R4}{R2 + R3 + R4}$ $\qquad Rout_{V1} = 131.66\Omega$ $\qquad Rin_{V2} := 51.944 \cdot 10^3 \Omega$

**A** $\quad e_{N.5534} := 4 \cdot 10^{-9} \cdot \frac{\sqrt{B_{20k}}}{\sqrt{B_1}} \cdot V$ $\qquad e_{N.5534} = 5.654 \times 10^{-7} V$

**B** $\quad i_{N.5534} := 0.6 \cdot 10^{-12} \cdot \frac{\sqrt{B_{20k}}}{\sqrt{B_1}} \cdot A$ $\qquad i_{N.5534} = 8.481 \times 10^{-11} A$

$Zin_{V2} := \left(\frac{1}{Rin_{V2}} + \frac{1}{Rout_{V1}}\right)^{-1}$ $\qquad Zin_{V2} = 131.327\Omega$ $\qquad e_{N.Zin.V2} := \sqrt{4 \cdot k \cdot T \cdot B_{20k} \cdot Zin_{V2}}$

➤ **MCD Worksheet IV:** SN evaluation of the DOSE MC pre-pre-amp - Page 4

**Step 5 Evaluation of the impedance of the feedback network of V2 at 1kHz:**

$C7 := 14.32 \cdot 10^{-9} \cdot F \qquad C8 := 50 \cdot 10^{-9} \cdot F \qquad R7 := 5.22 \cdot 10^{3} \Omega$

$R8 := 63.5 \cdot 10^{3} \Omega \qquad h := 1000Hz$

$$Z_{fb.V2} := \left[ \left| \frac{1}{220\Omega} + \frac{1}{\left[ \left( \frac{1}{R7} + 2j \cdot \pi \cdot h \cdot C7 \right)^{-1} + \left( \frac{1}{R8} + 2j \cdot \pi \cdot h \cdot C8 \right)^{-1} \right]} \right| \right]^{-1}$$

$Z_{fb.V2} = 215.42\Omega$

$e_{N.Z.fb.V2} := \sqrt{4 \cdot k \cdot T \cdot B_{20k} \cdot Z_{fb.V2}}$

**Step 6 Calculation of the total noise voltage and $SN_{ne}$ at the input of V1:**

$$e_{N.in.V2} := \sqrt{e_{N.ppa}^2 + e_{N.5534}^2 + e_{N.Zin.V2}^2 + e_{N.Z.fb.V2}^2 + i_{N.5534}^2 \cdot \left( Zin_{V2} + Z_{fb.V2} \right)^2}$$

$e_{N.in.V2} = 1.532 \times 10^{-6} V \qquad \frac{e_{N.in.V2} \cdot \sqrt{B_1}}{\sqrt{B_{20k}}} = 1.084 \times 10^{-8} V$

$e_{in.V1.eff} := \frac{e_{N.in.V2}}{G_{ppa}} \qquad e_{in.V1.eff} = 1.532 \times 10^{-7} V$

$\frac{e_{in.V1.eff} \cdot \sqrt{B_1}}{\sqrt{B_{20k}}} = 1.084 \times 10^{-9} V$

$SN_{ne.dose.43.eff} := 20 \cdot \log\left( \frac{e_{in.V1.eff}}{e_{in.nom}} \right) \qquad SN_{ne.dose.43.eff} = -70.273$

**Step 7 Evaluation of influence of the input voltage devider effects at the input of V2 (MM phono-amp) and at the input of DOSE's additional IEC-filter in [dB]:**

loss of voltage devider between ppa (V1) and MM phono-amp (V2):

$loss1 := 20 \cdot \log\left( \frac{Rin_{V2}}{Rout_{V1} + Rin_{V2}} \right) \qquad loss1 = -0.022$

loss of voltage devider at the input of the IEC 20Hz hp (values of components are taken from DOSE's book "Self on Audio", Chapter 5, Figure 8):

$loss2 := 20 \cdot \log\left( \frac{12 \cdot 10^{3} \Omega}{12 \cdot 10^{3} \Omega + 470\Omega + 68\Omega} \right) \qquad loss2 = -0.381$

➤ **MCD Worksheet IV:** SN evaluation of the DOSE MC pre-pre-amp - Page 5

**Step 8 Evaluation of final SNs with reference to B $_{20k}$ and 0.5mV$_{rms}$ /1kHz:**

$SN_{riaa.dose.43} := SN_{ne.dose.43.eff} + SN_r - loss1 - loss2$ $\qquad SN_{riaa.dose.43} = -73.515$

(–73.194)

$SN_{ariaa.dose.43} := SN_{ne.dose.43.eff} + SN_{ar} - loss1 - loss2$ $\qquad SN_{ariaa.dose.43} = -77.805$

(–77.484)

$SN_{sriaa.dose.43} := SN_{ne.dose.43.eff} + SN_{sr} - loss1 - loss2$ $\qquad SN_{sriaa.dose.43} = -77.723$

(–77.402)

$SN_{ariaa.dose.43.ppa} - SN_{ariaa.dose.43} = -1.294$ (5532: –1.615 dB)

To get the respective figures for a "5532" op-amp as input device the following noise voltage and current values have to be taken for **A** and **B**:

$$e_{N.5532} := 5 \cdot 10^{-9} V \cdot \frac{\sqrt{B_{20k}}}{\sqrt{B_1}} \qquad i_{N.5532} := 0.7 \cdot 10^{-12} A \cdot \frac{\sqrt{B_{20k}}}{\sqrt{B_1}}$$

➢ **MCD Worksheet V:** SN evaluation of the BUVO MM phono-amp + DOSE MC ppa - Page 1

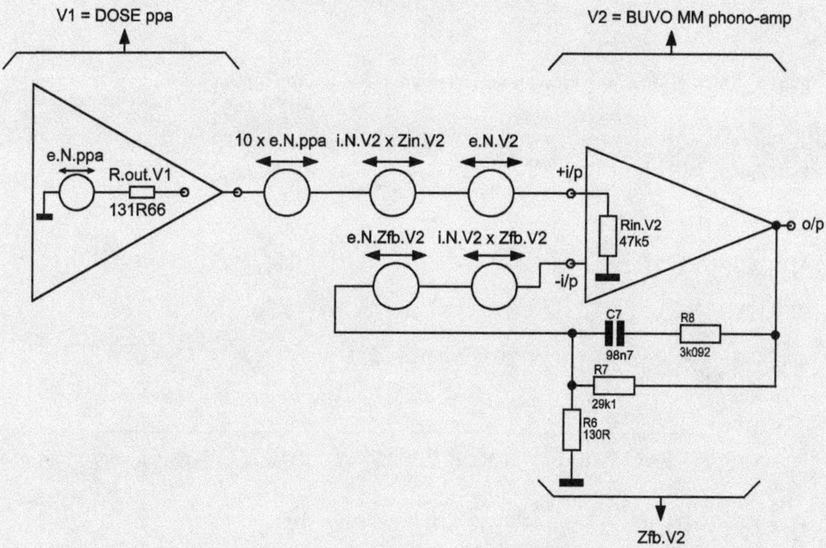


Figure 7.3 Equivalant circuit of the DOSE ppa connected to the BUVO phono-amp (with all meaningful noise sources)

**Step 1 Definition of all meaningful constants, components, etc. (see also Worksheet IV):**

$T := 300\cdot K$ $\quad k := 1.38065\cdot 10^{-23}\cdot V\cdot A\cdot s\cdot K^{-1}$ $\quad e_{in.nom} := 0.5\cdot 10^{-3} V$

$B_{20k} := 19980\cdot Hz$ $\quad B_1 := 1Hz$ $\quad h := 1000Hz$

$SN_r := -3.64566$ $\quad SN_{ar} := -7.93532$ $\quad SN_a := -2.04623$ $\quad SN_{sr} := -7.853$

$e_{N.in.tot} := 1.377\cdot 10^{-7} V$ $\quad \dfrac{e_{N.in.tot}\cdot\sqrt{B_1}}{\sqrt{B_{20k}}} = 9.742\times 10^{-10} V$

$G_{ppa} := 10$ $\quad e_{N.ppa} := G_{ppa}\cdot e_{N.in.tot}$

$e_{N.ppa} = 1.377\times 10^{-6} V$ $\quad \dfrac{e_{N.ppa}\cdot\sqrt{B_1}}{\sqrt{B_{20k}}} = 9.742\times 10^{-9} V$

$Rout_{V1} := 131.66\Omega$ $\quad Rin_{V2} := 47.5\cdot 10^3\Omega$

**Step 2 Evaluation of the impedance of the feedback network of V2 at 1kHz:**

$C7 := 98.7\cdot 10^{-9}\cdot F$ $\quad R7 := 29.1\cdot 10^3\Omega$ $\quad R8 := 3.092\cdot 10^3\Omega$ $\quad R6 := 130\Omega$

$$Z_{fb.V2} := \left|\left[\frac{1}{R6} + \left[\left(\frac{1}{R7}\right) + \left(R8 + \frac{1}{2j\cdot\pi\cdot h\cdot C7}\right)^{-1}\right]\right]^{-1}\right| \qquad Z_{fb.V2} = 125.281\,\Omega$$

$$e_{N.Z.fb.V2} := \sqrt{4\cdot k\cdot T\cdot B_{20k}\cdot Z_{fb.V2}}$$

➢ **MCD Worksheet V:** SN evaluation of the BUVO MM phono-amp + DOSE MC ppa - Page 2

**Step 3 Calculation of the total noise voltage and $SN_{ne}$ at the input of V1:**

$$e_{N.buvo} := \sqrt{2}\cdot 1.54\cdot 10^{-9}\cdot\frac{\sqrt{B_{20k}}}{\sqrt{B_1}}\cdot V \qquad e_{N.buvo} = 3.078\times 10^{-7}\,V$$

$$i_{N.buvo} := 0.23\cdot 10^{-12}\cdot\frac{\sqrt{B_{20k}}}{\sqrt{B_1}}\cdot A \qquad i_{N.buvo} = 3.251\times 10^{-11}\,A$$

$$Zin_{V2} := \left(\frac{1}{Rin_{V2}} + \frac{1}{Rout_{V1}}\right)^{-1} \qquad Zin_{V2} = 131.296\,\Omega \qquad e_{N.Zin.V2} := \sqrt{4\cdot k\cdot T\cdot B_{20k}\cdot Zin_{V2}}$$

$$e_{N.in.V2} := \sqrt{e_{N.ppa}^{\;2} + e_{N.buvo}^{\;2} + e_{N.Zin.V2}^{\;2} + e_{N.Z.fb.V2}^{\;2} + i_{N.buvo}^{\;2}\cdot\left(Zin_{V2} + Z_{fb.V2}\right)^2}$$

$$e_{N.in.V2} = 1.441\times 10^{-6}\,V \qquad \frac{e_{N.in.V2}\cdot\sqrt{B_1}}{\sqrt{B_{20k}}} = 1.019\times 10^{-8}\,V$$

$$e_{in.V1.eff} := \frac{e_{N.in.V2}}{G_{ppa}} \qquad e_{in.V1.eff} = 1.441\times 10^{-7}\,V \qquad \frac{e_{in.V1.eff}\cdot\sqrt{B_1}}{\sqrt{B_{20k}}} = 1.019\times 10^{-9}\,V$$

**Step 4 Evaluation of SNs with reference to $B_{20k}$ and $0.5mV_{rms}$ /1kHz nominal input voltage:**

loss of voltage devider between ppa (V1) and MM phono-amp (V2):

$$loss1 := 20\cdot\log\left(\frac{Rin_{V2}}{Rout_{V1} + Rin_{V2}}\right) \qquad loss1 = -0.024$$

$$SN_{ne.43} := 20\cdot\log\left(\frac{e_{in.V1.eff}}{e_{in.nom}}\right) - loss1 \qquad SN_{ne.43} = -70.783$$

$$SN_{riaa.43} := SN_{ne.43} + SN_r \qquad SN_{riaa.43} = -74.429$$

$$SN_{ariaa.43} := SN_{ne.43} + SN_{ar} \qquad SN_{ariaa.43} = -78.719$$

$$SN_{sriaa.43} := SN_{ne.43} + SN_{sr} \qquad SN_{sriaa.43} = -78.636$$

➢ **MCD Worksheet VI:** SN evaluation of the BUVO MC trafo-coupled phono-amp - Page 1

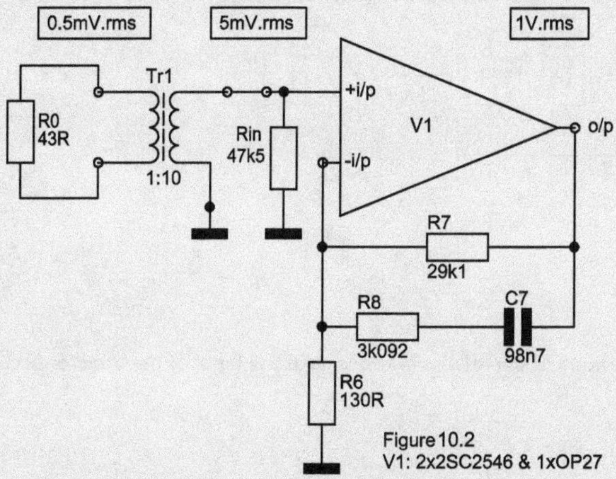


Figure 7.4 General components arrangement of the transformer driven BUVO MC phono-amp

**Step 1 Definition of all meaningful constants, components, etc.:**

$T := 300 \cdot K$ $\quad k := 1.38065 \cdot 10^{-23} \cdot V \cdot A \cdot s \cdot K^{-1}$ $\quad e_{in.nom} := 0.5 \cdot 10^{-3} V$

$B_{20k} := 19980 \cdot Hz$ $\quad B_1 := 1Hz$ $\quad h := 1000Hz$

$SN_r := -3.64566$ $\quad SN_{ar} := -7.93532$ $\quad SN_a := -2.04623$ $\quad SN_{sr} := -7.853$

$R0 := 43\Omega$ $\quad Rin_{V1} := 47.5 \cdot 10^3 \Omega$ $\quad R6 := 130\Omega$ $\quad R7 := 29.1 \cdot 10^3 \Omega$

$C7 := 98.7 \cdot 10^{-9} F$ $\quad R8 := 3092\Omega$

**Step 2 Calculation of Zin = transformer impedance || Rin (derivation from Jensen Transformer's AS040 and Figures 6.3a,b):**

Transformer values: $Rp := 3.0\Omega$ $\quad Rs := 950\Omega$ $\quad tr := 10$

$R0_{sec} := R0 \cdot tr^2$ $\quad R0_{sec} = 4.3 \times 10^3 \Omega$

$Rp_{sec} := Rp \cdot tr^2$ $\quad Rp_{sec} = 300\Omega$

$Z_{tr} := R0_{sec} + Rp_{sec} + Rs$ $\quad Z_{tr} = 5.55 \times 10^3 \Omega$ $\quad Zin := \frac{Rin_{V1} \cdot Z_{tr}}{Rin_{V1} + Z_{tr}}$

$Zin = 4.969 \times 10^3 \Omega$

$e_{N.Zin} := \sqrt{4 \cdot k \cdot T \cdot B_{20k} \cdot Zin}$ $\quad e_{N.Zin} = 1.283 \times 10^{-6} V$ $\quad \frac{e_{N.Zin} \cdot \sqrt{B_1}}{\sqrt{B_{20k}}} = 9.074 \times 10^{-9} V$

➤ **MCD Worksheet VI:** SN evaluation of the BUVO MC trafo-coupled phono-amp - Page 2

**Step 3 Evaluation of the impedance of the feedback network of V1 at 1kHz:**

$$Z_{fb.V1} := \left|\left[\frac{1}{R6} + \left[\left(\frac{1}{R7}\right) + \left(R8 + \frac{1}{2j\cdot\pi\cdot h\cdot C7}\right)^{-1}\right]\right]^{-1}\right| \qquad Z_{fb.V1} = 125.281\,\Omega$$

$$e_{N.Zfb.V1} := \sqrt{4\cdot k\cdot T\cdot B_{20k}\cdot Z_{fb.V1}} \qquad e_{N.Zfb.V1} = 2.036\times 10^{-7}\,V$$

$$\frac{e_{N.Zfb.V1}\cdot\sqrt{B_1}}{\sqrt{B_{20k}}} = 1.441\times 10^{-9}\,V$$

**Step 4 Calculation of the total noise voltage at the input of the transformer:**

$$e_{N.buvo} := \sqrt{2}\cdot 1.54\cdot 10^{-9}\cdot\frac{\sqrt{B_{20k}}}{\sqrt{B_1}}\cdot V \qquad e_{N.buvo} = 3.078\times 10^{-7}\,V$$

$$i_{N.buvo} := 0.23\cdot 10^{-12}\cdot\frac{\sqrt{B_{20k}}}{\sqrt{B_1}}\cdot A \qquad i_{N.buvo} = 3.251\times 10^{-11}\,A$$

$$e_{N.in.tr} := \frac{\sqrt{e_{N.buvo}^{2} + e_{N.Zin}^{2} + e_{N.Zfb.V1}^{2} + i_{N.buvo}^{2}\cdot\left(Zin + Z_{fb.V1}\right)^{2}}}{tr}$$

$$e_{N.in.tr} = 1.345\times 10^{-7}\,V \qquad \frac{e_{N.in.tr}\cdot\sqrt{B_1}}{\sqrt{B_{20k}}} = 9.514\times 10^{-10}\,V$$

**Step 5 Calculations of loss [dB] due to changed load of cartridge and the voltage devider situation between U0, U1, $Z_{tr}$ and Rin $_{V1}$ [ equation (3.278) ff ] :**

$$G_{loss.43} := \frac{Rin_{V1} + Rs + tr^{2}\cdot Rp + tr^{2}\cdot R0}{Rin_{V1}} \qquad G_{loss.43} = 1.117$$

$$G_{e.loss.43} := 20\cdot\log\left(G_{loss.43}\right) \qquad G_{e.loss.43} = 0.96$$

**Step 6 Evaluation of SNs [dB] with referrence to B $_{20k}$ and 0.5mV$_{rms}$ /1kHz nominal input voltage:**

$$SN_{ne.43} := 20\cdot\log\left(\frac{e_{N.in.tr}}{e_{in.nom}}\right) + G_{e.loss.43} \qquad SN_{ne.43} = -70.446$$

$$SN_{riaa.43} := SN_{ne.43} + SN_{r} \qquad SN_{riaa.43} = -74.092$$

$$SN_{ariaa.43} := SN_{ne.43} + SN_{ar} \qquad SN_{ariaa.43} = -78.381$$

$$SN_{sriaa.43} := SN_{ne.43} + SN_{sr} \qquad SN_{sriaa.43} = -78.299$$

## ➢ MCD Worksheet VII: SN evaluation of the BUVO MC solid-state phono-amp - Page 1

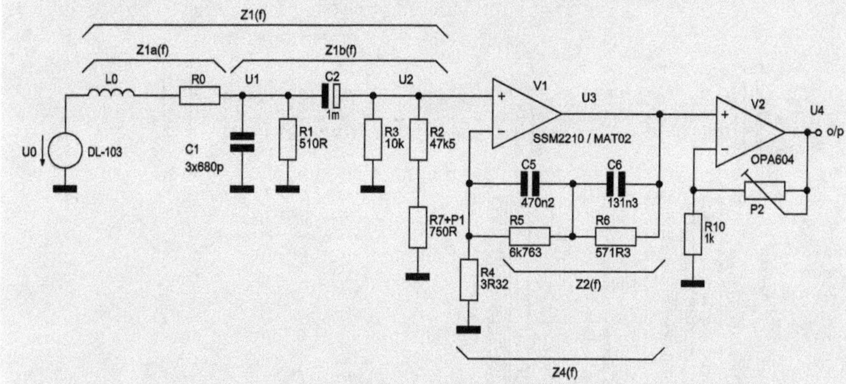


Figure 7.5 General components arrangement of the solid-state driven BUVO MC phono-amp

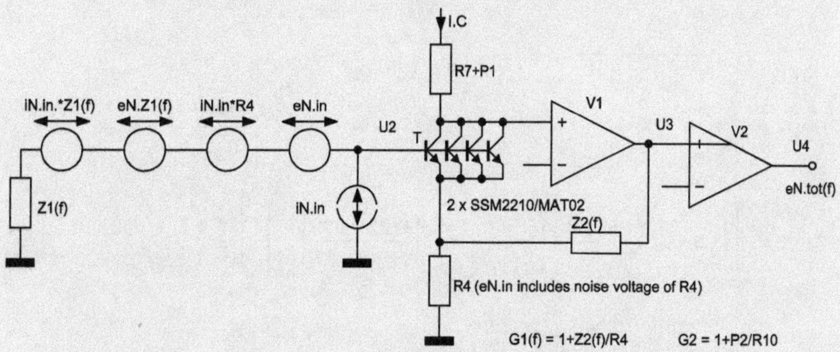


Figure 7.6 Equivalent noise model of the solid-state driven BUVO MC phono-amp

**Step 1 Definition of all meaningful constants, components, etc.:**

$k := 1.38065 \cdot 10^{-23} \cdot V \cdot A \cdot s \cdot K^{-1}$ $q := 1.6021765 \cdot 10^{-19} A \cdot s$ $T := 300 \cdot K$ $e_{in.nom} := 0.5 \cdot 10^{-3} V$

$B_{20k} := 19980 \cdot Hz$ $B_1 := 1Hz$ $h := 1000Hz$ $e_{out.nom} := 1V$

$SN_r := -3.64566$ $SN_{ar} := -7.93532$ $SN_a := -2.04623$ $SN_{sr} := -7.853$

$R0 := 43\Omega$ $R1 := 510\Omega$ $R2 := 47.5 \cdot 10^3 \Omega$ $R3 := 10 \cdot 10^3 \Omega$

$R7 := 750\Omega$ $R4 := 3.32\Omega$ $R5 := 6.763 \cdot 10^3 \Omega$ $R6 := 571.3\Omega$

$C1 := 2.04 \cdot 10^{-9} F$ $C2 := 10^{-3} F$ $C5 := 470.2 \cdot 10^{-9} F$ $C6 := 131.3 \cdot 10^{-9} F$

$I_C := 6.7 \cdot 10^{-3} A$ $h_{FE} := 680$ $r_{bb} := \frac{30}{4} \Omega$ $RA := 10 \cdot 10^3 \Omega$ $RB := 47.5 \cdot 10^3 \Omega$

$V_{DC.C} := R7 \cdot I_C$ $V_{DC.E} := 0.6V$ $NI := 31.62 \cdot 10^{-9} V$ $G_{1st} := 100$

➤ **MCD Worksheet VII:** SN evalua tion of the BUVO MC solid-state phono-amp - Page 2

**Step 2 Evaluation of the impedance of the input network impedances of V1:**

$f := 10, 20 .. 20000$

$$Z1(f) := \left| \left[ \left[ \left[ \frac{1}{R0} + (2j\cdot\pi\cdot f\cdot C1) + \frac{1}{R1} \right]^{-1} + \frac{1}{2j\cdot\pi\cdot f\cdot C2} \right]^{-1} + \left[ \frac{R3\cdot(R2+R7)}{R2+R3+R7} \right]^{-1} \right]^{-1} \right|$$

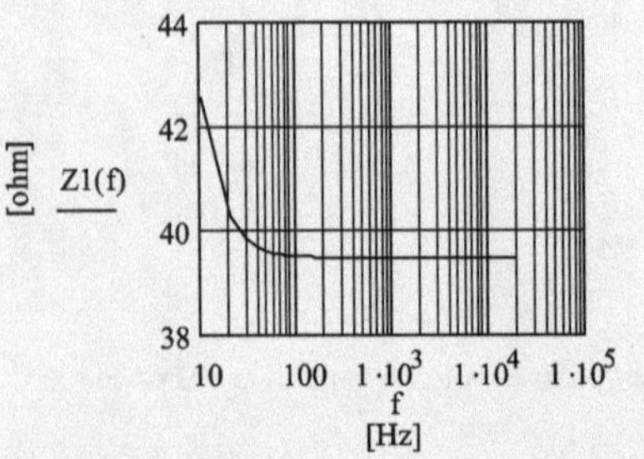


$Z1(h) = 39.468\,\Omega$

$Z1a(f) := R0$

Figure 7.7
Impedance if the input network Z1(f) of the BUVO solid-state design

$e_{N.Z1}(f) := \sqrt{4\cdot k\cdot T\cdot B_{20k}\cdot Z1(f)}$ $\qquad e_{N.Z1}(h) = 1.143\times 10^{-7}\,V$

$$\frac{e_{N.Z1}(h)\cdot\sqrt{B_1}}{\sqrt{B_{20k}}} = 8.086\times 10^{-10}\,V$$

$$Z1b(f) := \left| \left[ \left[ (2j\cdot\pi\cdot f\cdot C1) + \frac{1}{R1} \right] + \left[ \frac{1}{2j\cdot\pi\cdot f\cdot C2} + \frac{R3\cdot(R2+R7)}{R2+R3+R7} \right]^{-1} \right]^{-1} \right|$$

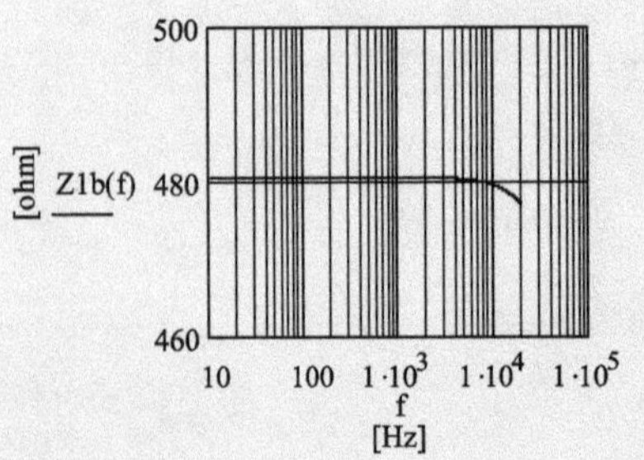


Figure 7.8
Input impedance Z1b(f) of the BUVO solid-state design

**Step 3 Evaluation of the impedance of the feedback network of V1 and gain $G_{V1}(h)$ [dB] at 1kHz:**

$$Z2(h) := \left| \left[ \left[ \left( \frac{1}{R5} + 2j\cdot\pi\cdot h\cdot C5 \right)^{-1} + \left( \frac{1}{R6} + 2j\cdot\pi\cdot h\cdot C6 \right)^{-1} \right]^{-1} \right]^{-1} \right|$$

$G_{V1}(h) := 20\cdot\log\left(1 + \frac{Z2(h)}{R4}\right)$ $\qquad Z2(h) = 738.864\,\Omega$

$G_{V1}(h) = 46.987$

➤ **MCD Worksheet VII:** SN evaluation of the BUVO MC solid-state phono-amp - **Page 3**

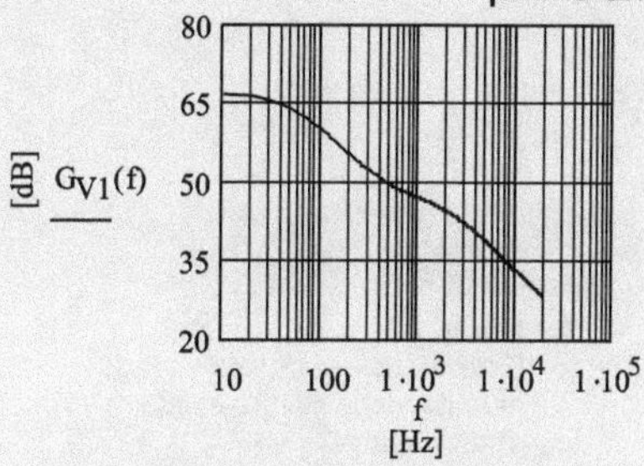


Figure 7.9
RIAA transfer of the BUVO solid-state MC phono-amp

**Note:** **With P2 adjustment of total gain (66dB) to get 1V$_{rms}$ rated output voltage with 0.5mV$_{rms}$ /1kHz rated input voltage! This will enable direct measurement reading in [-dBV].**

**Step 4** **Calculation of input noise current and voltage [equations (3.93), (3.94) and (3.59) ff]:**

$$i_N := \sqrt{\frac{2\cdot q\cdot I_C}{h_{FE}}\cdot B_{20k} + \frac{4\cdot k\cdot T}{\left(\frac{1}{RA}+\frac{1}{RB}\right)^{-1}}\cdot B_{20k}} \qquad i_N = 3.212\times 10^{-10}\,A$$

$$\frac{i_N\cdot\sqrt{B_1}}{\sqrt{B_{20k}}} = 2.272\times 10^{-12}\,A$$

$$e_N := \sqrt{\frac{2\cdot k^2\cdot T^2}{q\cdot I_C}\cdot B_{20k} + 4\cdot k\cdot T\cdot\left(r_{bb}+R4\right)\cdot B_{20k}} \qquad e_N = 6.496\times 10^{-8}\,V$$

$$\frac{e_N\cdot\sqrt{B_1}}{\sqrt{B_{20k}}} = 4.596\times 10^{-10}\,V$$

$$e_{N.R7ex.3d} := NI\cdot\sqrt{3}\cdot\frac{V_{DC.C}}{1V} \qquad e_{N.R7ex.3d} = 2.752\times 10^{-7}\,V \qquad e_{N.R7} := \sqrt{4\cdot k\cdot T\cdot R7\cdot B_{20k}}$$

$$e_{N.R4ex.3d} := NI\cdot\sqrt{3}\cdot\frac{V_{DC.E}}{1V} \qquad e_{N.R4ex.3d} = 3.286\times 10^{-8}\,V \qquad e_{N.R4} := \sqrt{4\cdot k\cdot T\cdot R4\cdot B_{20k}}$$

$$e_{N.ex.tot} := \sqrt{e_{N.R7ex.3d}^{\,2} + e_{N.R7}^{\,2} + e_{N.R4ex.3d}^{\,2} + e_{N.R4}^{\,2}} \qquad e_{N.ex.tot} = 5.711\times 10^{-7}\,V$$

$$e_{N.tot}(f) := \sqrt{e_N^{\,2} + i_N^{\,2}\cdot(Z1(f)+R4)^2 + e_{N.Z1}(f)^2 + \left(\frac{e_{N.ex.tot}}{G_{1st}}\right)^2}$$

$$e_{N.tot}(h) = 1.323\times 10^{-7}\,V \qquad \frac{e_{N.tot}(h)\cdot\sqrt{B_1}}{\sqrt{B_{20k}}} = 9.361\times 10^{-10}\,V$$

excess noise of R7 and R4 can beneglected in this case because of the high gain $G_{1st}$ of the 1st stage:

➤ **MCD Worksheet VII:** SN evaluation of the BUVO MC solid-state phono-amp - Page 4

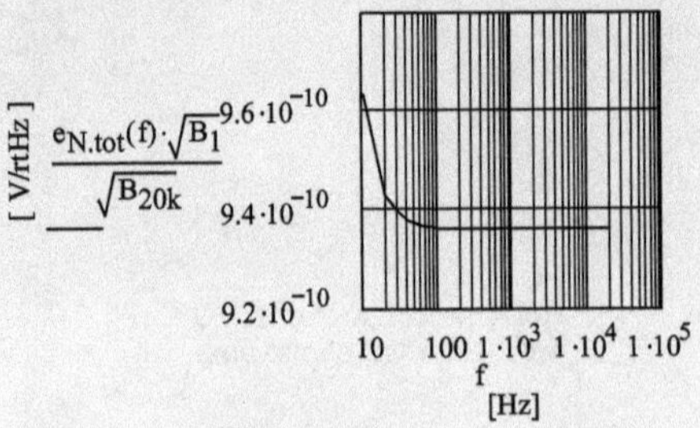


Figure 7.10
Equivalent input noise voltage density of the BUVO solid state phono-amp

**Step 5 Input load = resistor 43R; evaluation of SNs [dB] with referrence to B $_{20k}$ and 0.5mV$_{rms}$ /1kHz nominal input voltage:**

$$SN_{ne.43} := 20 \cdot \log\left[\frac{\sqrt{\frac{1}{B_{20k}}\left[\int_{20Hz}^{20000Hz}\left(e_{N.tot}(f)\right)^2 df\right]}}{e_{in.nom}}\right] \qquad SN_{ne.43} = -71.547$$

$$SN_{riaa.43} := SN_{ne.43} + SN_r \qquad SN_{riaa.43} = -75.193$$

$$SN_{ariaa.43} := SN_{ne.43} + SN_{ar} \qquad SN_{ariaa.43} = -79.483$$

$SN_{ariaa.43}$ without excess noise: -79.491

$$SN_{sriaa.43} := SN_{ne.43} + SN_{sr} \qquad SN_{sriaa.43} = -79.4$$

**Step 6 Input load = DL-103 cartridge**
**Calculations of loss1 [dB] due to changed load of cartridge and the voltage devider situation between U0, U1, R0 and Z1(bf):**

with $R_{load.1}$ = 1k the DL-103 output voltage at 8cm/s velocity becomes at 1kHz (data sheet spec):

$$R_{load.1} := 10^3\Omega \qquad U0 := 0.56 \cdot 10^{-3}V$$

the change from $R_{load.1}$ to $R_{load.2}$ = 480R reduces U0 to U1:

$$R_{load.2} := 480\Omega \qquad U1 := U0 \cdot \frac{\left(R0 + R_{load.1}\right) \cdot R_{load.2}}{\left(R0 + R_{load.2}\right) \cdot R_{load.1}} \qquad U1 = 5.361 \times 10^{-4}V$$

$$loss1 := 20 \cdot \log\left(\frac{U0}{U1}\right) \qquad loss1 = 0.38$$

$$SN_{ne.43.loss} := SN_{ne.43} + loss1 \qquad SN_{ne.43.loss} = -71.168$$

all other SNs got reduced accordingly!

**Step 7 Calculation course for 2 x LM394 as input devices:**

to be changed in Step 7: $I_C = 12.05 \cdot 10^{-3}A$ $\quad h_{FE} = 530$ $\quad r_{bb} = \frac{40}{4}\Omega$

results: see Table 6.2

➢ **MCD Worksheet VIII:** SN evaluation of a BFW16A MC solid-state phono-amp - Page 1

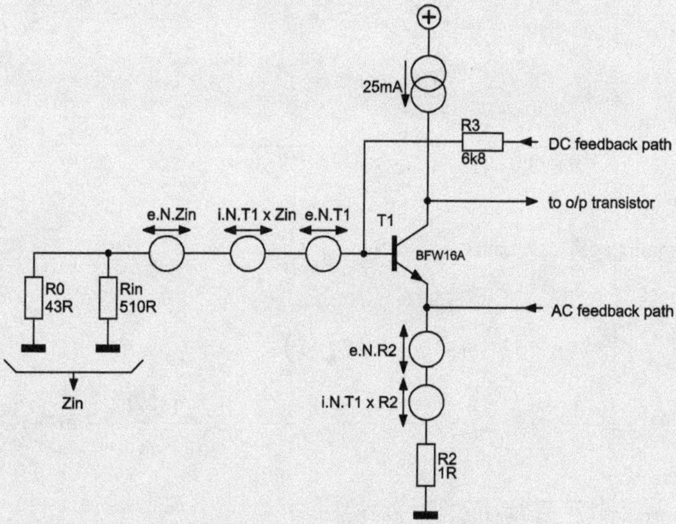


Figure 7.11 Equivalent circuit of the Nordholt / van Vierzen design (with all meaningful noise sources)

**Step 1 Definition of all meaningful constants, components, etc.:**

$T := 300 \cdot K$ $\quad k := 1.38065 \cdot 10^{-23} \cdot V \cdot A \cdot s \cdot K^{-1}$ $\quad q := 1.6021765 \cdot 10^{-19} \cdot A \cdot s$

$B_{20k} := 19980 \cdot Hz$ $\quad B_1 := 1Hz$ $\quad e_{in.nom} := 0.5 \cdot 10^{-3} V$

$SN_r := -3.64566$ $\quad SN_{ar} := -7.93532$ $\quad SN_a := -2.04623$ $\quad SN_{sr} := -7.853$

$R0 := 43\Omega$ $\quad Rin := 510\Omega$ $\quad R2 := 1\Omega$ $\quad I_C := 25 \cdot 10^{-3} A$

**Step 2 Assumptions:** according to N / vV BFW16A with $h_{FE} := 40$

insertion of Rin = 510R
$f_{ci}$ = 800Hz (according to M/C for rather old transistor types like the 2N930)
$f_{ce}$ = 300Hz (according to Fig. 5, JAES 04-1980, Northold / van Vierzen)

**Step 3 Evaluation of input noise voltage and SNs with reference to $B_{20k}$, 0.5mV$_{rms}$ / 1kHz and including any 1/f effect:**

with equations (3.94) and (3.5) and $f_{ci} := 800Hz$:

$$i_{n.T} := \sqrt{\frac{2 \cdot q \cdot I_C}{h_{FE}} \cdot B_{20k}} \qquad i_{n.T} = 2 \times 10^{-9} A \qquad \frac{i_{n.T} \cdot \sqrt{B_1}}{\sqrt{B_{20k}}} = 1.415 \times 10^{-11} A$$

$$i_{N.T} := \frac{i_{n.T}}{\sqrt{B_{20k}}} \cdot \sqrt{f_{ci} \cdot \ln\left(\frac{20000}{20}\right) + B_{20k}} \qquad i_{N.T} = 2.26 \times 10^{-9} A \qquad \frac{i_{N.T} \cdot \sqrt{B_1}}{\sqrt{B_{20k}}} = 1.599 \times 10^{-11} A$$

➢ **MCD Worksheet VIII:** SN evaluation of a BFW16A MC solid-state phono-amp - Page 2

$$Zin := \left(\frac{1}{R0} + \frac{1}{Rin}\right)^{-1} \qquad Zin = 39.656\Omega$$

$$e_{N.Zin} := \sqrt{4 \cdot k \cdot T \cdot B_{20k} \cdot Zin} \qquad e_{N.R2} := \sqrt{4 \cdot k \cdot T \cdot B_{20k} \cdot R2}$$

with equation (3.93), (3.4) and $f_{ce} := 300Hz$, $r_{bb} := 4\Omega$:

$$e_{n.T} := \sqrt{\frac{2 \cdot k^2 \cdot T^2}{q \cdot I_C} \cdot B_{20k} + 4 \cdot k \cdot T \cdot r_{bb} \cdot B_{20k} + \left(i_{N.T} \cdot r_{bb}\right)^2}$$

$$e_{n.T} = 3.971 \times 10^{-8} V \qquad \frac{e_{n.T} \cdot \sqrt{B_1}}{\sqrt{B_{20k}}} = 2.809 \times 10^{-10} V$$

$$e_{N.T} := \frac{e_{n.T}}{\sqrt{B_{20k}}} \cdot \sqrt{f_{ce} \cdot \ln\left(\frac{20000}{20}\right) + B_{20k}}$$

$$e_{N.T} = 4.172 \times 10^{-8} V \qquad \frac{e_{N.T} \cdot \sqrt{B_1}}{\sqrt{B_{20k}}} = 2.952 \times 10^{-10} V$$

$$e_{N.in.43} := \sqrt{e_{N.T}^2 + \left[i_{N.T} \cdot (Zin + R2)\right]^2 + e_{N.Zin}^2 + e_{N.R2}^2}$$

$$e_{N.in.43} = 1.538 \times 10^{-7} V \qquad \frac{e_{N.in.43} \cdot \sqrt{B_1}}{\sqrt{B_{20k}}} = 1.088 \times 10^{-9} V$$

$$SN_{ne.43} := 20 \cdot \log\left(\frac{e_{N.in.43}}{e_{in.nom}}\right) \qquad SN_{ne.43} = -70.242$$

$$SN_{riaa.43} := SN_{ne.43} + SN_r \qquad SN_{riaa.43} = -73.888$$

$$SN_{ariaa.43} := SN_{ne.43} + SN_{ar} \qquad SN_{ariaa.43} = -78.178$$

$$SN_{sriaa.43} := SN_{ne.43} + SN_{sr} \qquad SN_{sriaa.43} = -78.095$$

# Chapter 8
# RIAA Networks

## Intro

Masterpieces of phono-amps are the result of precision design calculation approaches followed by precision building-up of a prototype of the designed circuitry – before it goes to manufacturing. But this is not the end. It is also necessary to select the right sound-relevant passive components, like e.g. capacitors, resistors and inductors. Although, this is a listening based subjective selection, it cannot be part of this book. Nevertheless, any RIAA network has a great influence on SNs of phono-amps. The relating mathematical proof is illustrated with Eqs. (6.8 ... 6.17).

With reference to 0 dB/1 kHz the transfer of the RIAA decoding function $R(f)$ is given in Eq. (2.6) whereas Eq. (2.3) describes the basic RIAA transfer function RIAA($f$) with time constants $T_1 = 3180\,\mu s$, $T_2 = 318\,\mu s$, $T_3 = 75\,\mu s$. To transfer these equations into a practical circuitry design they allow several solutions of combinations of the following RIAA time constant based equations, e.g.

$$R_1(f) = \frac{1}{\sqrt{1+(2\pi f T_1)^2}} \tag{8.1}$$

$$R_2(f) = \sqrt{1+(2\pi f T_2)^2} \tag{8.2}$$

$$R_3(f) = \frac{1}{\sqrt{1+(2\pi f T_3)^2}} \tag{8.3}$$

To refer these equations to 0 dB/1 kHz multiplication with the respective inverse function with $f = 1000\,\text{Hz}$ must take place.

Now, in any case, to get the complete RIAA transfer RIAA($f$) all three equations have to be multiplied:

$$RIAA(f) = R_1(f)R_2(f)R_3(f) \tag{8.4}$$


DOI: 10.1007/978-3-540-76884-5, ©Springer 2008

Expressed in circuit design language we can select any combination of passive, active, passive-active, active-passive solution that lies between the following boundaries illustrated by Figs. 8.1 … 8.2.

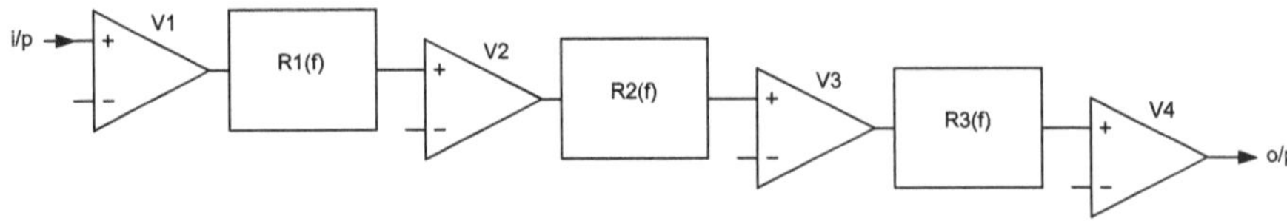


**Fig. 8.1** Total split of RIAA($f$) into three sections designed for a passive components solution separated by active amplifying stages

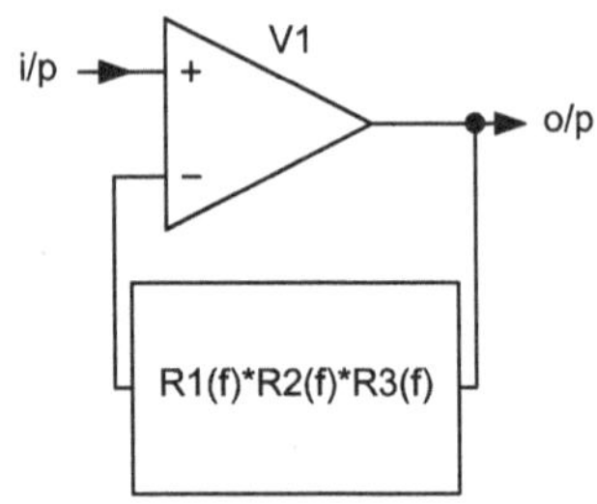


**Fig. 8.2** No split of RIAA($f$) with a fully active feedback mode solution

## Un-Balanced (ub) Solutions

To reduce the many possibilities to a practical amount of calculation efforts I'll go through only four of them – the most common ones for un-balanced (ub) purposes (balanced (b) RIAA amplification will be touched a bit further on in this chapter):

1. the fully passive solution – 1 step
2. the fully passive solution – 2 steps
3. the active-passive solution – 2 steps
4. the fully active solution – 1 step

## The Fully Passive 1-Step-Solution Types ($A_{ub}$) and ($B_{ub}$)

Basically, the fully passive 1-step-solutions types ($A_{ub}$) and ($B_{ub}$) lead to two different design approaches shown in Figs. 8.3 … 8.4.

Both passive solutions can be switched between active gain stages that produce enough gain to drive any following pre-amp. It is obvious that especially overload matters trigger the choice of the gain of the 1st stage because it has to handle an

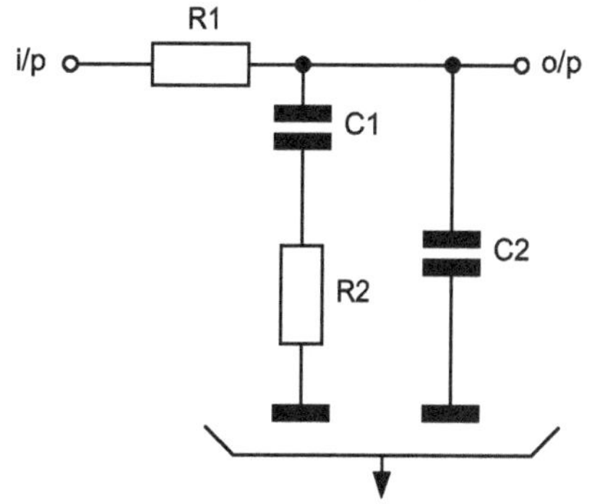


**Fig. 8.3** Passive RIAA($f$) 1-step-solution type ($A_{ub}$)

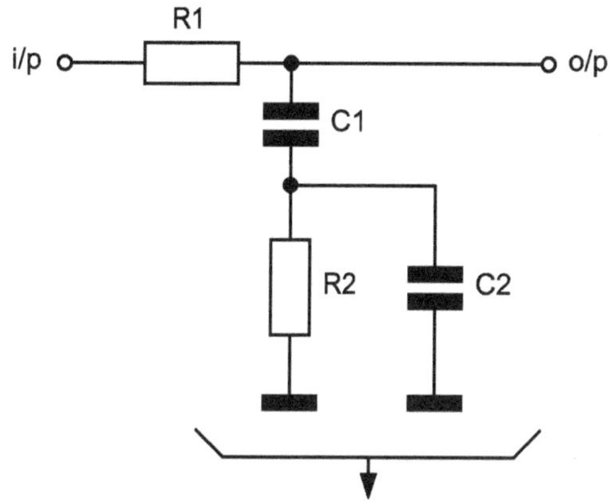


**Fig. 8.4** Passive RIAA($f$) 1-step-solution type ($B_{ub}$)

app. 20 dB higher signal level at 20 kHz than at 1 kHz. That's why these passive solutions are the ideal choice for high voltage gain stages like e.g. valves.

The basic equations to calculate the values for the components could be found e.g. in[1] and are given here in a more precise form[2] (although, calculated with the found equations only the deviation of the networks lies within +0.0005 dB/ −0.0045 dB for ($A_{ub}$) and +0 dB/−0.0016 dB for ($B_{ub}$) in $B_{20\,k}$):

**($A_{ub}$):**

$$r_1 = \frac{R_1}{R_2} = 6.877366 \qquad \text{(JLH: 6.88)} \tag{8.5}$$

$$R_1 \times C_1 = 2187.00207\,\mu s \qquad \text{(JLH: 2187}\,\mu s) \tag{8.6}$$

$$R_2 \times C_2 = 109.0534\,\mu s \qquad \text{(JLH: 109}\,\mu s) \tag{8.7}$$

[1] "The art of linear electronics", John Linsley Hood (JLH), Newnes, 2nd edition 1998

[2] see also calculations on the Worksheets I and II in the following Chap. 9

**($B_{ub}$):**

$$r_1 = \frac{R_1}{R_2} = 12.40316 \qquad \text{(JLH: 12.40)} \tag{8.8}$$

$$R_1 \times C_1 = 2937.001\,\mu\text{s} \qquad \text{(JLH: 2937}\,\mu\text{s)} \tag{8.9}$$

$$R_2 \times C_2 = 81.2053\,\mu\text{s} \qquad \text{(JLH: 81.21}\,\mu\text{s)} \tag{8.10}$$

Unfortunately, these networks and the respective calculation formulae are valid for ideal output and input situations only, say: zero ohm output impedance and infinite input impedance of the gain stages. The real situation looks very much different: the output impedances in e.g. valve gain stages are $\gg 0\,\text{R}$ and the input impedances of any kind of gain stage are built-up by a network of a certain resistance parallel to a sum of capacitances with input-C (BJT: $C_{be}$, $C_{bc}$, valve: $C_{gc}$, $C_{ga}$, FET: $C_{GS}$, $C_{GD}$) plus Miller-C plus stray-C.

Consequently, Figs. 8.3 ... 8.4 have to be changed into Figs. 8.5 ... 8.6 that mirror the real situation for RIAA transfer networks and their calculation needs:

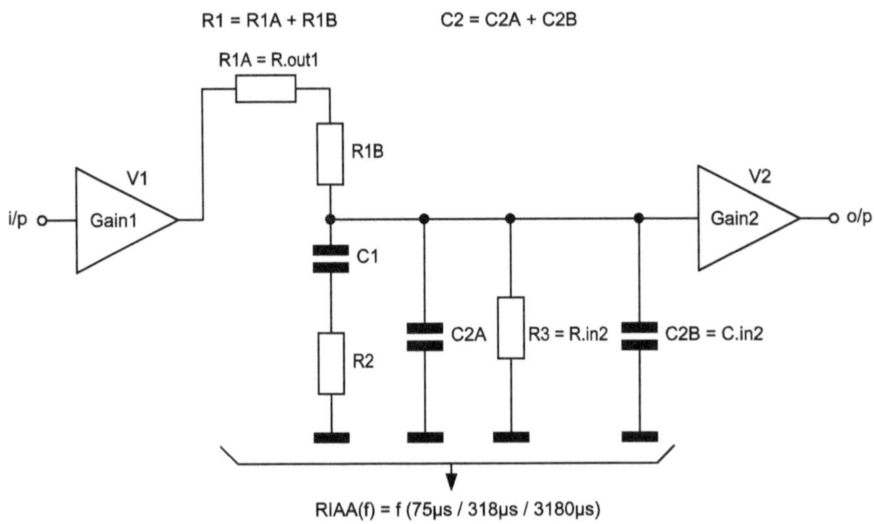


**Fig. 8.5** 1-step passive RIAA transfer network (type ($A_{ub}$)) between gain stage $V_1$ with output impedance $= R_{out1} = R_{1A}$ and gain stage $V_2$ with input impedance $= (R_3 = R_{in2})\,||\,(C_{2B} = C_{in2})$

The respective calculation formulae will change severely. They totally depend on the input impedance of the second stage, whereas, in both cases, the output impedance $R_{out1}$ of the 1st stage will be covered by the calculation of the value for

$$R_1 = R_{1A} + R_{1B} = R_{out1} + R_{1B} \tag{8.11}$$

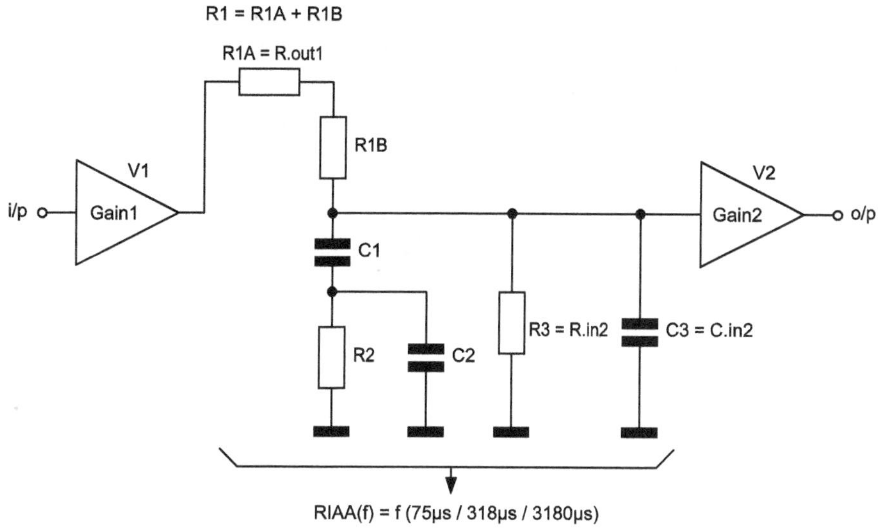


**Fig. 8.6** 1-step passive RIAA transfer network (type ($B_{ub}$)) between gain stage $V_1$ with output impedance $= R_{out1} = R_{1A}$ and gain stage $V_2$ with input impedance $= (R_3 = R_{in2}) \,||\, (C_3 = C_{in2})$

Thus, $R_{1B}$ is the real resistor to find:

$$R_{1B} = R_1 - R_{out1} \tag{8.12}$$

The input impedance of the 2nd stage is formed by resistor $R_3$ and capacitance $C_{2B}$ and/or $C_3$. In both design cases $R_3$ must be selected by the designer and is part of the whole biasing set-up of that gain stage. $C_{2B}$ and $C_3$ are formed by an input capacitance of the first active device $C_{in.dev}$ of that gain stage plus its Miller capacitance $C_{M.dev}$ plus stray capacitances, expressed as $C_{stray}$.

In the type ($A_{ub}$) design $C_{2B}$ is part of the calculated value of $C_2$. Thus, $C_{2A}$ is the real capacitor to find:

$$C_{2A} = C_2 - C_{2B} \tag{8.13}$$

$$C_{2B} = C_{in.dev} + C_{M.dev} + C_{stray} \tag{8.14}$$

In the type ($B_{ub}$) design $C_3$ is the same like $C_{2B}$ in the ($A_{ub}$) design:

$$C_3 = C_{in.dev} + C_{M.dev} + C_{stray} \tag{8.15}$$

- Before going further on it must be mentioned that later on in this chapter the unchanged Eqs. (8.5 … 8.10) will play a role in one of the fully active calculation approaches (see Figs. 8.15 … 8.16), but here, they have to be adapted to the above mentioned real situation.

Example calculation results will demonstrate the differences. Worksheets in the following Chap. 9 give calculation details. Table 8.1 … 8.2 show calculation results for various 2nd stage input resistances and network capacitors:

With a biasing resistor $R_3$

- $R_3 = 475\,\mathrm{k}$

and an input capacitance $C_{2B} = C_3$ that includes a calculated Miller-C[3] (there is no Miller-C as long as this gain stage produces a gain of only $\leq 1$ as a cathode or emitter or source follower) plus a data sheet given input-C plus a guessed stray-C

- $C_{2B} = C_3 = 100\,\mathrm{p}$

the respective formulae look as follows:

**Type ($A_{\mathbf{ub.real}}$)[4]:**

$$r_1 = \frac{R_1}{R_2} = 7.624252 \tag{8.16}$$

$$R_1 \times C_1 = 2425\,\mu\mathrm{s} \tag{8.17}$$

$$R_2 \times C_2 = 109.0534\,\mu\mathrm{s} \tag{8.18}$$

additionally, for all frequencies, the voltage reduction – as a result of the insertion loss of the voltage divider formed by $R_1$ and $R_3$ – should be made as close to 1 as possible, thus, keeping the need for a 2nd stage with rather high gain (automatically, that means a decrease in SN) as low as possible:

$$G_{\mathrm{il.A}} = \frac{R_3}{R_1 + R_3} = 0.9017 \tag{8.19}$$

**Type ($B_{\mathbf{ub.real}}$)[5]:**

$$r_1 = \frac{R_1}{R_2} = 12.72467 \tag{8.20}$$

$$R_1 \times C_1 = 3014.05\,\mu\mathrm{s} \tag{8.21}$$

$$R_2 \times C_2 = 81.1053\,\mu\mathrm{s} \tag{8.22}$$

additionally, like in the type ($A_{\mathrm{ub}}$) case, the insertion loss – as a result of the voltage divider formed by $R_1$ and $R_3$ – should be made as close to 1 as possible:

---

[3] see Chap. 3, Sect. 3.3, Eq. (3.144)

[4] see Worksheet III in the following Chap. 9

[5] see Worksheet IV in the following Chap. 9

$$G_{\text{il.B}} = \frac{R_3}{R_1 + R_3} = 0.974 \tag{8.23}$$

All calculation results of the worksheets will be achieved with the following method – demonstrated by the type ($A_{\text{ub}}$) design:

- to get a 0 dB/1 kHz related function of the frequency dependent RIAA transfer performing network its voltage divider type of gain $G_{\text{nw}}(f)$ must be multiplied with the inverse function $R(10^3)$ of Eq. (2.4) and with a factor that compensates the voltage divider insertion loss. Than, by subtraction of the ideal RIAA transfer function $R_0(f)$ (= Eq. (2.6)) and by a successive approximation (succ-apps) of certain variables we can bring the resulting deviation $D(f)$ to nearly 0 dB within set boundaries, like e.g. ±0.0005 dB. This can easily be monitored on a graph like the one shown in Fig. 8.7.

With reference to 0 dB/1 kHz the magnitude of the gain $G_0(f)$ of the RIAA network $G_{\text{nw}}(f)$ can be written with admittances. Thus, in MCD style it looks as follows:

$$G_0(f) = G_{\text{il.A}}^{-1} \times G_{\text{nw}}(f) \times R(10^3)^{-1} \tag{8.24}$$

$$G_{\text{nw}}(f) = \left| \frac{\left(2j\pi fC_2 + \frac{1}{R_3} + \frac{1}{R_2 + \frac{1}{2j\pi fC_1}}\right)^{-1}}{R_1 + \left(2j\pi fC_2 + \frac{1}{R_3} + \frac{1}{\frac{1}{2j\pi fC_1}}\right)^{-1}} \right| \tag{8.25}$$

As the result of all succ-apps efforts the deviation between the ideal RIAA transfer function $R_0(f)$ and the real transfer function $G_0(f)$ can be monitored on the graph of Fig. 8.7 (ideally, in $B_{20\,\text{k}}$ it should result in a straight horizontal line at 0 dB) by applying the following equation:

$$D(f) = 20\log(G_0(f)) - 20\log(R_0(f)) \tag{8.26}$$

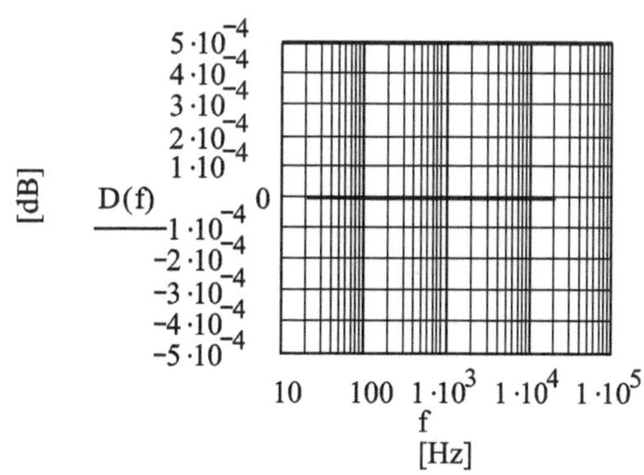


**Fig. 8.7** Graph showing the trace of the deviation $D(f)$ between the ideal RIAA transfer function ($= R_0(f)$) and a calculated one ($= G_0(f)$) referenced to 0 dB/1 kHz

Hence, with the pre-defined values by the designer

- $C_1 = 47\,\mathrm{n}$
- $C_{2B} = 100\,\mathrm{p}$
- $R_3 = 475\,\mathrm{k}$
- $R_{1A} = 73\,\mathrm{R}$ (e.g. PCC88 with $R_c = 1\,\mathrm{k}$)[6]

and – to get the trace in Fig. 8.7 to "0 dB" – succ-apps with the below values

- $T_1 = R_1 \times C_1 \approx 2425\,\mu\mathrm{s}$
- $T_2 = R_2 \times C_1 \approx 109\,\mu\mathrm{s}$
- $r_1 = R_1/R_2 \approx 6.9\ldots7.7$

the other components of Fig. 8.5 become the following calculated values:

- $C_2 = 16\,\mathrm{n}118$
- $C_{2A} = C_2 - C_{2B} = 16\,\mathrm{n}018$
- $R_1 = 51\,\mathrm{k}585$
- $R_{1B} = R_1 - R_{1A} = 51\,\mathrm{k}512$
- $R_2 = 6\,\mathrm{k}766$

and the insertion loss becomes

- $G_{\mathrm{il.A}} = 0.9017$

With Worksheet III, given in the following Chap. 9, it is possible to calculate Table 8.1 … 8.2 results.

Equivalent tables for a type ($B_{\mathrm{ub}}$) 1-step passive network can be generated with Worksheet IV of Chap. 9.

### The Fully Passive 2-Step-Solution Type ($AB_{\mathrm{ub}}$)

Basically, the fully passive 2-step-solution type ($AB_{\mathrm{ub}}$) leads to a rather simple design approach with a tiny disadvantage over the previous approach: we need one more gain stage. But from a noise point of view this will also be an advantage. The following Fig. 8.8 illustrates the network/gain stage set-up.

For that solution approach the basic rules of the game are:

$$T_1 = 75\,\mu\mathrm{s} = R_1 \times C_1 \tag{8.27}$$

$$T_2 = 318\,\mu\mathrm{s} = R_3 \times C_2 \tag{8.28}$$

$$T_3 = 3180\,\mu\mathrm{s} = C_2 \times (R_2 + R_3) \tag{8.29}$$

$$R_2 = 9 \times R_3 \tag{8.30}$$

[6] Eq. (3.185)

**Table 8.1** Calculation results for a type ($A_{ub}$) 1-step passive RIAA network that is located between two gain stages with gains >1 and different input impedances ($R_3||C_{2B}$) of the 2nd stage

| 1/A | B | C | D | E | F | G | H | I | J | K | L | M | N | O |
|---|---|---|---|---|---|---|---|---|---|---|---|---|---|---|
| 2 | Pre-defined by designer | | | | Variables for succ-apps | | | | | Calculation results | | | | |
| 3 | $R_3$ | $C_1$ | $C_{2B}$ | $R_{1A}$ | $r_1$ | $T_1$ | $T_2^*$ | $R_1$ | $R_{1B}$ | $R_2$ | $G_{il.A}$ | $T_3^{**}$ | $C_2$ | $C_{2A}$ |
| 4 | [kΩ] | [nF] | [nF] | [Ω] | | [μs] | [μs] | [Ω] | [Ω] | [Ω] | | $R_2 \times C_1$ [μs] | [nF] | [nF] |
| 5 | infinite | 33 | 0.1 | 73 | 6.8773660 | 2187.00207 | 109.0534 | 66,273 | 66,200 | 9,636 | 1 | 318 | 11.317 | 11.217 |
| 6 | 1,000 | 47 | 0.1 | 73 | 7.2130010 | 2293.73432 | 109.0534 | 48,803 | 48,730 | 6,766 | 0,9533 | 318 | 16.118 | 16.018 |
| 7 | 825 | 47 | 0.1 | 73 | 7.2884504 | 2317.72729 | 109.0534 | 49,313 | 49,240 | 6,766 | 0,9434 | 318 | 16.118 | 16.018 |
| 8 | 475 | 68 | 0.1 | 73 | 7.3768440 | 2345.85639 | 109.0534 | 34,498 | 34,425 | 4,676 | 0,9320 | 318 | 23.320 | 23.220 |
| 9 | 221 | 100 | 0.1 | 73 | 7.6326909 | 2427.19571 | 109.0534 | 24,272 | 24,199 | 3,180 | 0,9009 | 318 | 34.294 | 34.194 |
| 10 | 100 | 220 | 0.1 | 73 | 7.6365040 | 2428.40827 | 109.0534 | 11,038 | 10,965 | 1,445 | 0,9009 | 318 | 75.446 | 75.346 |

* advise: don't change; ** should be always 318 μs

**Table 8.2** Calculation results for a type ($A_{ub}$) 1-step passive RIAA network that is located between two gain stages with gains >1 with different values for $C_1$ and a fixed input impedance ($R_3 = 475$ k)||($C_{2B} = 100$ p) of the 2nd stage

| 1/A | B | C | D | E | F | G | H | I | J | K | L | M | N | O |
|---|---|---|---|---|---|---|---|---|---|---|---|---|---|---|
| 2 | Pre-defined by designer | | | | Variables for succ-apps | | | | | Calculation results | | | | |
| 3 | $R_3$ | $C_1$ | $C_{2B}$ | $R_{1A}$ | $r_1$ | $T_1$ | $T_2^*$ | $R_1$ | $R_{1B}$ | $R_2$ | $G_{il.A}$ | $T_3^{**}$ | $C_2$ | $C_{2A}$ |
| 4 | [kΩ] | [nF] | [nF] | [Ω] | | [μs] | [μs] | [Ω] | [Ω] | [Ω] | | $R_2 \times C_1$ [μs] | [nF] | [nF] |
| 5 | 475 | 100 | 0.1 | 73 | 7.2092970 | 2292.55600 | 109.0534 | 22,926 | 22,853 | 3,180 | 0.9542 | 318 | 34.294 | 34.194 |
| 6 | 475 | 82 | 0.1 | 73 | 7.2864940 | 2317.10500 | 109.0534 | 28,257 | 28,184 | 3,878 | 0.9443 | 318 | 28.121 | 28.021 |
| 7 | 475 | 68 | 0.1 | 73 | 7.3768440 | 2345.83639 | 109.0534 | 34,498 | 34,425 | 4,676 | 0.9320 | 318 | 23.320 | 23.220 |
| 8 | 475 | 56 | 0.1 | 73 | 7.4934635 | 2382.92200 | 109.0534 | 42,552 | 42,479 | 5,679 | 0.9178 | 318 | 19.204 | 19.104 |
| 9 | 475 | 47 | 0.1 | 73 | 7.6242520 | 2424.51000 | 109.0534 | 51,585 | 51,512 | 6,766 | 0.9017 | 318 | 16.118 | 16.018 |
| 10 | 475 | 33 | 0.1 | 73 | 7.9924910 | 2541.61000 | 109.0534 | 77,019 | 76,946 | 9,636 | 0.8606 | 318 | 11.317 | 11.217 |
| 11 | 475 | 22 | 0.1 | 73 | 8.6976260 | 2765.84500 | 109.0534 | 125,720 | 125,647 | 14,455 | 0.7905 | 318 | 7.545 | 7.445 |
| 12 | 475 | 10 | 0.1 | 73 | 12.7457000 | 4053.16000 | 109.0534 | 405,313 | 405,240 | 31,800 | 0.5397 | 318 | 3.429 | 3.329 |

* advise: don't change; ** should be always 318 μs

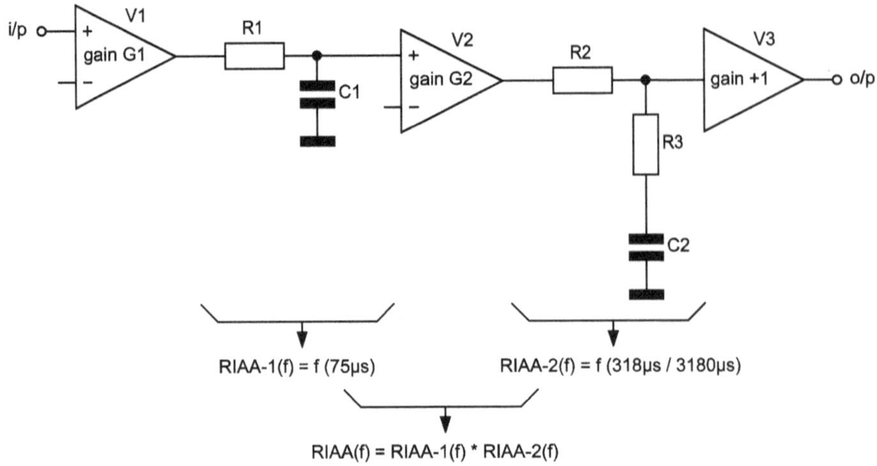


**Fig. 8.8** RIAA equalization with an ideal 2-step passive transfer network type ($AB_{ub}$)

and:

- total gain of the phono-amp is concentrated in $V_1$ and $V_2$ only
- to avoid Miller-C problems that might influence RIAA-2$(f)$ buffer gain of $V_3$ is only $\leq 1$, thus, no Miller-C touches that network negatively
- Miller-C influence on RIAA-1$(f)$ is covered by $C_1$ – which includes it mathematically
- RIAA-1$(f)$ must be located between $V_1$ and $V_2$, RIAA-2$(f)$ between $V_2$ and $V_3$ – not the other way around. Reasons for that are:

  1. RIAA-2$(f)$ is a frequency dependent voltage divider with a loss of app. 1/10 vs. 20 Hz at frequencies >1 kHz. If we would move this network between $V_1$ and $V_2$ this loss must be compensated by the gain of $V_2$, thus, amplifying the noise voltage of that – in most cases high resistive – network plus the noise voltage of the 1st stage with a factor of 10 at least, hence, increasing the total noise voltage. SN automatically becomes worse when compared with the recommended case.
  2. the overload problem for frequencies >1 kHz is concentrated on the 1st stage alone. At the output of the lp-filter RIAA-1$(f)$ all frequencies >1 kHz get equalized. Frequencies <1 kHz won't play a role.
  3. especially with valve driven gain stages that are not configured in the low-output-impedance SRPP mode RIAA-2$(f)$ might change the load of the anode, thus more and more decreasing its gain with growing frequency >1 kHz.

Although, the design of the 2-step solution looks easier to handle than the one of the 1-step solution, it looks and is still challenging. This will be demonstrated with the next figure:

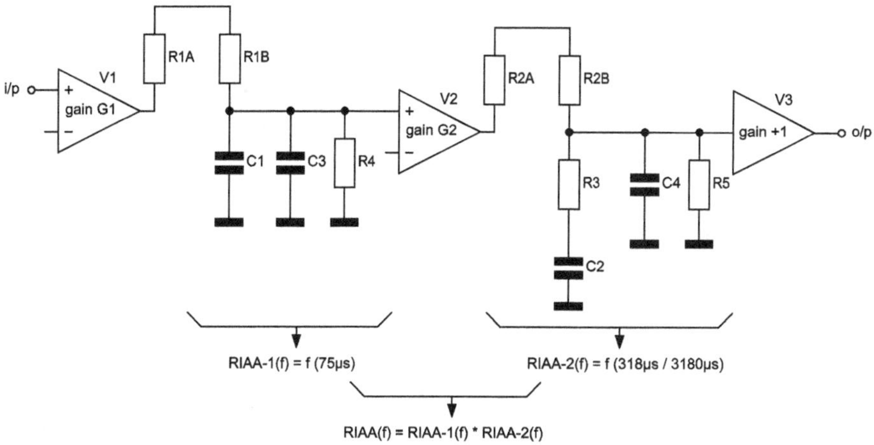


**Fig. 8.9** RIAA equalization with a 2-step passive transfer network type ($AB_{ub}$) in a real circuit environment with resistive output impedances $R_{1A}$, $R_{2A}$, resistive input impedances and biasing resistors $R_4$, $R_5$ and gain stage input capacitances $C_3$, $C_4$

The calculation formulae for Fig. 8.9 look as follows:

$$T_1 = 75\,\mu\text{s} = (C_1 + C_3) \times \left(\frac{1}{R_{1A} + R_{1B}} + \frac{1}{R_4}\right)^{-1} \tag{8.31}$$

$C_3$ represents the Miller-C $C_M$ plus any other data-sheet given input-C $C_{in}$ of the 2nd stage plus a guessing for stray-C $C_{stray}$, hence

$$C_3 = C_{M.2nd} + C_{in.2nd} + C_{stray.2nd} \tag{8.32}$$

the insertion $G_{il.1}$ loss of RIAA-1$(f)$ becomes:

$$G_{il.1} = \frac{R_4}{R_{1A} + R_{1B} + R_4} \tag{8.33}$$

With a direct coupling solution of $V_1$ and $V_2$ we could make $R_4$ nearly infinite, hence, $G_{il.1}$ becomes

$$G_{il.1} = 1 \tag{8.34}$$

Assumed that

- $C_4$ is very small compared with $C_2$ (app. a factor of 1000 smaller) and has a value of app. 10 p (that's why it is recommended to use a buffer with gain of only +1 for this stage)
- $C_4$ does not create a lp-filter effect with an additional time constant $(R_{2A} + R_{2B}) \times C_4$ in $B_{20\,k}$
- $R_5$ can be made nearly infinite because we use a direct stage coupling solution of the two stages

than, $T_2$, $T_3$ become:

$$T_2 = 318\,\mu s = R_3 \times C_2 \tag{8.35}$$
$$T_3 = 3180\,\mu s = C_2 \times (R_{2A} + R_{2B} + R_3) \tag{8.36}$$
$$R_{2A} + R_{2B} = 9 \times R_3 \tag{8.37}$$

the insertion $G_{il.2}$ loss of RIAA-2$(f)$ becomes:

$$G_{il.1} = \frac{R_5}{R_{2A} + R_{2B} + R_5} \tag{8.38}$$

As long as $R_5$ is infinite $G_{il.2}$ becomes:

$$G_{il.2} = 1 \tag{8.39}$$

Otherwise, if the above mentioned assumptions are not valid and a direct coupling of the stages is not possible and the 3rd stage needs gain (consequently with creation of Miller-C), than, the calculation of RIAA-2$(f)$ goes the same way that is described under the previous point for the fully passive 1-step solution of a type ($A_{ub}$) network.

When comparing the 2-step passive network with the 1-step passive solutions of the previous paragraphs one of the many advantages of this type of RIAA equalization is the fact that, from a noise point of view, the rather noisy part of the network can be switched at the end – in front of a buffer – of the amp chain, thus, leaving the noise making elements that count to $V_1$, RIAA-1$(f)$ and $V_2$ alone. With the calculation rules for "contribution allowed"[7] these three elements can easily be noise optimised, whether it's a valve driven or any other solid-state driven part of the amp chain.

### The Active-Passive 2-Step-Solution Types ($C_{ub}$) and ($D_{ub}$)

The active-passive 2-step-solutions types ($C_{ub}$) and ($D_{ub}$) set on an active management of the 318 μs/3180 μs time constants followed by a passive 75 μs time constant solution formed by one simple R-C 1st order lp-filter.

In contrast to the fully passive 2-step solution described in the previous paragraph the 318 μs/3180 μs active gain stage must be located at the 1st place of the amp chain.

Main reasons for that are:

1. if we would set the active part at the 2nd place in the amp chain with reference to 0 dB at 1 kHz the whole noise of the 1st stage <1 kHz would be more and more amplified with decreasing frequency – from 1 at 1 kHz up to app. 10 times at 20 Hz, thus, creating unnecessary SN worsening

[7] Chap. 3, Sect. 3.2 Noise in BJTs

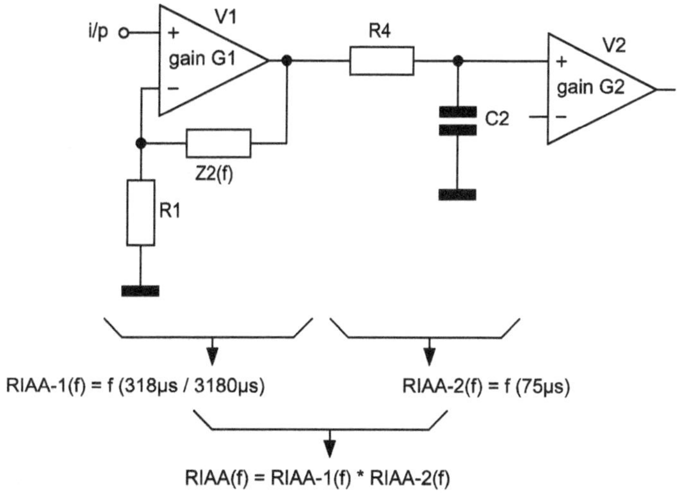


**Fig. 8.10** Phono-amp with 2-step active-passive RIAA equalization

2. the whole noise created by the 1st stage will be filtered by the 75 µs-lp-filter

$Z_2(f)$ can be formed by a R-C network. Together with $R_1$ the gain $G_1$ at 1 kHz of that stage can be set. Basically, two different types can make it:

**Type ($C_{ub}$):**

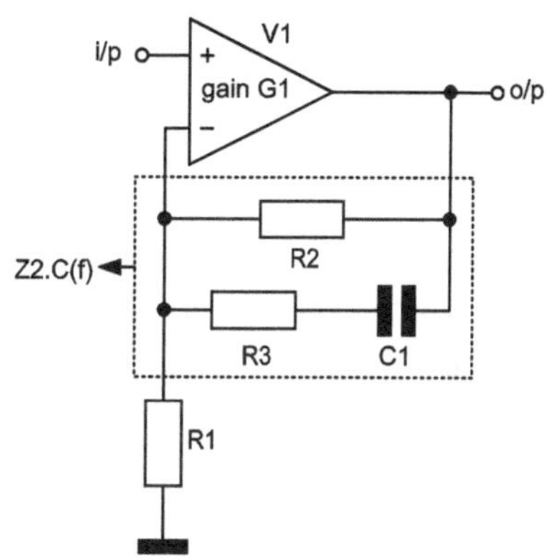


**Fig. 8.11** Type ($C_{ub}$) feedback network of a 1st gain stage for 318 µs/3180 µs equalization

**Type ($D_{ub}$):**

Both types of network work equally well but need different design calculations. It is obvious that the minimum gain $G_1$ of the 1st stage must generate at least a gain of 10 to enable the nearly +20 dB 20 Hz boost of the RIAA transfer. Never-

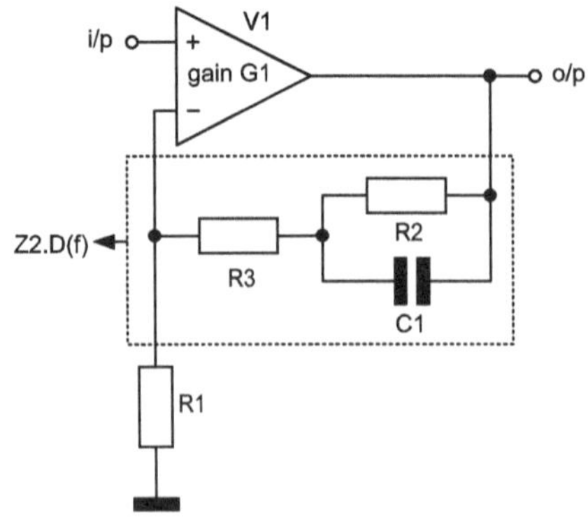


**Fig. 8.12** Type ($D_{ub}$) feedback network of a 1st gain stage for 318 μs/3180 μs equalization

theless, to overcome the overload problem generated by frequencies >1 kHz $G_1$ should be chosen as low as possible. A good compromise could be 20 … 30, thus, amplifying a 5 mV$_{rms}$ 1 kHz signal up to 100 mV$_{rms}$ … 150 mV$_{rms}$. This gives a sufficient 1st stage overload capability for the high frequencies[8]: app. 16 dB at 20 kHz with a gain of 30, assumed we take an op-amp with a ±15 V power supply.

The calculation rules for the components of the type ($C_{ub}$) network look as follows[9], whereas the derivation of the calculation formulae and an example calculation will be demonstrated on a separate MCD Worksheet[10]:

- choose $R_1$ – the most noise-sensitive component of the network
- choose gain $G1_1$ at 1 kHz – the most headroom-sensitive figure of the network
- calculate the DC gain $G1_0$ of the 1st stage as a function of $G1_1$

$$G1_0 = G1_1 \times 8.953721 = \frac{G1_1}{0.111685} \tag{8.40}$$

- calculate $R_2$

$$R_2 = R_1 \times G1_0 - R_1 \tag{8.41}$$

- calculate $R_1$

$$R_1 = \frac{R_2}{9}\left(1 - \frac{10}{G1_0}\right) \tag{8.42}$$

---

[8] Because of the massive loss of headroom with growing frequency of up to additional 20 dB at 20 kH it should be pointed out that the types ($C_{ub}$) and ($D_{ub}$) phono amps might have a certain disadvantage for music listeners with record signals far beyond the 0 dB level of a vinyl record. It is worth reading the respective letter from D/S: Letter in EW 05-2001. On the other hand, I never noticed any clipping when listening to purely electronic generated disco sounds with my MM phono-amp (Fig. 6.3a)

[9] see also "AN346" National Semiconductor Application Note

[10] Worksheet V, Chap. 9

- calculate $C_1$

$$C_1 = \frac{3180\,\mu\text{s}}{R_2 + R_3} \tag{8.43}$$

- cross-check for $T_2$: it should become 318 µs

$$T_2 = C_1 \frac{R_1R_3 + R_2R_3 + R_1R_2}{R_1 + R_2} \tag{8.44}$$

The calculation rules for the components of the type ($D_{\text{ub}}$) network look as follows, whereas the derivation of the calculation formulae and an example calculation will be demonstrated on a separate Worksheet[11]:

- choose $R_1$ – the most noise-sensitive component of the network
- choose gain $G1_1$ at 1 kHz – the most headroom-sensitive figure of the network
- calculate the DC gain $G1_0$ of the 1st stage as a function of $G1_1$

$$\begin{aligned} G1_0 &= G1_1 \times 8.953721 \\ &= \frac{G1_1}{0.111685} \end{aligned} \tag{8.45}$$

- calculate $a$

$$a = R_1 \times (G1_0 - 1) = R_2 + R_3 \tag{8.46}$$

- calculate $R_3$

$$R_3 = \frac{(a - 9 \times R_1)}{10} \tag{8.47}$$

- calculate $R_2$

$$R_2 = a - R_3 \tag{8.48}$$

- calculate $C_1$

$$C_1 = \frac{3180\,\mu\text{s}}{R_2} \tag{8.49}$$

- cross-check for $T_2$: it should become 318 µs

$$T_2 = C_1 \times R_2 \times \frac{R_1 + R_3}{R_1 + R_2 + R_3} \tag{8.50}$$

### The Fully Active 1-Step Solution Types ($E_{\text{ub}}$), ($F_{\text{ub}} - A$), ($F_{\text{ub}} - B$)

The fully active 1-step solution phono-amp configuration manages the RIAA transfer inside the feedback loop of only one amplifying stage à la Fig. 8.2.

[11] Worksheet VI, Chap. 9

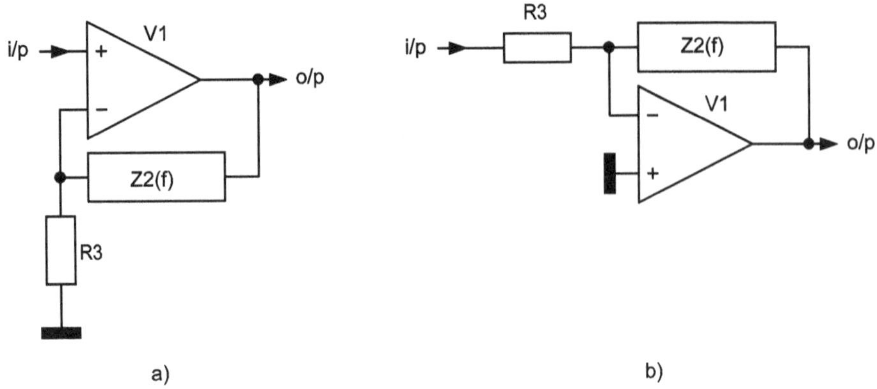


**Fig. 8.13** Basic 1-stage active RIAA network configuration in a) series (= non-inverting) mode and b) shunt (= inverting) mode

For the feedback path $Z_2(f)$ many different network configurations were developed in the past. The following three are the most common ones:

**Type ($E_{ub}$):**

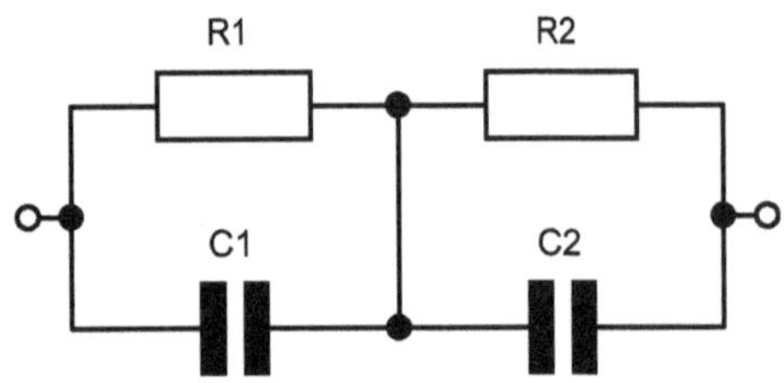


**Fig. 8.14** Type ($E_{ub}$) RIAA network feedback path configuration

**Type ($F_{ub} - A$):**

the active version of the 1-step passive network type ($A_{ub}$) of Fig. 8.3:

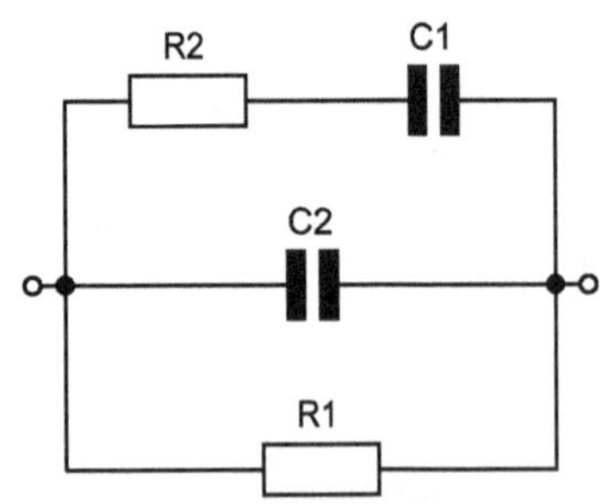


**Fig. 8.15** Type ($F_{ub} - A$) RIAA network feedback path configuration

**Type ($F_{ub} - B$):**

the active version of the Fig. 8.4 type ($B_{ub}$) 1-step passive solution:

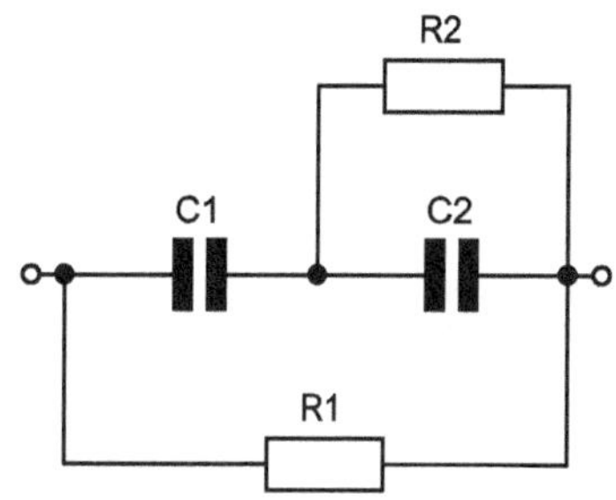


**Fig. 8.16** Type ($F_{ub} - B$) RIAA network feedback path configuration

The calculation methods for these three RIAA networks in shunt and series mode configured phono-amps will be the content of the following paragraphs.

## Shunt Mode

The advantage of the shunt mode configuration is the fact that the calculation rules for the feedback components look rather simple and no additional gain adjusting component $R_3$ influences them on the feedback side of the amp – as long as the open-loop gain of the amplifier is very high, thus making the feedback factor $\beta \gg 100$.

The disadvantage of that kind of configuration is the fact that *SN* will become severely worse when comparing it with the series mode amplifier arrangement. Calculated deterioration will then be app. 13 dB[12].

Assumed that the open-loop gain $c$ of the amp is nearly infinite ($c \geq 10^6$) the formulae to calculate the three different feedback networks for shunt mode configured phono-amps look as follows:

**Type ($E_{ub}$)[13]:**

with

$$T_1 = 3180\,\mu\text{s} \quad T_2 = 75\,\mu\text{s} \quad T_3 = 318\,\mu\text{s} \tag{8.51}$$

the proportional factors $c_1$ and $r_1$ become:

$$c_1 = 3.6 \tag{8.52}$$

$$r_1 = 11.7\overline{7} \tag{8.53}$$

[12] "Low-noise Audio Amplifiers", H. P. Walker, WW 05-1972

[13] Derivation see Worksheet X of Chap. 9

thus, the passive components can be calculated as follows:

$$C_1 = c_1 \times C_2 \tag{8.54}$$

$$R_1 = r_1 \times R_2 \tag{8.55}$$

$$T_1 = R_1 \times C_1 \tag{8.56}$$

$$T_2 = R_2 \times C_2 \tag{8.57}$$

$$T_3 = (C_1 + C_2) \times \left(\frac{R_1 \times R_2}{R_1 + R_2}\right) \tag{8.58}$$

gain $G1_{0.E}$ at DC becomes:

$$G1_{0.E} = -\frac{R_1 + R_2}{R_3} \tag{8.59}$$

gain $G1_{1.E}$ at 1 kHz becomes:

$$G1_{1.E} = G1_{0.E} \times R_{1000} = G1_{0.E} \times 0.10103 \tag{8.60}$$

**Type ($F_{\mathbf{ub}} - A$):**

The formulae to calculate this type of network are the same like the ones for the type ($A_{\text{ub}}$) network:

Eqs. (8.5 … 8.7).

Additionally, DC and 1 kHz gain $G1_{0.F.A}$ and $G1_{1.F.A}$ become:

$$G1_{0.F.A} = -\frac{R_1}{R_3} \tag{8.61}$$

$$G1_{1.F.A} = G1_{0.F.A} \times R_{1000} = G1_{0.F.A} \times 0.10103 \tag{8.62}$$

**Type ($F_{\mathbf{ub}} - B$):**

The formulae to calculate this type of network are the same like the ones for the type ($B_{\text{ub}}$) network:

Eqs. (8.8 … 8.10).

Additionally, DC and 1 kHz gain $G1_{0.F.B}$ and $G1_{1.F.B}$ become:

$$G1_{0.F.B} = -\frac{R_1}{R_3} \tag{8.63}$$

$$G1_{1.F.B} = G1_{0.F.B} \times R_{1000} = G1_{0.F.B} \times 0.10103 \tag{8.64}$$

**Series Mode**

4th Time Constant $T_4$

Compared with shunt mode configured amps in the series mode case a 4th time constant $T_4$ plays an additional role. It is the result of the fact that the gain of a series configured amp always tends to become only 1 at higher frequencies. $T_4$ is a function of the amp's DC gain $G1_0$, its feedback factor $\beta$ and all other effected time constants of the feedback impedance.

To simplify things a bit we assume a phono-amp with open-loop gain $c \geq 10^6$ and $\beta \geq 500$. Thus, as a function of the DC gain $T_4$ becomes[14]:

$$T_4 = \frac{750\,\mu\text{s}}{G1_0} \tag{8.65}$$

In contrast to that simple looking equation lower values for $c$ and $\beta$ require the full calculation formula for $T_4$ (calculation example: see Worksheet IX):

$$T_4 = G1_0^{-1}\left(\frac{\beta+1}{\beta+\frac{2862}{3105}}\right)\left(\frac{\beta+1}{\beta+\frac{243}{3105}}\right)\left(\frac{T_1R \times T_2R}{T_3R}\right) \tag{8.66}$$

**Type ($E_{ub}$):**

For phono-amps configured in series mode three Worksheets [VII ... IX] explain the derivation of the respective succ-apps approaches and specific formulae calculations to get the values for the type ($E_{ub}$) feedback network. A fourth Worksheet [X] explains the derivation of two important proportional factors that will very much ease the calculations of Worksheets VII ... IX.

I've chosen to present two different calculation methods. To demonstrate them with an existing example the RIAA feedback network of the BUVO MC phono-amp of Fig. 6.4 will be calculated.

**Type ($E_{ub}$) Succ-Apps Method**

The first method works with succ-apps and is illustrated on Worksheet VII (as version v2.0). The difference to v1.0 will be explained a bit later – but no difference of the results of the two calculation approaches could be observed.

---

[14] "RIAA-Iisierung", J. W. van Dael/J. A. Kruithof, Elektor 11-1973
Dutch and German version only

With the assumption of $c \geq 10^6$, with the DC gain $G1_{0.E}$ and the gain $G1_{1.E}$ at 1 kHz[15]:

$$G1_{0.E} = \frac{R_1 + R_2 + R_3}{R_3} \tag{8.67}$$

$$G1_{1.E} = G1_{0.E} \times R_{1000} = G1_{0.E} \times 0.10103 \tag{8.68}$$

theoretically, the deviation $D(f)$ from the exact RIAA transfer can be made marginal (Fig. 8.17)[16]. Practically, the tolerances of the components (recommended $\pm 1\%$) always trigger higher deviations than calculated.

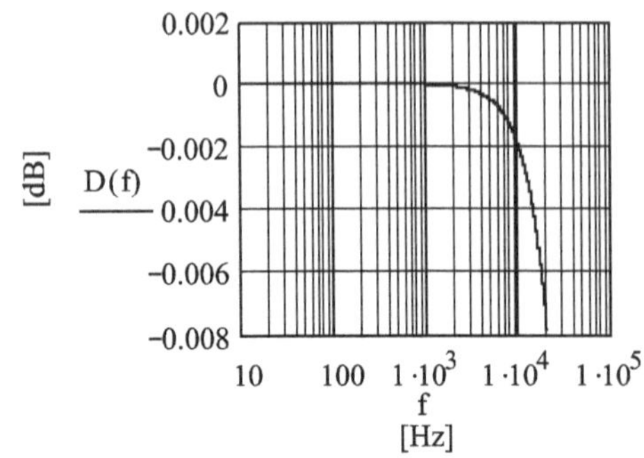


**Fig. 8.17** With succ-apps determined deviation $D(f)$ from the exact RIAA transfer of the RIAA network of the BUVO MC phono-amp of Fig. 6.4

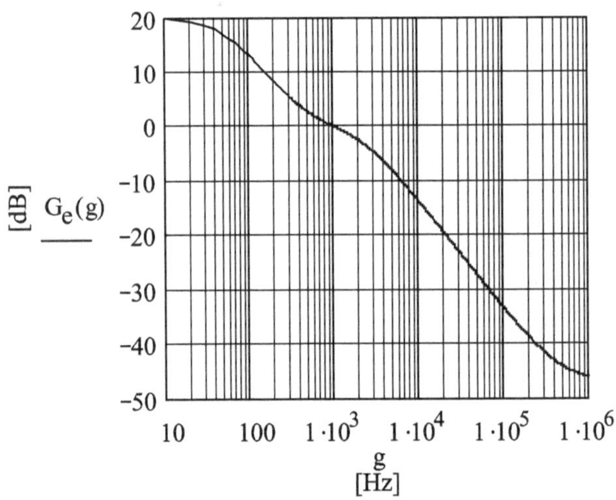


**Fig. 8.18** Plot of the 0 dB/1 kHz referenced RIAA transfer of the BUVO MC phono-amp that shows the $f_4$ corner frequency

[15] see Worksheets VII ... IX, Chap. 9

[16] taken from Worksheet VII, Chap. 9

The respective plot $G_e(g)$[17] of the RIAA transfer of the BUVO MC phono-amp looks as shown in Fig. 8.18. It is referenced to 0 dB at 1 kHz and clearly demonstrates the $T_4$ triggered $f_4$ corner frequency >400 kHz.

## Type ($E_{ub}$) Specific Formulae Method

The second method is demonstrated on Worksheet IX of Chap. 9. I call it "specific formulae approach". I've included this calculation method of Messrs Deal and Kruithof[14] because it's a memory and tribute to the old – Windows- and MAC-free – days of the early 70-ies of last century. At that time complex calculations could only be performed on huge computers. The preparations were very extensive and to get calculation time was really hard won. Nevertheless, the calculation results are nearly as close to the exact RIAA transfer as the results from a succ-apps approach à la Worksheet VII. The possibility to calculate networks for phono-amps with $c < 10^6$ and $\beta$ down to 1 is one of the advantages of this calculation method.

With reference to 0 dB at 1 kHz the deviation $D(g)$[18] from the exact RIAA transfer looks as follows:

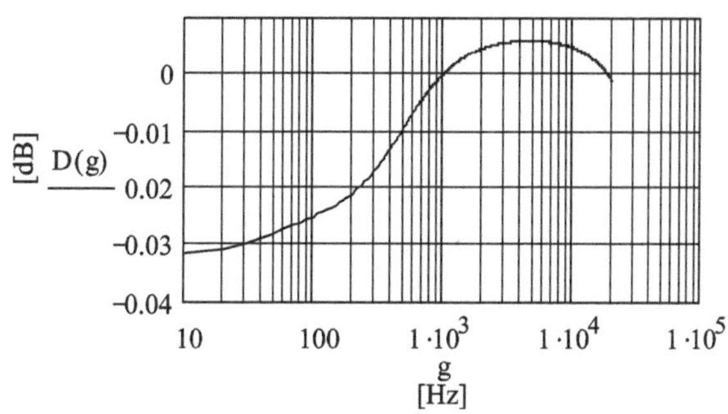


**Fig. 8.19** Deviation $D(g)$ from the exact RIAA transfer

To demonstrate what it means for $T_4$ to calculate a series mode RIAA network of a MM phono-amp I've included Worksheet VIII. Because the DC and 1 kHz gain is nearly 20 dB lower than in the MC case the corner frequency $f_4$ got nearly halved (Fig. 8.20)[19].

In addition, comparing it with the MC case the deviation $D(f)$ from the exact RIAA transfer >1 kHz becomes a calculated order of magnitude worse (Fig. 8.21)[20]. Switched into the amp chain after the RIAA stage a buffered passive 1st order lp-filter with $f_c = f_4$ would totally eliminate this disadvantage.

[17] taken from Worksheet VII, Chap. 9

[18] taken from Worksheet IX, Chap. 9

[19] taken from Worksheet VIII, Chap. 9

[20] taken from Worksheet VIII, Chap. 9

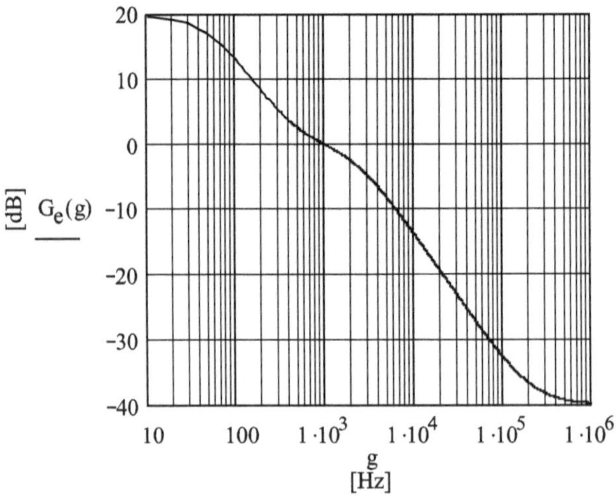


**Fig. 8.20** Plot of a 0 dB/1 kHz referenced RIAA transfer of a MM phono-amp that shows the $f_4$ corner frequency

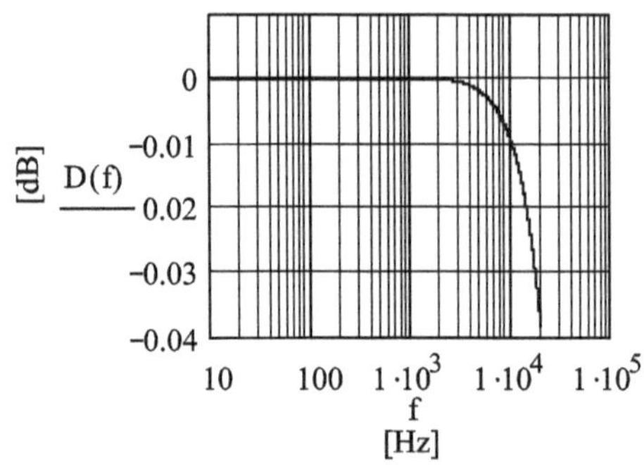


**Fig. 8.21** With succ-apps determined deviation $D(f)$ from the exact RIAA transfer of a RIAA network of a MM phono-amp

### Difference Between v1.0 and v2.0

- The differences between v1.0 and v2.0 are the following ones[21]:
  - In v1.0 "guessing according to Eqs. (8.52 … 8.55)" starts with the calculation of $R_1$ and $C_1$ followed by the application of succ-apps to get $R_2$ and $C_2$.
  - In v2.0 "guessing according to Eqs. (8.52 … 8.55)" starts with the calculation of $R_2$ and $C_2$ followed by the application of succ-apps to get $R_1$ and $C_1$.

### Type ($F_{ub} - A$) Succ-Apps Method

Attached to a series configured amp this type of network can be calculated by the help of Eqs. (8.5 … 8.7). They form the starting basis for the application of the succ-

[21] see Worksheets VII and VIII, Chap. 9

apps method à la type ($E_{ub}$) that is demonstrated on Worksheet VII. $T_4$ calculation rules are outlined above.

With the assumption of $c \geq 10^6$, with DC gain $G1_{0.F.A}$ and gain $G1_{1.F.A}$ at 1 kHz:

$$G1_{0.F.A} = \frac{R_1 + R_3}{R_3} \tag{8.69}$$

$$G1_{1.F.A} = G1_{0.F.A} \times R_{1000} = G1_{0.F.A} \times 0.10103 \tag{8.70}$$

and with $Z2_{F.A}(f)$ and $G1_{F.A}(f)$:

$$Z2_{F.A}(f) = \left( \frac{1}{R_1} + 2j\pi f C_2 + \left( R_2 + \frac{1}{2j\pi f C_1} \right)^{-1} \right)^{-1} \tag{8.71}$$

$$G1_{F.A} = \left| \frac{Z2_{F.A}(f)}{R_3} + 1 \right| \tag{8.72}$$

all components of the feedback network can be determined by bringing the deviation function $D(f)$ and respective trace as close to 0 dB as possible.

### Type ($F_{ub} - B$) Succ-Apps Method

Attached to a series configured amp this type of network can be calculated by the help of Eqs. (8.8 … 8.10). They form the starting basis for the application of the succ-apps method à la type (E) that is demonstrated on Worksheet VII. $T_4$ calculation rules are outlined above.

With the assumption of $c \geq 10^6$, with DC gain $G1_0$ and gain $G1_1$ at 1 kHz:

$$G1_{0.F.B} = \frac{R_1 + R_3}{R_3} \tag{8.73}$$

$$G1_{1.F.B} = G1_{0.F.B} \times R_{1000} = G1_{0.F.B} \times 0.10103 \tag{8.74}$$

and with $Z2_{F.B}(f)$ and $G1_{F.B}(f)$:

$$Z2_{F.B}(f) = \left( \frac{1}{R_1} + \left[ (2j\pi f C_1)^{-1} + \left( \frac{1}{R_2} + 2j\pi f C_2 \right)^{-1} \right]^{-1} \right)^{-1} \tag{8.75}$$

$$G1_{F.B} = \left| \frac{Z2_{F.B}(f)}{R_3} + 1 \right| \tag{8.76}$$

all components of the feedback network can be determined by bringing the deviation function $D(f)$ and respective trace as close to 0 dB as possible.

## Balanced (b) Solutions

For balanced (b) purposes I'll only concentrate on two different circuit solutions. Once they have been understood all other kinds of network as insertion into a balanced amp-chain set-up should become easy to do.

## The Fully Passive 2-Step Solution Type ($AB_b$)

The change of a 2-step passive un-balanced solution into a 2-step balanced solution will be demonstrated with the following Figs.:

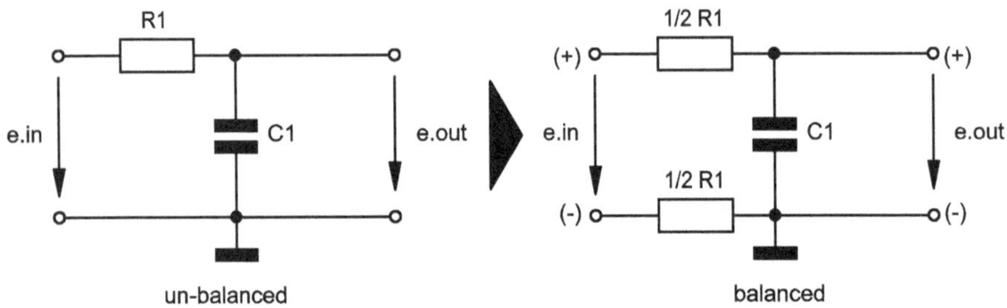


**Fig. 8.22** Type ($AB_b$) RIAA network step 1 (75 μs): change from an un-balanced to a balanced configuration

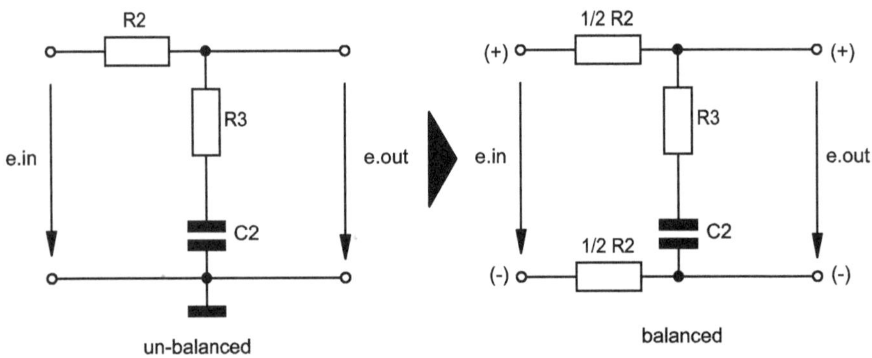


**Fig. 8.23** Type ($AB_b$) RIAA network step 2 (318 μs/3180 μs): change from an un-balanced to a balanced configuration

- All calculation formulae that are given in "The fully passive 2-step-solution Type ($AB_{ub}$)" paragraph are fully valid to calculate the components of this kind of network (Eqs. (8.27 … 8.39)).

- Calculation results of $R_1$ and $R_4$ should be included into the circuitry as indicated in Figs. 8.22 ... 8.23.
- (+) and (−) indicate the opposite amp chains in the balanced mode.

### The Active/Passive 2-Step Solution Type ($C_b$) and ($D_b$)

We've seen in Sect. 3.6 of Chap. 3 that, basically, we can use two different in-amp topologies to form a balanced input section of an amp chain (Figs. 3.69 ... 3.70). The following paragraphs will discuss these types of input topologies in conjunction with adequate RIAA networks.

### Balanced In/Un-Balanced Out

Figure 3.71 gives more details on a circuitry of the type 1 topology that includes RIAA feedback impedances $Z_1(f)$ and $Z_2(f)$. This type of configuration allows to form a balanced input section followed by an un-balanced output stage.

- The feedback networks ($2 \times Z_2(f)$) can both be formed by either the type ($C_{ub}$) or ($D_{ub}$) networks of Figs. 8.11 ... 8.12.
- Equations (8.40 ... 8.50) will give the values for the network components.
- To simplify calculations in these equations $R_1$ should to be formed as follows:

$$R_1 = 0.5 \times \left( \frac{1}{RG} + \frac{1}{R_1 a + R_1 b} \right)^{-1} \tag{8.77}$$

$$R_{1a} = R_{1b} \approx 100 \times RG \tag{8.78}$$

  thus, compared with the un-balanced situation, doubling the gain of the 1st stage.
- $Z_2(f)$ can also be formed by a type ($E_{ub}$) or ($F_{ub} - A$) or ($F_{ub} - B$) feedback network. Hence, the 2nd stage becomes unnecessary. The respective calculation course is given in "The fully active 1-step solution" paragraph and in Worksheets VII ... X. $R_1$ should be set à la Eqs. (8.77 ... 8.78).
- To avoid significant differences between the ideal RIAA transfer and the actual one $Z_2(f)$ components should be $\leq 1\%$ types as well as being exactly paired.

### Balanced In/Balanced Out

If we would take the un-balanced RIAA transfer forming situation of Fig. 8.10 and transform it into a balanced solution we'll end up with a circuitry shown in the following figure which demonstrates the second possibility to form a balanced RIAA transfer solution à la in-amp type 2 topology.

- The two $318\,\mu s/3180\,\mu s$ forming impedances $Z_2(f)$ can both be formed of type ($C_{ub}$) or ($D_{ub}$) feedback networks (Figs. 8.11 ... 8.12).

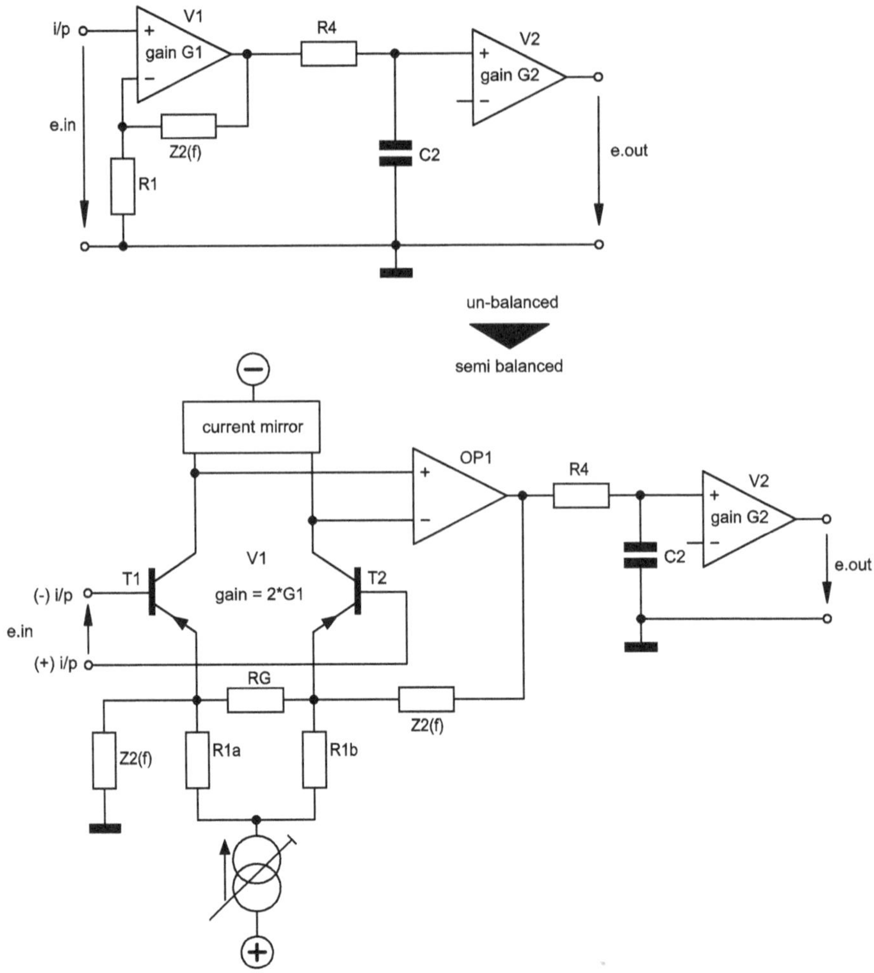


**Fig. 8.24** Type ($C_b$ or $D_b$) RIAA network (active 318 μs/3180 μs + passive 75 μs): change from an un-balanced to a semi balanced configuration with in-amp topology type 1

- The calculation approach is given with Eqs. (8.40 ... 8.50).
- As indicated in Fig. 8.25 calculation results of $R_1$ and $R_4$ should be included into the circuitry.
- The handling of the 75 μs forming network $R_4/C_2$ is the same like the one in the Type ($AB_b$) case of Fig. 8.22.
- $G_1$ and $G_3$ must have equal gains – as well as $G_2$ and $G_4$.
- To avoid significant differences between the ideal RIAA transfer and the actual one $Z_2(f)$ components should be $\leq 1\%$ types as well as being exactly paired.

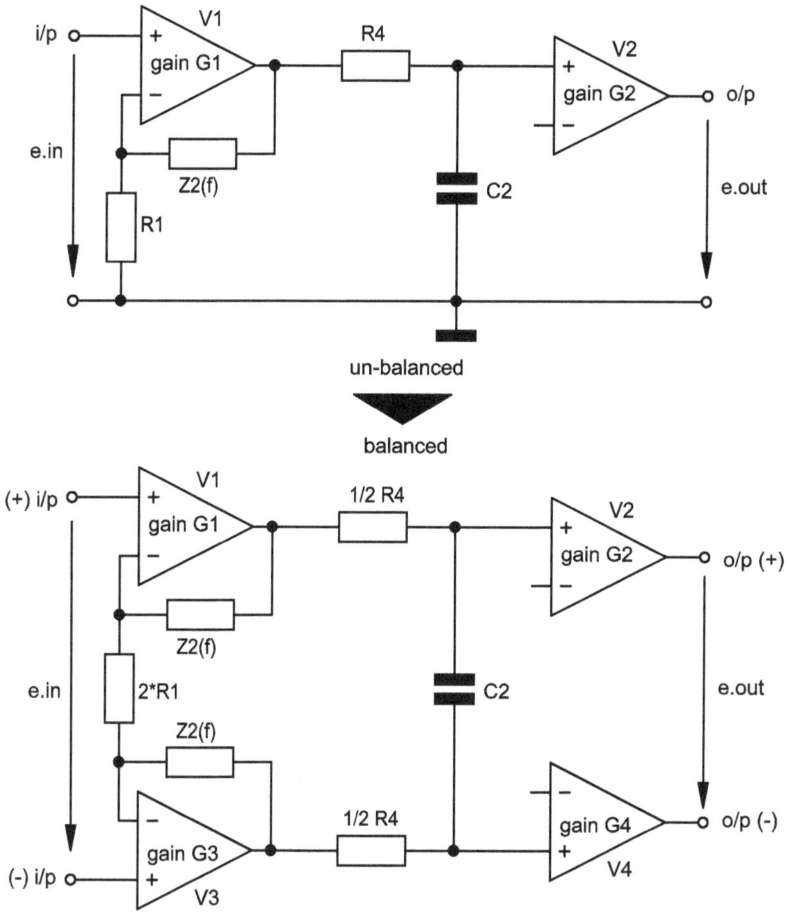


**Fig. 8.25** Type ($C_b$ or $D_b$) RIAA network (active 318 µs/3180 µs + passive 75 µs): change from an un-balanced to a balanced configuration with in-amp topology type 2

# Chapter 9
# RIAA Networks – Mathematical Calculation Course

The following pages show all relevant Mathcad worksheets to enable the reader to follow the calculation results for RIAA networks given in the previous chapter:

- Worksheet I: 1-step passive type ($A_{ub}$) solution in an ideal environment
- Worksheet II: 1-step passive type ($B_{ub}$) solution in an ideal environment
- Worksheet III: 1-step passive type ($A_{ub}$) solution in a real environment
- Worksheet IV: 1-step passive type ($B_{ub}$) solution in a real environment
- Worksheet V: 2-step active-passive type ($C_{ub}$) solution
- Worksheet VI: 2-step active-passive type ($D_{ub}$) solution
- Worksheet VII: Fully active type ($E_{ub}$) succ-apps solution for MC phono-amps
- Worksheet VIII: Fully active type ($E_{ub}$) succ-apps solution for MM phono-amps
- Worksheet IX: Fully active type ($E_{ub}$) formulae solution for MC phono-amps
- Worksheet X: Derivation of proportional factors to calculate type ($E_{ub}$) RIAA networks

B. Vogel, *The Sound of Silence*
DOI: 10.1007/978-3-540-76884-5, ©Springer 2008

➢ **MCD Worksheet I:** Calculation of components for an ideal type ($A_{ub}$) 1-step passive RIAA network (buvo vs. jhl) - Page 1

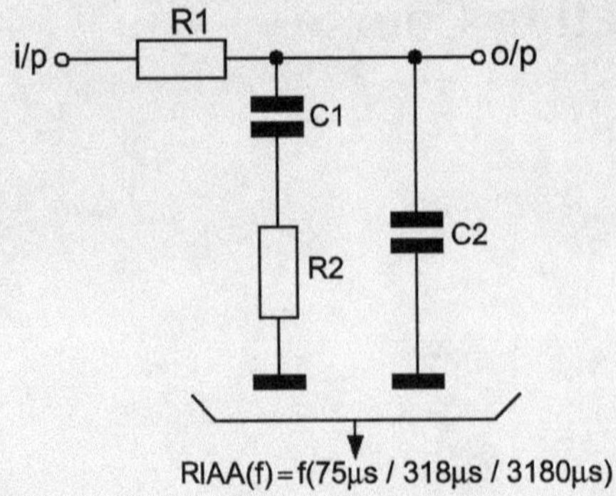


Figure 9.1 = Figure 8.3

**1 pre-defined value plus 3 values for successive approximation (equations (8.5 ... 8.7):**

$C1_{buvo} := 33 \cdot 10^{-9} F$ $\quad T1_{buvo} := 2187.00207 \cdot 10^{-6} s$

$r1_{buvo} := 6.877366$ $\quad T2_{buvo} := 109.0534 \cdot 10^{-6} s$

$C1_{jlh} := C1_{buvo}$ $\quad T1_{jlh} := T1_{buvo}$

$r1_{jlh} := 6.88$ $\quad T2_{jlh} := 109 \cdot 10^{-6} s$

**calculation results:**

$$R1_{buvo} := \frac{T1_{buvo}}{C1_{buvo}}$$

$$R1_{buvo} = 66.273 \times 10^{3} \Omega$$

$$R2_{buvo} := \frac{R1_{buvo}}{r1_{buvo}}$$

$$R2_{buvo} = 9.636 \times 10^{3} \Omega$$

$$C2_{buvo} := \frac{T2_{buvo}}{R2_{buvo}}$$

$$C2_{buvo} = 11.317 \times 10^{-9} F$$

$$R1_{jlh} := \frac{T1_{jlh}}{C1_{jlh}}$$

$$R1_{jlh} = 66.273 \times 10^{3} \Omega$$

$$R2_{jlh} := \frac{R1_{jlh}}{r1_{jlh}}$$

$$R2_{jlh} = 9.633 \times 10^{3} \Omega$$

$$C2_{jlh} := \frac{T2_{jlh}}{R2_{buvo}}$$

$$C2_{jlh} = 11.311 \times 10^{-9} F$$

➢ **MCD Worksheet I:** Calculation of components for an ideal type ($A_{ub}$) 1-step passive RIAA network (buvo vs. jhl) - Page 2

**monitoring of deviation in 10Hz-steps in the frequency range of** $f := 20, 30 .. 20000$ :

$$R\left(10^3\right) = R_{1000} \qquad R_{1000} := \frac{\sqrt{1+\left(2\cdot\pi\cdot 10^3\cdot 318\cdot 10^{-6}\right)^2}}{\sqrt{1+\left(2\cdot\pi\cdot 10^3\cdot 3180\cdot 10^{-6}\right)^2}\cdot\sqrt{1+\left(2\cdot\pi\cdot 10^3\cdot 75\cdot 10^{-6}\right)^2}}$$

$$R_0(f) := \frac{\sqrt{1+\left(2\cdot\pi\cdot f\cdot 318\cdot 10^{-6}\right)^2}}{\sqrt{1+\left(2\cdot\pi\cdot f\cdot 3180\cdot 10^{-6}\right)^2}\cdot\sqrt{1+\left(2\cdot\pi\cdot f\cdot 75\cdot 10^{-6}\right)^2}}\cdot\left(R_{1000}\right)^{-1}$$

$$G_{nw.buvo}(f) := \left|\frac{\left(2j\cdot\pi\cdot f\cdot C2_{buvo} + \dfrac{1}{R2_{buvo} + \dfrac{1}{2j\cdot\pi\cdot f\cdot C1_{buvo}}}\right)^{-1}}{R1_{buvo} + \left(2j\cdot\pi\cdot f\cdot C2_{buvo} + \dfrac{1}{R2_{buvo} + \dfrac{1}{2j\cdot\pi\cdot f\cdot C1_{buvo}}}\right)^{-1}}\right|$$

$$G_{nw.jlh}(f) := \left|\frac{\left(2j\cdot\pi\cdot f\cdot C2_{jlh} + \dfrac{1}{R2_{jlh} + \dfrac{1}{2j\cdot\pi\cdot f\cdot C1_{jlh}}}\right)^{-1}}{R1_{jlh} + \left(2j\cdot\pi\cdot f\cdot C2_{jlh} + \dfrac{1}{R2_{jlh} + \dfrac{1}{2j\cdot\pi\cdot f\cdot C1_{jlh}}}\right)^{-1}}\right|$$

$$G_{0.buvo}(f) := G_{nw.buvo}(f)\cdot R_{1000}{}^{-1} \qquad D_{buvo}(f) := 20\cdot\log\left(R_0(f)\right) - 20\cdot\log\left(G_{0.buvo}(f)\right)$$

$$G_{0.jlh}(f) := G_{nw.jlh}(f)\cdot R_{1000}{}^{-1} \qquad D_{jlh}(f) := 20\cdot\log\left(R_0(f)\right) - 20\cdot\log\left(G_{0.jlh}(f)\right)$$

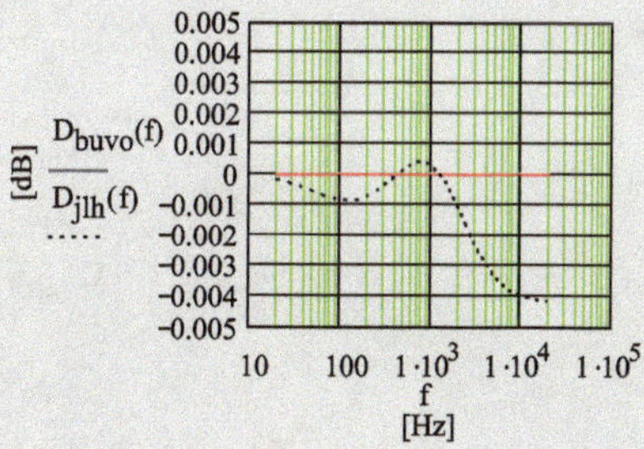


Figure 9.2 BUVO (*red*) and JLH (*black-dotted*) deviations

➢ **MCD Worksheet II:** Calculation of components for an ideal type ($B_{ub}$) 1-step passive RIAA network (buvo vs. jlh) - Page 1

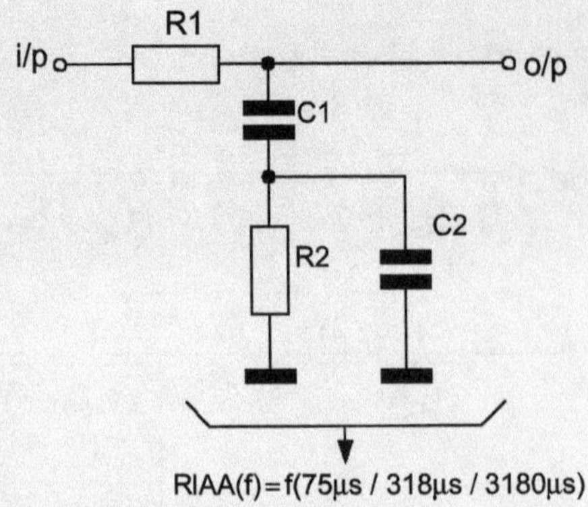


Figure 9.3 = Figure 8.4

**1 pre-defined value and 3 values for successive approximation (equations (8.8 ... 8.10):**

$R2_{buvo} := 1 \cdot 10^{3} \Omega$ $R2_{jlh} := R2_{buvo}$

$T1_{buvo} := 2937.001 \cdot 10^{-6} s$ $T1_{jlh} := 2937 \cdot 10^{-6} s$

$T2_{buvo} := 81.2053 \cdot 10^{-6} s$ $T2_{jlh} := 81.21 \cdot 10^{-6} s$

$r1_{buvo} := 12.40316$ $r1_{jlh} := 12.40$

**calculation results:**

$R1_{buvo} := r1_{buvo} \cdot R2_{buvo}$

$R1_{buvo} = 12.403 \times 10^{3} \Omega$

$C1_{buvo} := \dfrac{T1_{buvo}}{R1_{buvo}}$

$C1_{buvo} = 236.79457 \times 10^{-9} F$

$C2_{buvo} := \dfrac{T2_{buvo}}{R2_{buvo}}$

$C2_{buvo} = 81.2053 \times 10^{-9} F$

$R1_{jlh} := r1_{jlh} \cdot R2_{jlh}$

$R1_{jlh} = 12.4 \times 10^{3} \Omega$

$C1_{jlh} := \dfrac{T1_{jlh}}{R1_{jlh}}$

$C1_{jlh} = 236.85484 \times 10^{-9} F$

$C2_{jlh} := \dfrac{T2_{buvo}}{R2_{buvo}}$

$C2_{jlh} = 81.2053 \times 10^{-9} F$

➢ **MCD Worksheet II:** Calculation of components for an ideal type ($B_{ub}$) 1-step passive RIAA network (buvo vs. jlh) - Page 2

**monitoring of deviation in 10Hz-steps in the frequency range of** $f := 20, 30 .. 20000$ **:**

$$R(10^3) = R_{1000} \qquad R_{1000} := \frac{\sqrt{1+\left(2\cdot\pi\cdot10^3\cdot318\cdot10^{-6}\right)^2}}{\sqrt{\left[1+\left(2\cdot\pi\cdot10^3\cdot3180\cdot10^{-6}\right)^2\right]}\cdot\sqrt{1+\left(2\cdot\pi\cdot10^3\cdot75\cdot10^{-6}\right)^2}}$$

$$R_0(f) := \frac{\sqrt{1+\left(2\cdot\pi\cdot f\cdot318\cdot10^{-6}\right)^2}}{\sqrt{1+\left(2\cdot\pi\cdot f\cdot3180\cdot10^{-6}\right)^2}\cdot\sqrt{1+\left(2\cdot\pi\cdot f\cdot75\cdot10^{-6}\right)^2}}\cdot\left(R_{1000}\right)^{-1}$$

$$G_{nw.buvo}(f) := \left|\frac{\frac{1}{2j\cdot\pi\cdot f\cdot C1_{buvo}}+\left(\frac{1}{R2_{buvo}}+2j\cdot\pi\cdot f\cdot C2_{buvo}\right)^{-1}}{R1_{buvo}+\frac{1}{2j\cdot\pi\cdot f\cdot C1_{buvo}}+\left(\frac{1}{R2_{buvo}}+2j\cdot\pi\cdot f\cdot C2_{buvo}\right)^{-1}}\right|$$

$$G_{nw.jlh}(f) := \left|\frac{\frac{1}{2j\cdot\pi\cdot f\cdot C1_{jlh}}+\left(\frac{1}{R2_{jlh}}+2j\cdot\pi\cdot f\cdot C2_{jlh}\right)^{-1}}{R1_{jlh}+\frac{1}{2j\cdot\pi\cdot f\cdot C1_{jlh}}+\left(\frac{1}{R2_{jlh}}+2j\cdot\pi\cdot f\cdot C2_{jlh}\right)^{-1}}\right|$$

$$G_{0.buvo}(f) := G_{nw.buvo}(f)\cdot R_{1000}{}^{-1} \qquad G_{0.jlh}(f) := G_{nw.jlh}(f)\cdot R_{1000}{}^{-1}$$

$$D_{buvo}(f) := 20\cdot\log\left(R_0(f)\right) - 20\cdot\log\left(G_{0.buvo}(f)\right) \qquad D_{jlh}(f) := 20\cdot\log\left(R_0(f)\right) - 20\cdot\log\left(G_{0.jlh}(f)\right)$$

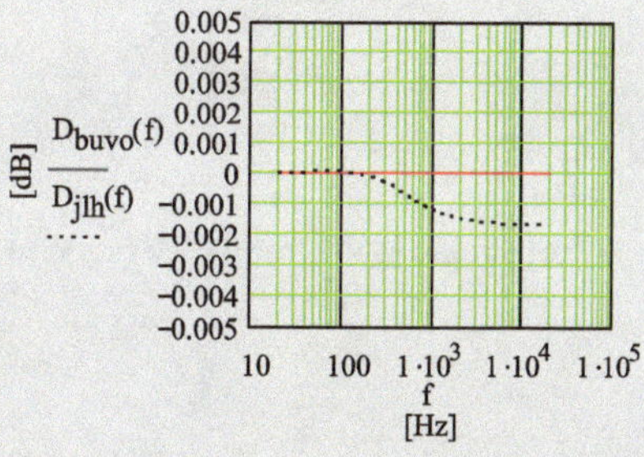


Figure 9.4 BUVO (*red*) and JLH (*black-dotted*) deviations

➢ **MCD Worksheet III:** Calculation of components for an ideal type ($A_{ub}$) 1-step passive RIAA network - Page 1

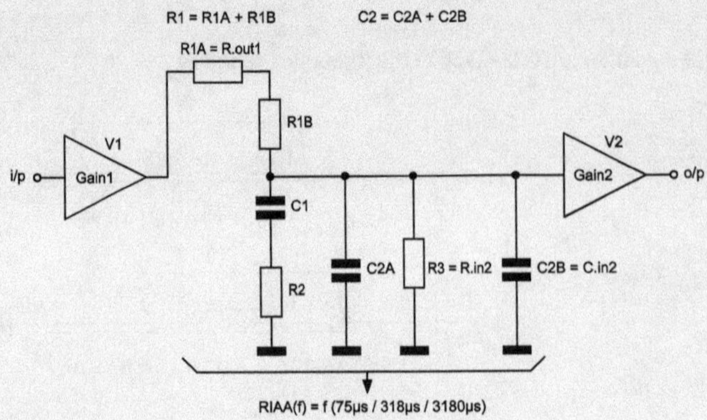


Figure 9.5 = Figure 8.5

**4 pre-defined values:**

$C1 := 47 \cdot 10^{-9} F$

$C2B := 100 \cdot 10^{-12} F$

$R1A := 73\Omega$

$R3 := 475 \cdot 10^{3}\Omega$

**3 values for succ-apps:**

$T1 := 2424.51 \cdot 10^{-6} s$

$T2 := 109.0534 \cdot 10^{-6} s$

$r1 := 7.624252$

**calculation results:**

$R1 := \frac{T1}{C1}$

$R1 = 51.585 \times 10^{3}\,\Omega$

$R2 := \frac{R1}{r1}$

$R2 = 6.766 \times 10^{3}\,\Omega$

$C2 := \frac{T2}{R2}$

$C2 = 16.118 \times 10^{-9}\,F$

$R1B := R1 - R1A$

$R1B = 51.512 \times 10^{3}\,\Omega$

$G_{il.A} := \frac{R3}{R1 + R3}$

$G_{il.A} = 902.038 \times 10^{-3}$

$C2A := C2 - C2B$

$C2A = 16.018 \times 10^{-9}\,F$

**monitoring of deviation in 10Hz-steps in the frequency range of** $f := 20, 30 .. 20000$ **:**

$R(10^{3}) = R_{1000}$

$$R_{1000} := \frac{\sqrt{1 + \left(2 \cdot \pi \cdot 10^{3} \cdot 318 \cdot 10^{-6}\right)^{2}}}{\sqrt{1 + \left(2 \cdot \pi \cdot 10^{3} \cdot 3180 \cdot 10^{-6}\right)^{2}} \cdot \sqrt{1 + \left(2 \cdot \pi \cdot 10^{3} \cdot 75 \cdot 10^{-6}\right)^{2}}}$$

➤ **MCD Worksheet III:** Calculation of components for areal type ($A_{ub}$) 1-step passive RIAA network - Page 2

$$R_0(f) := \frac{\sqrt{1+\left(2\cdot\pi\cdot f\cdot 318\cdot 10^{-6}\right)^2}}{\sqrt{1+\left(2\cdot\pi\cdot f\cdot 3180\cdot 10^{-6}\right)^2}\cdot\sqrt{1+\left(2\cdot\pi\cdot f\cdot 75\cdot 10^{-6}\right)^2}}\cdot\left(R_{1000}\right)^{-1}$$

$$G_{nw}(f) := \left|\frac{\left(2j\cdot\pi\cdot f\cdot C2+\frac{1}{R3}+\frac{1}{R2+\frac{1}{2j\cdot\pi\cdot f\cdot C1}}\right)^{-1}}{R1+\left(2j\cdot\pi\cdot f\cdot C2+\frac{1}{R3}+\frac{1}{R2+\frac{1}{2j\cdot\pi\cdot f\cdot C1}}\right)^{-1}}\right|$$

$$G_0(f) := G_{il.A}{}^{-1}\cdot G_{nw}(f)\cdot R_{1000}{}^{-1}$$

$$D(f) := 20\cdot\log\left(R_0(f)\right) - 20\cdot\log\left(G_0(f)\right)$$

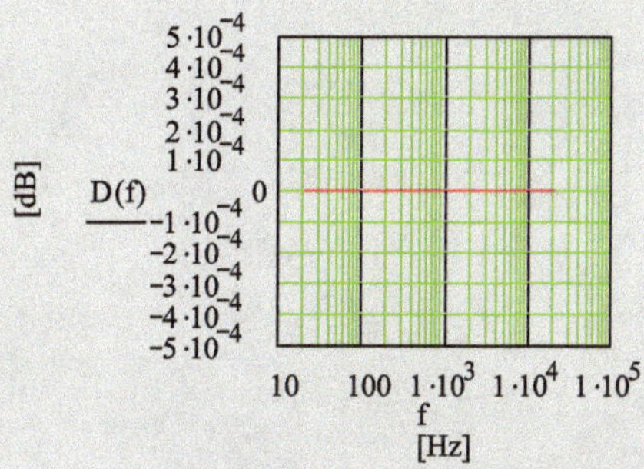


Figure 9.6 Deviation after succ-apps

➤ **MCD Worksheet IV:** Calculation of components for a real type ($B_{ub}$) 1-step passive RIAA network - Page 1

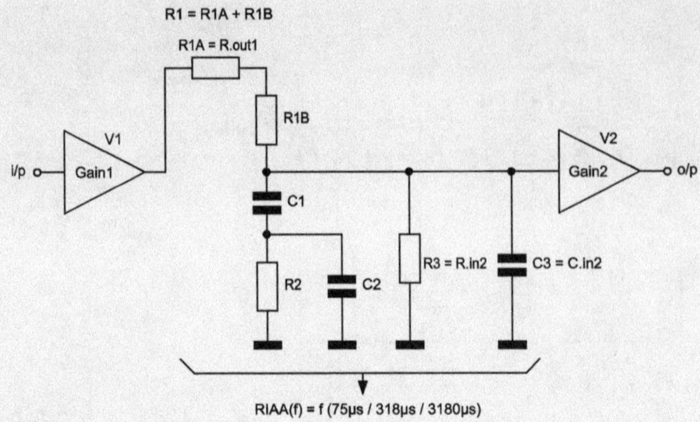


Figure 9.7 = Figure 8.6

**4 pre-defined values:**

$C3 := 100 \cdot 10^{-12} F$

$R1A := 73\Omega$

$R2 := 1 \cdot 10^{3}\Omega$

$R3 := 475 \cdot 10^{3}\Omega$

**3 values for succ-apps:**

$T1 := 3014.405 \cdot 10^{-6} s$

$T2 := 81.1053 \cdot 10^{-6} s$

$r1 := 12.72467$

**calculation results:**

$R1 := r1 \cdot R2$ $\qquad C1 := \frac{T1}{R1}$ $\qquad C2 := \frac{T2}{R2}$

$R1 = 12.725 \times 10^{3}\Omega$ $\qquad C1 = 236.89455 \times 10^{-9} F$ $\qquad C2 = 81.1053 \times 10^{-9} F$

$R1B := R1 - R1A$ $\qquad G_{il.B} := \frac{R3}{R1 + R3}$

$R1B = 12.652 \times 10^{3}\Omega$ $\qquad G_{il.B} = 0.974$

**monitoring of deviation in 10Hz-steps in the frequency range of** $f := 20, 30 .. 20000$ **:**

$$R\left(10^{3}\right) = R_{1000}$$

$$R_{1000} := \frac{\sqrt{1 + \left(2 \cdot \pi \cdot 10^{3} \cdot 318 \cdot 10^{-6}\right)^{2}}}{\sqrt{\left[1 + \left(2 \cdot \pi \cdot 10^{3} \cdot 3180 \cdot 10^{-6}\right)^{2}\right]} \cdot \sqrt{1 + \left(2 \cdot \pi \cdot 10^{3} \cdot 75 \cdot 10^{-6}\right)^{2}}}$$

➤ **MCD Worksheet IV:** Calculation of components for a real type ($B_{ub}$) 1-step passive RIAA network - Page 2

$$G_{nw}(f) := \left| \frac{\left[\frac{1}{R3} + 2j\cdot\pi\cdot f\cdot C3 + \left[\frac{1}{2j\cdot\pi\cdot f\cdot C1} + \left(\frac{1}{R2} + 2j\cdot\pi\cdot f\cdot C2\right)^{-1}\right]^{-1}\right]^{-1}}{R1 + \left[\frac{1}{R3} + 2j\cdot\pi\cdot f\cdot C3 + \left[\frac{1}{2j\cdot\pi\cdot f\cdot C1} + \left(\frac{1}{R2} + 2j\cdot\pi\cdot f\cdot C2\right)^{-1}\right]^{-1}\right]^{-1}} \right|$$

$$R_0(f) := \frac{\sqrt{1 + \left(2\cdot\pi\cdot f\cdot 318\cdot 10^{-6}\right)^2}}{\sqrt{1 + \left(2\cdot\pi\cdot f\cdot 3180\cdot 10^{-6}\right)^2}\cdot\sqrt{1 + \left(2\cdot\pi\cdot f\cdot 75\cdot 10^{-6}\right)^2}}\cdot\left(R_{1000}\right)^{-1}$$

$$G_0(f) := G_{i1.B}{}^{-1}\cdot G_{nw}(f)\cdot R_{1000}{}^{-1}$$

$$D(f) := 20\cdot\log\left(R_0(f)\right) - 20\cdot\log\left(G_0(f)\right)$$

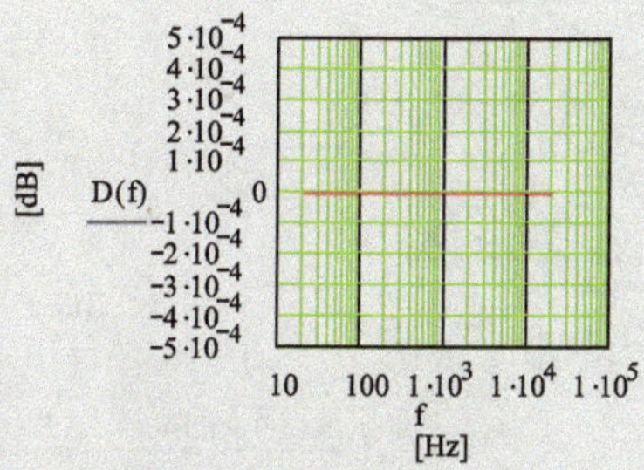


Figure 9.8 Deviation after succ-apps

➢ **MCD Worksheet V:** Calculation of components for a type ($C_{ub}$) 2-step active-passive RIAA network - Page 1

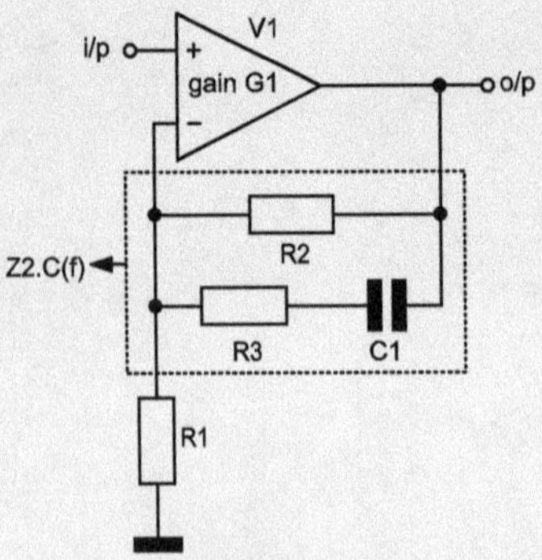


Figure 9.9 = Figure 8.11

**derivation of calculation rules:**

$$G1(f) = \frac{Z2_C(f) + R1}{R1} \quad \text{with:} \quad p = 2j \cdot \pi \cdot f \tag{1}$$

$$Z2_C(p) = \frac{R2 \cdot \left(R3 + \frac{1}{pC1}\right)}{R2 + R3 + \frac{1}{pC1}} = \frac{p \cdot C1 \cdot R3 \cdot R2 + R2}{p \cdot C1 \cdot (R3 + R2) + 1} = R2 \cdot \frac{(1 + p \cdot C1 \cdot R3)}{[\,1 + p \cdot C1 \cdot (R3 + R2)\,]} \tag{2}$$

$$G1(p) = \frac{Z2_C(p)}{R1} + 1 = \frac{R2}{R1} \cdot \frac{(1 + p \cdot C \cdot R3)}{[\,1 + p \cdot C \cdot (R3 + R2)\,]} + 1$$

$$G1(p) = \frac{(R2 + R2 \cdot p \cdot C \cdot R3 + R1 + R1 \cdot p \cdot C \cdot R3 + R1 \cdot p \cdot C \cdot R2)}{R1 \cdot (1 + p \cdot C \cdot R3 + p \cdot C \cdot R2)} \tag{3}$$

$$G1(p) = \frac{R1 + R2}{R1} \cdot \frac{\left(1 + p \cdot C \cdot \frac{R1 \cdot R3 + R1 \cdot R2 + R3 \cdot R2}{R3 + R2}\right)}{[\,1 + p \cdot C \cdot (R3 + R2)\,]} \tag{4}$$

$$G1_0 = \frac{R1 + R2}{R1} \tag{5}$$

$$T1 = C \cdot (R3 + R2) = 3180\mu s \tag{6}$$

$$T2 = C \cdot \frac{R1 \cdot R3 + R1 \cdot R2 + R3 \cdot R2}{R1 + R2} = 318\mu s \tag{7}$$

$$G1_0 = \frac{R2 + R1}{R1} \tag{8}$$

$$R2 = R1 \cdot G1_0 - R1 \tag{9}$$

$$\frac{T1}{T2} = 10 = \frac{R3 + R2}{\frac{R3 \cdot (R1+R2) + R1 \cdot R2}{R1+R2}} = \frac{R3 + R2}{R3 + \frac{R1 \cdot R2}{R1 + R2}} \tag{10}$$

$$10 \cdot R3 + \frac{10 \cdot R2}{G1_0} = R3 + R2 \tag{11}$$

$$R3 = \frac{R2}{9} \cdot \left(1 - \frac{10}{G1_0}\right) \tag{12}$$

$$C1 = \frac{T1}{R2 + R3} = \frac{3180\mu s}{R2 + R3} \tag{13}$$

**cross-check:**

$$T2 = C1 \cdot \frac{R1 \cdot R3 + R1 \cdot R2 + R3 \cdot R2}{R1 + R2} \tag{14}$$

➢ **MCD Worksheet V:** Calculation of components for a type ($C_{ub}$) 2-step active-passive RIAA network - **Page 2**

**calculation example:**

1. choose R1
2. choose G1 at 1kHz = $G1_1$
3. calculate $G1_0 = G1_1 * 8.953721$ or $G1_1/0.111685$
4. calculate $R2 = R1 * G1_0 - R1$
5. calculate $R3 = (R2 / 9) * (1 - 10 / G1_0)$
6. calculate $C1 = 3180\,\mu s / (R3 + R2)$
7. or: $C1 = 318\mu s / (R3+(R2 \| R1))$
   if C1 does not hit a value that is easy to get succ-apps with $G1_1$ should be used as long as the resulting change of C1 hits a desired value

**chosen:**

$R1 := 130\Omega$ $\qquad G1_1 := 24.81083958$

**calculation og DC gain $G1_0$ with equations (4) and (5):**

$$G1_0 := \frac{G1_1}{\dfrac{\sqrt{1+\left(2\cdot\pi\cdot 1000\text{Hz}\cdot 318\cdot 10^{-6}\text{s}\right)^2}}{\sqrt{1+\left(2\cdot\pi\cdot 1000\text{Hz}\cdot 3180\cdot 10^{-6}\text{s}\right)^2}}} \qquad \frac{\sqrt{1+\left(2\cdot\pi\cdot 1000\text{Hz}\cdot 318\cdot 10^{-6}\text{s}\right)^2}}{\sqrt{1+\left(2\cdot\pi\cdot 1000\text{Hz}\cdot 3180\cdot 10^{-6}\text{s}\right)^2}} = 0.111685$$

$G1_0 = 222.149$

**calculation of components feedback network with equations (11) ... (13):**

$$R2 := R1\cdot G1_0 - R1 \qquad R2 = 28.749\times 10^3\Omega$$

$$R3 := \frac{R2}{9}\cdot\left(1-\frac{10}{G1_0}\right) \qquad R3 = 3.1\times 10^3\Omega$$

succ-apps with G1 $_1$ will lead to a useful value for C1:

$$C1 := \frac{3180\cdot 10^{-6}\text{s}}{R3+R2} \qquad C1 = 100\times 10^{-9}\text{F}$$

**cross-check:**

$$T2 := C1\cdot\frac{R3\cdot R1 + R1\cdot R2 + R3\cdot R2}{R1+R2} \qquad T2 = 318\times 10^{-6}\text{s}$$

gain at $f := 1000\text{Hz}$:

$$Z2_C := \left|\frac{R2\cdot\left(R3+\dfrac{1}{2j\cdot\pi\cdot f\cdot C1}\right)}{R2+R3+\dfrac{1}{2j\cdot\pi\cdot f\cdot C1}}\right| \qquad Z2_C = 3.107\times 10^3\Omega$$

$$G1_{1000} := \frac{R1+Z2_C}{R1} \qquad G1_{1000} = 24.899$$

➤ **MCD Worksheet VI:** Calculation of components for a type ($D_{ub}$) 2-step active-passive RIAA network - Page 1

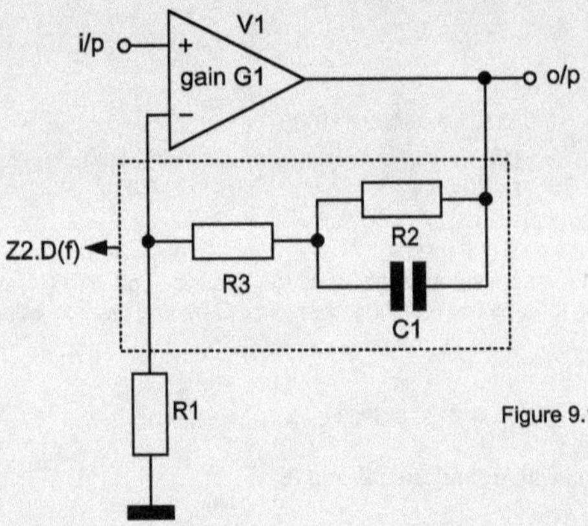


Figure 9.10 = Figure 8.12

**derivation of calculation rules:**

$$G1(f) = \frac{Z2_D(f) + R1}{R1} \qquad \text{with:} \quad p = 2j \cdot \pi \cdot f \tag{1}$$

$$Z2_D(p) = \frac{\frac{1}{p \cdot C1} \cdot R2}{R2 + \frac{1}{p \cdot C1}} + R3 = \frac{(R2 + R3 \cdot R2 \cdot p \cdot C1 + R3)}{(R2 \cdot p \cdot C1 + 1)} = \frac{(R2+R3) \cdot \left(1 + p \cdot C1 \cdot \frac{R2 \cdot R3}{R2 + R3}\right)}{(1 + p \cdot C1 \cdot R2)} \tag{2}$$

$$G1(p) = \frac{Z2_D(p)}{R1} + 1 = \frac{\frac{(R2+R3) \cdot \left(1+p \cdot C1 \cdot \frac{R2 \cdot R3}{R2+R3}\right)}{(1+p \cdot C1 \cdot R2)}}{R1} + 1$$

$$G1(p) = \frac{(R2+R3) \cdot \left(1 + p \cdot C1 \cdot \frac{R2 \cdot R3}{R2+R3}\right) + R1 \cdot (1 + p \cdot C1 \cdot R2)}{R1 \cdot (1 + p \cdot C1 \cdot R2)} \tag{3}$$

$$G1(p) = \frac{(R1 + R2 + R3)}{R1} \cdot \frac{\left(1 + p \cdot C1 \cdot \frac{R2 \cdot R1 + R2 \cdot R3}{R1 + R2 + R3}\right)}{(1 + p \cdot C1 \cdot R2)} \quad (4) \qquad G1_0 = \frac{R1 + R2 + R3}{R1} \quad (5)$$

$$T1 = C1 \cdot R2 = 3180\mu s \tag{6}$$

$$\frac{T1}{T2} = 10 = \frac{R1 + R2 + R3}{R1 + R3} \quad (7) \qquad 9 \cdot R1 = R2 - 9 \cdot R3 \quad (8)$$

$$a = R2 + R3 \qquad R2 = a - R3 \qquad a = 9 \cdot R1 + 10 \cdot R3 \tag{9}$$

$$R3 = \frac{a - 9 \cdot R1}{10} \qquad R3 = \frac{R2}{9} - R1 \qquad R2 = 9 \cdot (R1 + R3) \tag{10}$$

$$C1 = \frac{3180\mu s}{R2} \tag{11}$$

**cross-check:**

$$T2 = C1 \cdot R2 \cdot \frac{R1 + R3}{R1 + R2 + R3} = 318\mu s \tag{12}$$

➢ **MCD Worksheet VI:** Calculation of components for a type ($D_{ub}$) 2-step active-passive RIAA network - Page 2

**calculation example:**

1. choose R1
2. choose G1 at 1kHz = $G1_1$
3. calculate $G1_0 = G1_1/0.111685$ or $G1_1*8.953721$ see Worksheet V
4. calculate a = R2+R3 = R1*($G1_0$-1) with equation (5)
5. calculate R3 = (a-9*R1)/10 with equations (9) and (10)
6. calculate R2 = a-R3 with equations (9)
7. calculate C1 with equation (11)
   if C1 does not hit a value that is easy to get succ-apps with $G1_1$ should be used as long as the resulting change of C1 hits a desired value

**chosen:**

$R1 := 130\Omega$ $G1_1 := 24.8815$

**calculation course:**

$$G1_0 := \frac{G1_1}{\dfrac{\sqrt{1+\left(2\cdot\pi\cdot 1000\text{Hz}\cdot 318\cdot 10^{-6}\text{s}\right)^2}}{\sqrt{1+\left(2\cdot\pi\cdot 1000\text{Hz}\cdot 3180\cdot 10^{-6}\text{s}\right)^2}}} \qquad G1_0 = 222.782$$

$$a := R1\cdot\left(G1_0 - 1\right) \qquad a = 28.832\times 10^3\Omega$$

$$R3 := \frac{(a - 9\cdot R1)}{10} \qquad R3 = 2.766\times 10^3\Omega$$

$$R2 := a - R3 \qquad R2 = 26.065\times 10^3\Omega$$

succ-apps with $G1_1$ will lead to a useful values for C1:

$$C1 := \frac{3180\cdot 10^{-6}\text{s}}{R2} \qquad C1 = 122\times 10^{-9}\text{F}$$

**cross-check:**

$$T2 := C1\cdot R2\cdot\frac{R1 + R3}{R1 + R2 + R3} \qquad T2 = 318\times 10^{-6}\text{s}$$

gain at $f := 1000\text{Hz}$:

$$Z2_D := \left|R3 + \left(2j\cdot\pi\cdot f\cdot C1 + \frac{1}{R2}\right)^{-1}\right| \qquad Z2_D = 3.116\times 10^3\Omega$$

$$G1_{1000} := 1 + \frac{Z2_D}{R1} \qquad G1_{1000} = 24.969$$

➤ **MCD Worksheet VII:** Calculation v2.0 of the BUVO MC RIAA feedback network type ($E_{ub}$) with succ-apps - Page 1

**Calculation v2.0 of the RIAA feedback network of the BUVO MC phono-amp by application of the succ-apps methode:**

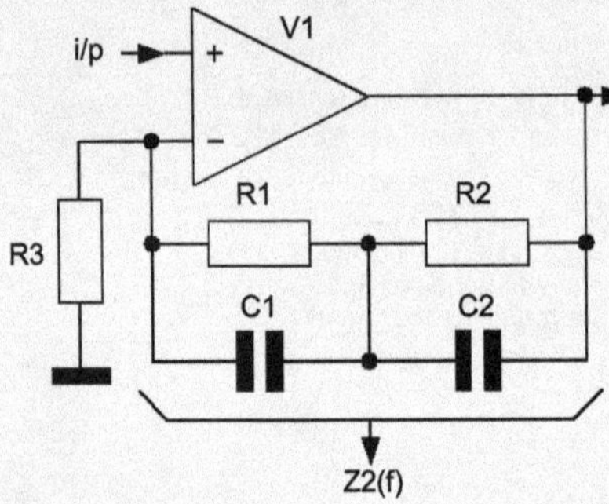


Figure 9.11 = Figures 8.13 + 8.14

**choose:**

gain at 1kHz and R3: $G1_1 := 223.46$ $R3 := 3.32\Omega$

**given equations (2.4 ...2. 6):**

$$R_{1000} := \frac{\sqrt{1+\left(2\cdot\pi\cdot10^3\text{Hz}\cdot318\cdot10^{-6}\text{s}\right)^2}}{\sqrt{1+\left(2\cdot\pi\cdot10^3\text{Hz}\cdot3180\cdot10^{-6}\text{s}\right)^2}\cdot\sqrt{1+\left(2\cdot\pi\cdot10^3\text{Hz}\cdot75\cdot10^{-6}\text{s}\right)^2}} \qquad R_{1000} = 101.03\times10^{-3}$$

$$R_0(f) := R_{1000}^{-1}\cdot\frac{\sqrt{1+\left(2\cdot\pi\cdot f\cdot318\cdot10^{-6}\text{s}\right)^2}}{\sqrt{1+\left(2\cdot\pi\cdot f\cdot3180\cdot10^{-6}\text{s}\right)^2}\cdot\sqrt{1+\left(2\cdot\pi\cdot f\cdot75\cdot10^{-6}\text{s}\right)^2}}$$

**calculate gain $G1_0$ at DC and x = R1+R2:**

$$G1_0 := \frac{G1_1}{R_{1000}} \qquad G1_0 = 2.212\times10^3$$

$$x = R1 + R2 \qquad x := G1_0\cdot R3 - R3 \qquad x = 7.34\times10^3\Omega$$

**guessings according to equations (8.52 ... 55):**

$$R2 = R1\cdot11.778^{-1} \qquad x = \left(11.778^{-1}+1\right)\cdot R1$$

$$R1 := \frac{x}{1+11.778^{-1}} \qquad R1 = 6.766\times10^3\Omega$$

$$T1 := 3180\cdot10^{-6}\text{s} \qquad C1 := \frac{T1}{R1} \qquad C1 = 470.031\times10^{-9}\text{F}$$

**application of succ-apps starting with c1 = 3.6 and r1 = 11.778:**

$$c1 := 3.5841 \qquad C2 := \frac{C1}{c1} \qquad C2 = 131.143\times10^{-9}\text{F}$$

➤ **MCD Worksheet VII:** Calculation v2.0 of the BUVO MC RIAA feedback network type ($E_{ub}$) with succ-apps - Page 2

$r1 := 11.82995$ $\quad R2 := \frac{R1}{r1}$ $\quad R2 = 571.897 \times 10^{0}\Omega$

**monitoring of the deviation D(f) between phono-amp gain G1(f) and ideal RIAA transfer $R_0$(f):**

$$Z2(f) := \left(\frac{1}{R1} + 2j\cdot\pi\cdot f\cdot C1\right)^{-1} + \left(\frac{1}{R2} + 2j\cdot\pi\cdot f\cdot C2\right)^{-1} \qquad G1(f) := \left|\frac{Z2(f)}{R3} + 1\right|$$

$h := 1000Hz$ $\quad f := 10, 20 .. 20000$ $\quad D(f) := 20\cdot\log(R_0(f)) - 20\cdot\log\left(G1(f)\cdot G1(h)^{-1}\right)$

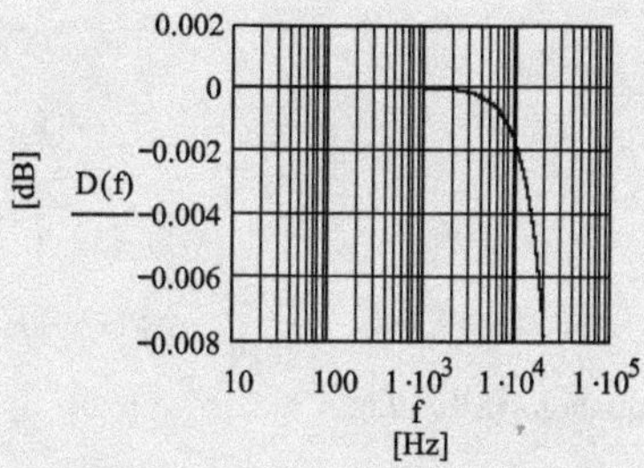


Figure 9.12 = Figure 8.17

**Result and conclusions:**

The deviation D(f) from the exact RIAA transfer can be made close to 0dB with the exception of frequencies >1kHz because the 4th time constant T4 forces the gain with growing frequency more and more close to 1. Following the amp stage V1 a 1st order lp with corner frequency of f4 would flatten the trace for D(f) towards 0dB.

With reference to 0dB at 1kHz the phono-amps gain looks as follows (f4 can be well observed):

$T4 := 750\cdot10^{-6}s\cdot\frac{1}{G1_0}$ $\quad T4 = 339.087 \times 10^{-9}s$ $\quad f4 := \frac{1}{2\cdot\pi\cdot T4}$ $\quad f4 = 469.364 \times 10^{3}Hz$

with reference to 0dB at 1kHzin the frequency range $\quad g := 10Hz, 20Hz .. 1000000Hz$

the RIAA transfer $G_e(g)$ becomes: $\quad G_e(g) := 20\cdot\log(G1(g)) - 20\cdot\log(G1(h))$

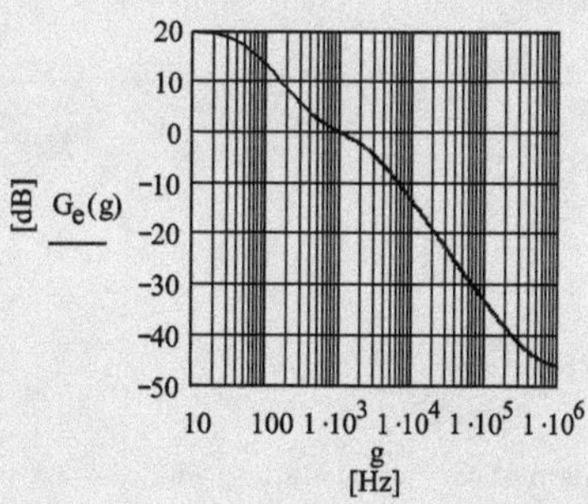


Figure 9.13 = Figure 8.18

**check of the two other RIAA time constants:**

$T2 := R2\cdot C2$ $\quad T2 = 75.000489 \times 10^{-6}s$ $\quad T3 := \frac{R1\cdot R2}{R1 + R2}\cdot(C1 + C2)$ $\quad T3 = 317.012 \times 10^{-6}s$

➢ **MCD Worksheet VIII:** Calculation v2.0 of a MM RIAA feedback network type ($E_{ub}$) with succ-apps - Page 1

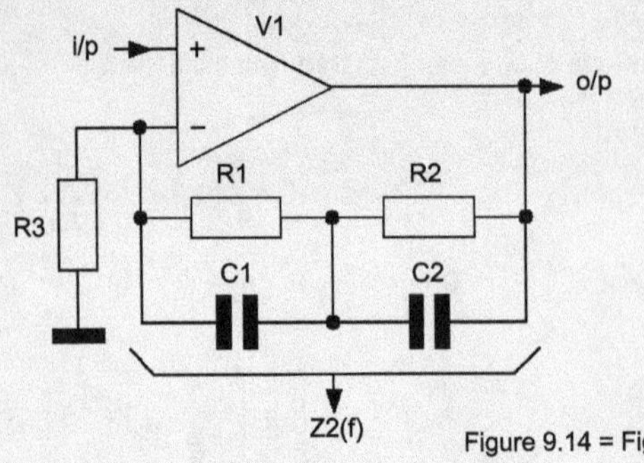


Figure 9.14 = Figures 8.13 + 8.14

**choose:**

gain at 1kHz and R3: $G1_1 := 100$ $R3 := 130\Omega$

**given equations (2.4 ... 2.6):**

$$R_{1000} := \frac{\sqrt{1+\left(2\cdot\pi\cdot10^3\text{Hz}\cdot318\cdot10^{-6}\text{s}\right)^2}}{\sqrt{1+\left(2\cdot\pi\cdot10^3\text{Hz}\cdot3180\cdot10^{-6}\text{s}\right)^2}\cdot\sqrt{1+\left(2\cdot\pi\cdot10^3\text{Hz}\cdot75\cdot10^{-6}\text{s}\right)^2}} \qquad R_{1000} = 101.03\times10^{-3}$$

$$R_0(f) := R_{1000}{}^{-1}\cdot\frac{\sqrt{1+\left(2\cdot\pi\cdot f\cdot318\cdot10^{-6}\text{s}\right)^2}}{\sqrt{1+\left(2\cdot\pi\cdot f\cdot3180\cdot10^{-6}\text{s}\right)^2}\cdot\sqrt{1+\left(2\cdot\pi\cdot f\cdot75\cdot10^{-6}\text{s}\right)^2}}$$

**calculate gain $G1_0$ at DC and x = R1+R2:**

$G1_0 := \frac{G1_1}{R_{1000}}$ $G1_0 = 989.808\times10^0$

$x = R1 + R2$ $x := G1_0\cdot R3 - R3$ $x = 128.545\times10^3\Omega$

**guessings according to equations (8.52 ... 8.55):**

$R2 = R1\cdot11.778^{-1}$ $x = \left(11.778^{-1}+1\right)\cdot R1$

$R1 := \frac{x}{1+11.778^{-1}}$ $R1 = 118.485\times10^3\Omega$

$T1 := 3180\cdot10^{-6}\text{s}$ $C1 := \frac{T1}{R1}$ $C1 = 26.839\times10^{-9}\text{F}$

**application of succ-apps starting with c1 = 3.6 and r1 = 11.778 from equ. (8.52 and 8.54):**

$c1 := 3.5641$ $C2 := \frac{C1}{c1}$ $C2 = 7.53\times10^{-9}\text{F}$

$r1 := 11.89547$ $R2 := \frac{R1}{r1}$ $R2 = 9.961\times10^3\Omega$

➢ **MCD Worksheet VIII:** Calculation v2.0 of a MM RIAA feedback network type ($E_{ub}$) with succ-apps - Page 2

**monitoring of the deviation D(f) between phono-amp gain G1(f) and ideal RIAA transfer $R_0$(f):**

$$Z2(f) := \left(\frac{1}{R1} + 2j\cdot\pi\cdot f\cdot C1\right)^{-1} + \left(\frac{1}{R2} + 2j\cdot\pi\cdot f\cdot C2\right)^{-1} \qquad G1(f) := \left|\frac{Z2(f)}{R3} + 1\right|$$

$$h := 1000Hz \qquad f := 10, 20..20000 \qquad D(f) := 20\cdot\log(R_0(f)) - 20\cdot\log\left(G1(f)\cdot G1(h)^{-1}\right)$$

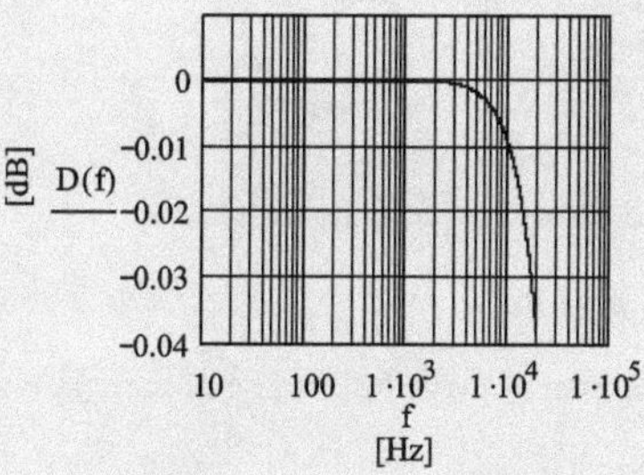


Figure 9.15 = Figure 8.21

**Result and conclusions:**

The deviation D(f) from the exact RIAA transfer can be made close to 0dB with the exception of frequencies >1kHz because with growing frequency the 4th time constant T4 forces the gain more and more close to 1. Following the amp stage V1 a 1st order lp with corner frequency of f4 would flatten the trace for D(f) towards 0dB.

With reference to 0dB at 1kHz the phono-amps gain looks as follows (f4 can be well observed):

$$T4 := 750\cdot 10^{-6}s\cdot\frac{1}{G1_0} \qquad T4 = 757.723\times 10^{-9}s \qquad f4 := \frac{1}{2\cdot\pi\cdot T4} \qquad f4 = 210.044\times 10^{3}Hz$$

with reference to 0dB at 1kHz in the frequency range: $g := 10Hz, 20Hz..1000000Hz$

the RIAA transfer $G_e(g)$ becomes: $G_e(g) := 20\cdot\log(G1(g)) - 20\cdot\log(G1(h))$

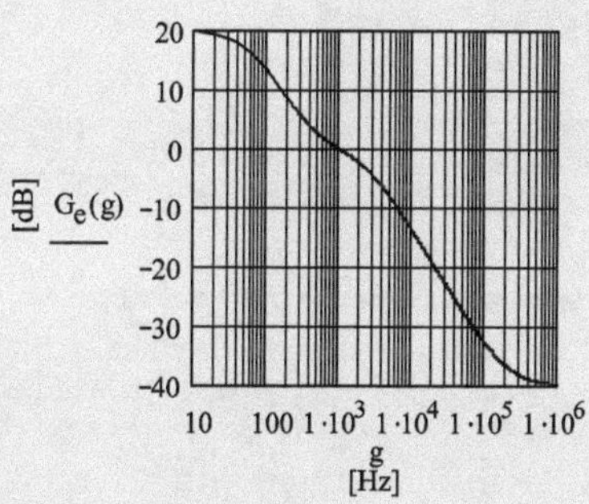


Figure 9.16 = Figure 8.20

**check of the two other RIAA time constants:**

$$T2 := R2\cdot C2 \qquad T2 = 75.005936\times 10^{-6}s \qquad T3 := \frac{R1\cdot R2}{R1 + R2}\cdot(C1 + C2) \qquad T3 = 315.788\times 10^{-6}s$$

➢ **MCD Worksheet IX:** Calculation of the RIAA feedback network type ($E_{ub}$) of the BUVO MC phono-amp with a specific formulae methode - Page 1

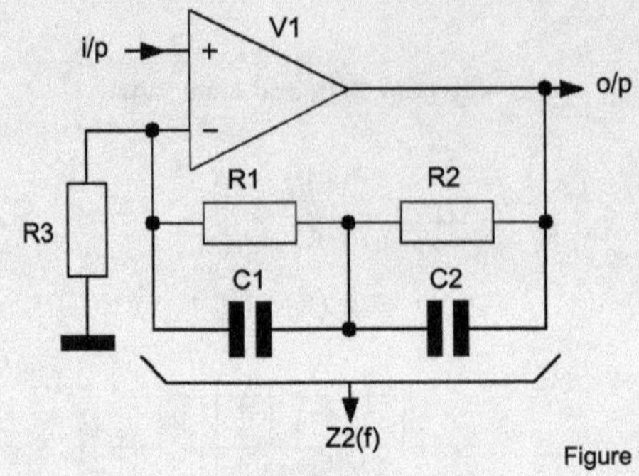


Figure 9.17 = Figures 8.13 + 8.14

**defined RIAA time constants:**

$T1R := 3180 \cdot 10^{-6} s$ $T2R := 75 \cdot 10^{-6} s$ $T3R := 318 \cdot 10^{-6} s$

**choose gain at 1kHz, R3 and open-loop gain c of V1:**

$G1_1 := 223.46$ $R3 := 3.32\Omega$ $c := 100 \cdot 25 \cdot 10^3$

c = open-loop gain of amp (5534.min = 25k)

**calculation of DC gain $G1_0$:**

$$R_{1000} := \frac{\sqrt{1 + \left(2 \cdot \pi \cdot 10^3 Hz \cdot T3R\right)^2}}{\sqrt{1 + \left(2 \cdot \pi \cdot 10^3 Hz \cdot T1R\right)^2} \cdot \sqrt{1 + \left(2 \cdot \pi \cdot 10^3 Hz \cdot T2R\right)^2}} \qquad R_{1000} = 101.03 \times 10^{-3}$$

$$G1_0 := \frac{G1_1}{R_{1000}} \qquad G1_0 = 2.212 \times 10^3$$

**calculation of DC impedance $Z2_0$ of Z2:**

$$Z2_0 := \left(G1_0 - 1\right) \cdot R3 \qquad Z2_0 = 7.34 \times 10^3 \Omega$$

**calculation of feedback factor:**

$$\beta := \frac{c}{G1_0} \qquad \beta = 1.13 \times 10^3$$

**application of the specific formulae (n = new) to get R1, R2, C1, C2:**

$$p = 2j \cdot \pi \cdot f \qquad Z2(p) = (R1 + R2) \cdot \frac{(1 + p \cdot T3n)}{(1 + p \cdot T1n) \cdot (1 + p \cdot T2n)}$$

$$T1R = T1n + \Delta 1 \qquad T2R = T2n + \Delta 2 \qquad T3R = T3n + \Delta 3$$

$$H(p) = G1_0 \cdot \frac{\beta}{(\beta + 1)} \cdot \frac{\left[1 + p \cdot (T3n + \Delta 3)\right] \cdot (1 + p \cdot T4)}{\left[1 + p \cdot (T1n + \Delta 1)\right] \cdot \left[1 + p \cdot (T2n + \Delta 2)\right]} \qquad G1_0 \cdot \frac{\beta}{(\beta + 1)} = 2209.8696$$

➢ **MCD Worksheet IX:** Calculation of the RIAA feedback network type ($E_{ub}$) of the BUVO MC phono-amp with a specific formulae methode - Page 2

if $G1_0 \geq 100$ and $\beta \geq 1$ and $V_0 \neq \infty$ than

$$T1n := \frac{\beta+1}{\beta+\frac{243}{3105}}\cdot T1R \qquad T1n = 3.182593\times 10^{-3}\,s$$

$$f1 := (T1n\cdot 2\cdot\pi)^{-1} \qquad f1 = 50.008\times 10^{0}\,Hz$$

$$T2n := \frac{\beta+1}{\beta+\frac{2862}{3105}}\cdot T2R \qquad T2n = 7.500519\times 10^{-5}\,s$$

$$f2 := (T2n\cdot 2\cdot\pi)^{-1} \qquad f2 = 2.122\times 10^{3}\,Hz$$

$$T3n := T3R\cdot\left(1-10\cdot G1_0^{-1}\cdot\frac{\beta+1}{\beta+\frac{243}{3105}}\right) - T2R\cdot G1_0^{-1}\cdot\left(1-10\cdot\frac{\beta+1}{\beta+\frac{243}{3105}}\right)$$

$$T3n = 3.168666\times 10^{-4}\,s \qquad f3 := (T3n\cdot 2\cdot\pi)^{-1} \qquad f3 = 502.278\times 10^{0}\,Hz$$

$$T4 := G1_0^{-1}\cdot\left(\frac{\beta+1}{\beta+\frac{2862}{3105}}\right)\cdot\left(\frac{\beta+1}{\beta+\frac{243}{3105}}\right)\cdot\left(\frac{T1R\cdot T2R}{T3R}\right) \qquad T4 = 339.387\times 10^{-9}\,s$$

$$f4 := (T4\cdot 2\cdot\pi)^{-1} \qquad f4 = 468.949\times 10^{3}\,Hz$$

$$x := \frac{T1n-T3n}{T3n-T2n} \qquad x = 11.849\times 10^{0}$$

$$y := \left(\frac{T2n}{T1n}\cdot\frac{T1n-T3n}{T3n-T2n}\right)^{-1} \qquad y = 3.581\times 10^{0}$$

$$R2 := \frac{Z2_0}{1+x} \qquad R2 = 571.262\times 10^{0}\,\Omega$$

$$R1 := x\cdot R2 \qquad R1 = 6.769\times 10^{3}\,\Omega$$

$$C1 := \frac{T1n}{R1} \qquad C1 = 470.194\times 10^{-9}\,F$$

$$C2 := \frac{C1}{y} \qquad C2 = 131.297\times 10^{-9}\,F$$

**calculation of RIAA transfer G1(f):**

$$p(f) := 2j\cdot\pi\cdot f$$

$$G1(f) := G1_0\cdot\frac{\beta}{(\beta+1)}\cdot\left|\frac{[1+p(f)\cdot(T3n)]\cdot(1+p(f)\cdot T4)}{[1+p(f)\cdot(T1n)]\cdot[1+p(f)\cdot(T2n)]}\right|$$

**calculation of gain $G1_1$ at** $h := 1000Hz$ **and the RIAA transfer $G_e(f)$ with reference to 0dB at 1kHz and with** $f := 10Hz, 20Hz .. 10^{6}Hz$:

$$G1_1(h) := G1_0\cdot\frac{\beta}{(\beta+1)}\cdot\left|\frac{[1+p(h)\cdot(T3n)]\cdot(1+p(h)\cdot T4)}{[1+p(h)\cdot(T1n)]\cdot[1+p(h)\cdot(T2n)]}\right| \qquad G1_1(h) = 222.443\times 10^{0}$$

➤ **MCD Worksheet IX:** Calculation of the RIAA feedback network type ($E_{ub}$) of the BUVO MC phono-amp with a specific formulae methode - Page 3

$$G_e(f) := 20\cdot\log(G1(f)) - 20\cdot\log\left(G1_1(h)\right)$$

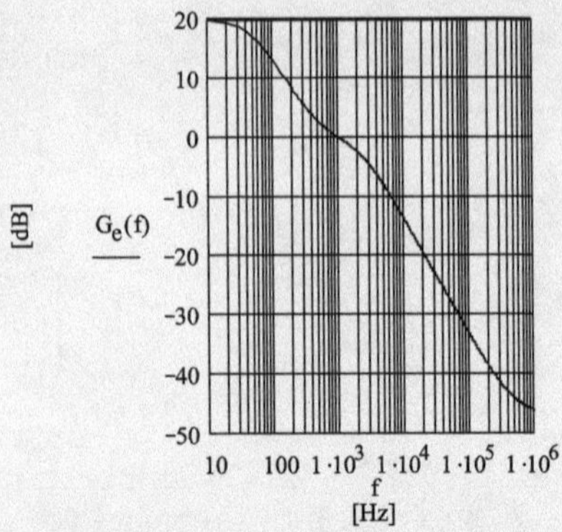


Figure 9.18 = Figure 8.18

**calculation of the magnitude of the impedance $Z2_{20k}$ of the feedback network at** hr:= 20000Hz:

$$Z2_{20k} := \left|\left(\frac{1}{R1} + 2j\cdot\pi\cdot hr\cdot C1\right)^{-1} + \left(\frac{1}{R2} + 2j\cdot\pi\cdot hr\cdot C2\right)^{-1}\right| \qquad Z2_{20k} = 77.124\times10^{0}\Omega$$

**calculation of the deviation D(g) between the ideal RIAA transfer $R_0$ (g) and the calculated G1(g) with reference to 0dB at 1kHz and with** g := 10Hz, 20Hz.. 20000Hz:

$$R_{1000} := \frac{\sqrt{1+(2\cdot\pi\cdot h\cdot T3R)^2}}{\sqrt{1+(2\cdot\pi\cdot h\cdot T1R)^2}\cdot\sqrt{1+(2\cdot\pi\cdot h\cdot T2R)^2}} \qquad R_{1000} = 101.03\times10^{-3}$$

$$R_0(g) := R_{1000}{}^{-1}\cdot\frac{\sqrt{1+(2\cdot\pi\cdot g\cdot T3R)^2}}{\sqrt{1+(2\cdot\pi\cdot g\cdot T1R)^2}\cdot\sqrt{1+(2\cdot\pi\cdot g\cdot T2R)^2}}$$

$$D(g) := 20\cdot\log\left(R_0(g)\right) - 20\cdot\log\left(G1(g)\cdot G1_1(h)^{-1}\right)$$

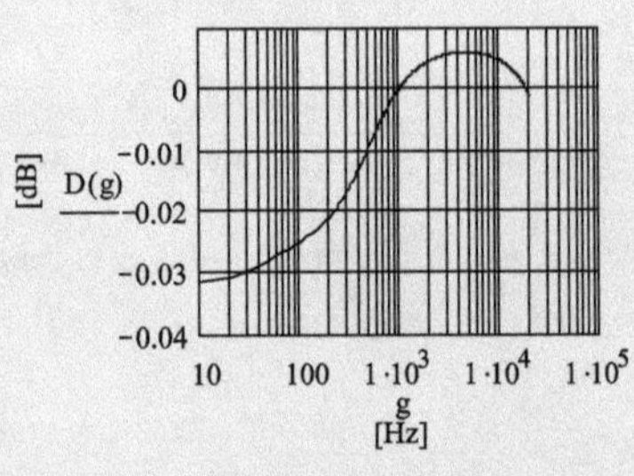


**calculation of the maximal deviation $D_{max}$:**

$D(20Hz) = -0.031$

$D(4700Hz) = 0.006$

$D_{max} := D(4700Hz) - D(20Hz)$

$D_{max} = 0.037$

Figure 9.19 = Figure 8.19

➢ **MCD Worksheet X:** Type ($E_{ub}$) RIAA feedback network: calculation of the proportional factors c1 and r1 - Page 1

**Derivation of the proportional factors r1 = 11.778 and c1 = 3.6 to calculate all components of a Figure 8.14 RIAA feedback network in a shunt mode configured op-amp with infinite open-loop gain:**

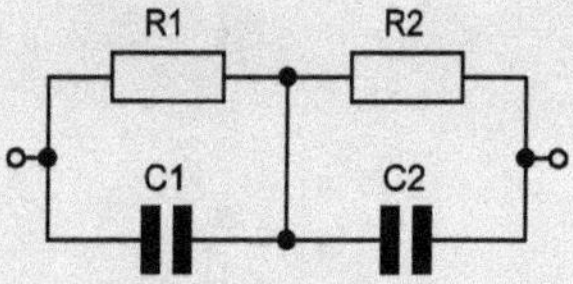


Figure 9.20 = Figure 8.14

with p = 2jπf the impedance Z2(p) becomes:

$$Z2(p) = \frac{R1\cdot\frac{1}{p\cdot C1}}{R1+\frac{1}{p\cdot C1}} + \frac{R2\cdot\frac{1}{p\cdot C2}}{R2+\frac{1}{p\cdot C2}}$$

by application of MCD tool "simplify" Z2(p) becomes:

$$Z2(p) = \frac{R1\cdot R2\cdot p\cdot C2 + R1 + R2\cdot R1\cdot p\cdot C1 + R2}{(R1\cdot p\cdot C1 + 1)\cdot(R2\cdot p\cdot C2 + 1)}$$

thus, the denominator represents the two RIAA time constants T1, T2:

$$T1 = C1\cdot R1 = 3180\mu s \qquad T2 = C2\cdot R2 = 75\mu s$$

reshaping the numerator $(p\cdot C1\cdot R1\cdot R2 + p\cdot C2\cdot R1\cdot R2 + R1 + R2)$ creates time constant T3:

$$p\cdot C1\cdot R1\cdot R2 + p\cdot C2\cdot R1\cdot R2 + R1 + R2 = (R1+R2)\cdot\left[1 + p\cdot(C1+C2)\cdot\left(\frac{R1\cdot R2}{R1+R2}\right)\right]$$

$$T3 = (C1+C2)\cdot\left(\frac{R1\cdot R2}{R1+R2}\right) = 318\mu s \qquad (1)$$

**determination of factor c1:**

$$R1 = \frac{T1}{C1} \quad \text{and} \quad R2 = \frac{T2}{C2}$$ put into equation (1) T3 becomes:

$$T3 = (C1+C2)\cdot\frac{\frac{T1\cdot T2}{C1\cdot C2}}{\frac{T1}{C1}+\frac{T2}{C2}}$$ with "simplify" T3 becomes:

$$T3 = (C1+C2)\cdot T1\cdot\frac{T2}{(T1\cdot C2 + T2\cdot C1)}$$

reshaping: $T1\cdot T3\cdot C2 + T2\cdot T3\cdot C1 = T1\cdot T2\cdot C1 + T1\cdot T2\cdot C2$

$$\frac{T1\cdot T3\cdot C2}{T1\cdot T2} + \frac{T2\cdot T3\cdot C1}{T1\cdot T2} = C1 + C2 \qquad \frac{T3}{T2}\cdot C2 + \frac{T3}{T1}\cdot C1 = C1 + C2 \qquad \left(\frac{T3}{T2}-1\right)\cdot C2 = \left(1-\frac{T3}{T1}\right)\cdot C1$$

➤ **MCD Worksheet X:** Type ($E_{ub}$) RIAA feedback network: calculation of the proportional factors c1 and r1 - Page 2

consequently:

$$\frac{C1}{C2} = \frac{\frac{T3}{T2} - 1}{1 - \frac{T3}{T1}} = \frac{\frac{T3-T2}{T2}}{\left(\frac{T1-T3}{T1}\right)} = \frac{T1}{T2} \cdot \frac{T3-T2}{T1-T3} \quad (2)$$

$T1 := 3180 \cdot 10^{-6} s$ $\quad T2 := 75 \cdot 10^{-6} s$ $\quad T3 := 318 \cdot 10^{-6} s$

$$c1 := \frac{T1}{T2} \cdot \frac{T3-T2}{T1-T3} \qquad c1 = 3.6 \times 10^{0} \quad (3)$$

**determination of factor r1:**

$C1 = \frac{T1}{R1}$ $\quad C2 = \frac{T2}{R2}$ $\quad$ put into equation (1) T3 becomes:

$T3 = \left(\frac{T1}{R1} + \frac{T2}{R2}\right) \cdot \frac{R1 \cdot R2}{(R1 + R2)}$ $\quad$ with "simplify" T3 becomes:

$$T3 = \frac{(T1 \cdot R2 + T2 \cdot R1)}{(R1 + R2)}$$

reshaping:

$$T3 \cdot R1 + T3 \cdot R2 = T2 \cdot R1 + T1 \cdot R2$$

$$R1 \cdot (T3 - T2) = R2 \cdot (T1 - T3)$$

consequently:

$$\frac{R1}{R2} = \frac{T1-T3}{T3-T2} \qquad r1 := \frac{T1-T3}{T3-T2} \quad (4)$$

$r1 = 11.777778$ (= 11.7 repetend) (5)

**example with C1 = 18n||18n = 36n:**

$C1 := 36 \cdot 10^{-9} F$ $\quad R1 := \frac{T1}{C1}$

$C2 := \frac{C1}{c1}$ $\quad R1 = 88.333 \times 10^{3} \Omega$

$C2 = 10 \times 10^{-9} F$ $\quad R2 := \frac{R1}{r1}$

$R2 = 7.5 \times 10^{3} \Omega$

**chross-check:**

$T2 := R2 \cdot C2$ $\quad T3 := \left(\frac{T1}{R1} + \frac{T2}{R2}\right) \cdot \frac{R1 \cdot R2}{(R1 + R2)}$

$T2 = 75 \times 10^{-6} s$ $\quad T3 = 318 \times 10^{-6} s$

# Part III
# Noise Measurement System

# Chapter 10
# System Overview

In Chap. 4 "Noise in MM cartridges" I've given a first insight into the measurement system (MS) to measure noise performances of MM cartridges. Figure 4.7 gives the details. Specifically, MM cartridge related parts of the shown measurement arrangement were covered by Chap. 4. The following topics will be subject of Chaps. 11 ... 12:

**Noise Measurement System Overview**

- Measurement amp with variable gain 0 dB ... +100/106.02 dB
- Measurement filters:
  - bp 20 Hz ... 20 kHz
  - lp 30 Hz
  - hp 355 Hz
  - A-weighting filter NAB
  - A-weighting filter CCIR
  - RIAA transfer equalizing network (decoder)
  - Anti RIAA transfer equalizing network (encoder)
- Indicators
  - digital with PC and CLIO40[1] or CLIO6.55[2]
  - analogue with electronic circuitry around a AD356

The need to develop a noise MS was based on the fact that – in most cases – access to one of the very expensive measurement instruments like those from Audio Precision (AP) still remain a dream. From an expense point of view the MS I wanted to develop should cost at least a 10th of the price of an AP system – based on year 2000

[1] 16-bit DOS version for HR2000 ISA PC-card

[2] 16-bit WIN98 version for HR2000 ISA PC-card; a 18-bit PCI card can be run with WIN2k ff and is called CLIO7.xy

B. Vogel, *The Sound of Silence*
DOI: 10.1007/978-3-540-76884-5, ©Springer 2008

prices. From a precision point of view I wanted to get noise, frequency response and phase measurement results close to $\pm 0.5\,\mathrm{dB}/\pm 0.1\,\mathrm{dB}/\pm 1^{\circ}$. After a search process I found out that the 16 bit CLIO MS nearly fulfilled all my requirements. I only had to develop a very low-noise variable-gain measurement amp, a handful of special filters and an analogue + digital dB-meter driving rectifying and logarithmic circuit.

Including the device under test (d.u.t.) the whole measurement set-up is shown in Fig. 10.1. It enables the user to carry out a double-check method for his measurements. The 1st check path goes from d.u.t. via the Test Terminal (TT) to the CLIO MS and indication screen whereas the 2nd check path goes via measurement amp, filter section and rectifier/log-converter to several indication meters. Alternatively, for very low-noise d.u.t. that need amplification of the measured noise signals before they got passed to the CLIO system, a 3rd check path goes via measurement amp and TT to the CLIO system.

Rather often, PC cards for measurement purposes suffer from the enormous amount of different interference signals inside a computer. They are based on different clock signals and the mains. Figure 10.2 shows a FFT plot of the disturbance situation of my WIN98 driven 500 MHz PC. It was made by shorting the balanced input of the TT, by setting the gain selector of the CLIO card on "auto" and an averaging rate of 50.

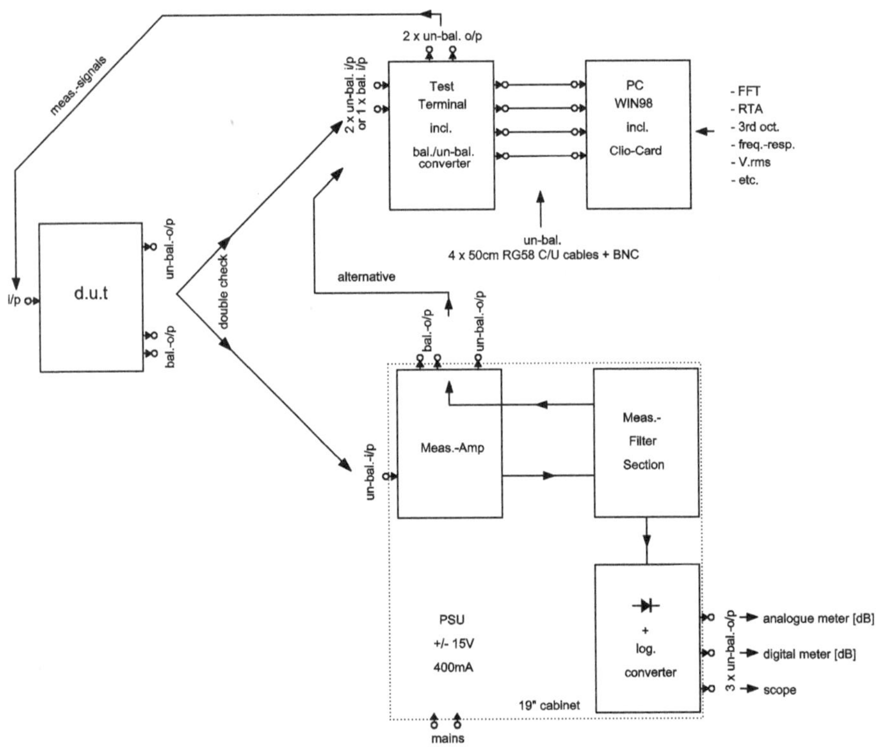


**Fig. 10.1** Overview of measurement set-up

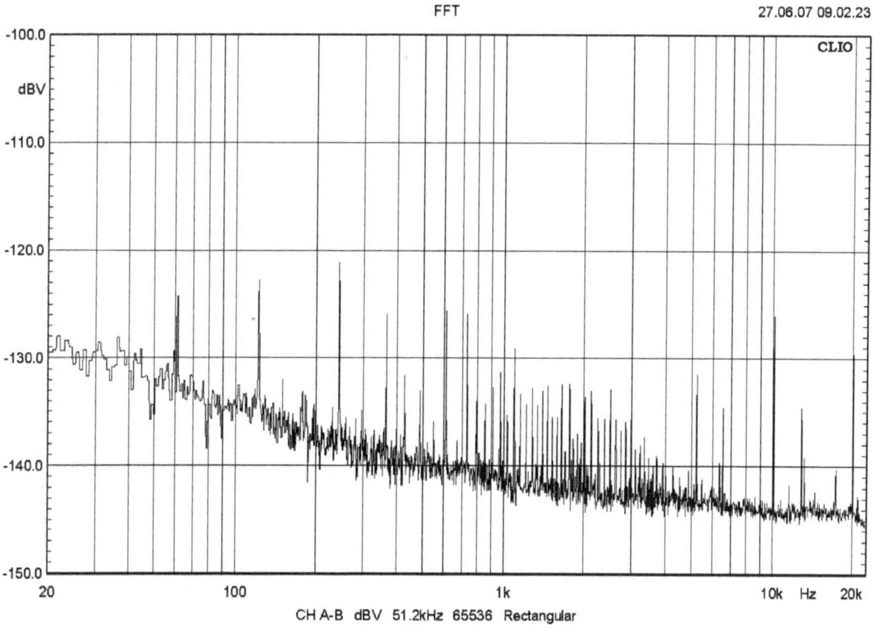


**Fig. 10.2** FFT analysis of the noise voltage of the CLIO card with balanced input shorted, averaging set to 50 and $SN_{\text{ne.clio.b}} = -99.66\,\text{dBV}$

Although the 150 Hz line is rather small ($-132\,\text{dBV}$) it is the only disturbance that will appear in nearly all future measurements. All the other spikes (they are all smaller than $-120\,\text{dBV}$) will disappear because for noise measurements we'll take into account CLIO's input gain stage contribution allowed effect[3].

The Fig. 10.2 non-equalized $SN_{\text{ne}}$ of the CLIO card becomes $-99.76\,\text{dBV}$ in $B_{20\,\text{k}}$. The insertion loss of the 2 transformers inside the TT is 0.1 dB, thus,

$$SN_{\text{ne.clio.b}} = -99.66\,\text{dBV} \tag{10.1}$$

To measure low-noise $< SN_{\text{ne.clio.b}}$ without PC interferences the balanced input signal of the CLIO card needs to be amplified by a certain gain rate $G_{\text{MS}}$ [dB] first. The gain rate $G_{\text{MS}}$ must have a size that ensures a nearly PC interferences free measurement result. To get the real SN of the d.u.t. subtraction of $G_{\text{MS}}$ from the CLIO result must follow:

$$SN_{\text{d.u.t.}} = SN_{\text{clio.measured}} - G_{\text{MS}} \tag{10.2}$$

Example:

- $SN_{\text{clio.measured}} = -27.55\,\text{dBV}$
- $G_{\text{MS}} = +60.00\,\text{dB}$
- $SN_{\text{d.u.t.}} = -87.55\,\text{dBV}$

[3] calculation of ca effects: see Chap. 3.2

To outwit the noise floor of the CLIO card and any hum interferences a balanced configured measurement set-up should be used as well as a measurement amp gain $G_{MS}$ of at least +40 dB should be taken. Best results will be achieved with a gain of +60 dB.

To connect the CLIO MS inputs and outputs to the outer world I also had to develop the special test terminal (TT) that enables the connection of the CLIO built-in direct coupled Cinch plugs with BNC ones and that also allows two different balanced input configurations:

- via the CLIO built-in direct coupled lines
- via potential-free transformer coupled lines

Figure 10.3 shows the TT that is built in a separate shielded insertion module.

The frequency and phase response performance of the TT is shown in Fig. 10.4.

The input signals came in via measurement amp with its gain $G_{MS}$ set to 6.02 dB and wihout any filters. The colours of the traces are the following ones:

left y-ordinate:
- green (upper trace at 10 Hz): frequency response direct coupled
- red (lower trace at 10 Hz): frequency response trafo coupled + potential-free

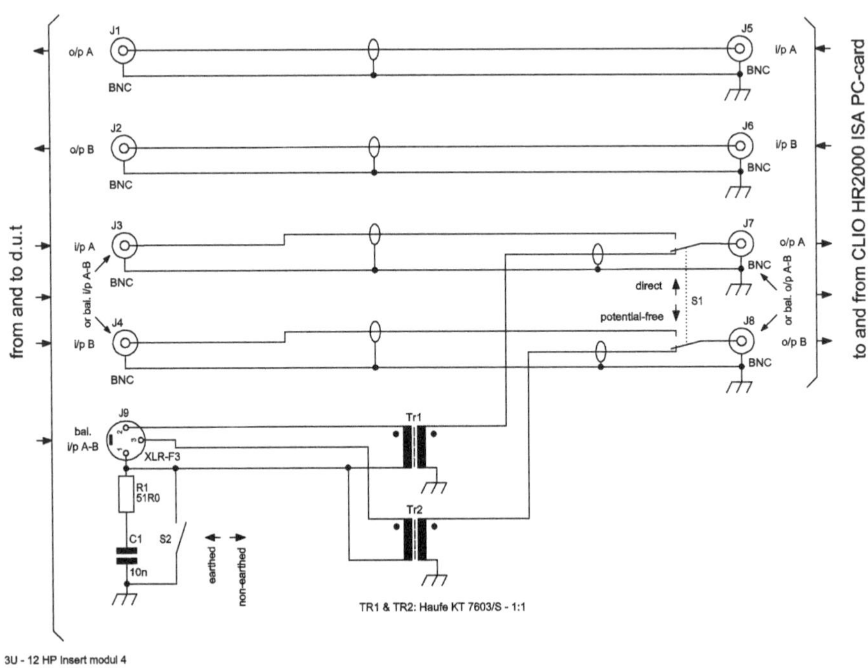


**Fig. 10.3** Test Terminal for the CLIO HR2000 ISA PC-card

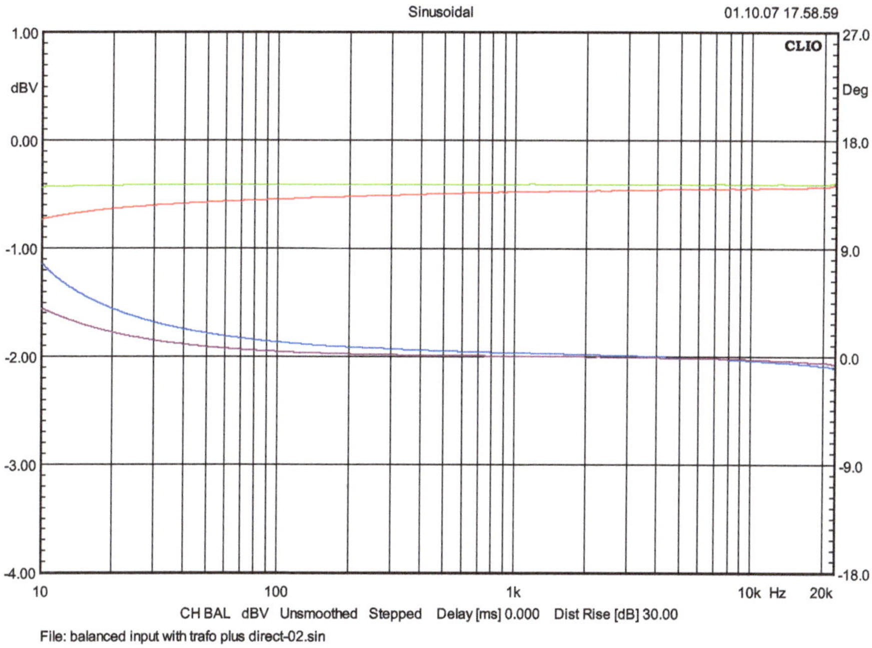


**Fig. 10.4** Test Terminal frequency (*left ordinate* – dBV) and phase (*right ordinate* – Deg) responses for two different balanced input configurations

right y-ordinate:

- violet (lower trace at 10 Hz): phase response direct coupled
- blue (upper trace at 10 Hz): phase response trafo coupled + potential-free

For noise measurements the phase response is not necessarily flat in $B_{20\,k}$.

With balanced direct coupled inputs the frequency and phase responses with reference to 0.00 dB at 1 kHz look as follows:

- 20 Hz: $\pm 0.00$ dB/$+2.00°$
- 1 kHz: $\pm 0.00$ dB/$\pm 0.00°$
- 20 kHz: $-0.02$ dB/$-0.52°$

With balanced trafo coupled inputs the frequency and phase responses with reference to 0.00 dB at 1 kHz look as follows:

- 20 Hz: $-0.15$ dB/$+3.69°$
- 1 kHz: $\pm 0.00$ dB/$\pm 0.00°$
- 20 kHz: $-0.04$ dB/$-1.12°$

Phase margins at >1 kHz partly come from inclusion of the measurement amp. Phase margins <1 kHz show difference in direct (lower trace) and trafo (upper trace) coupling.

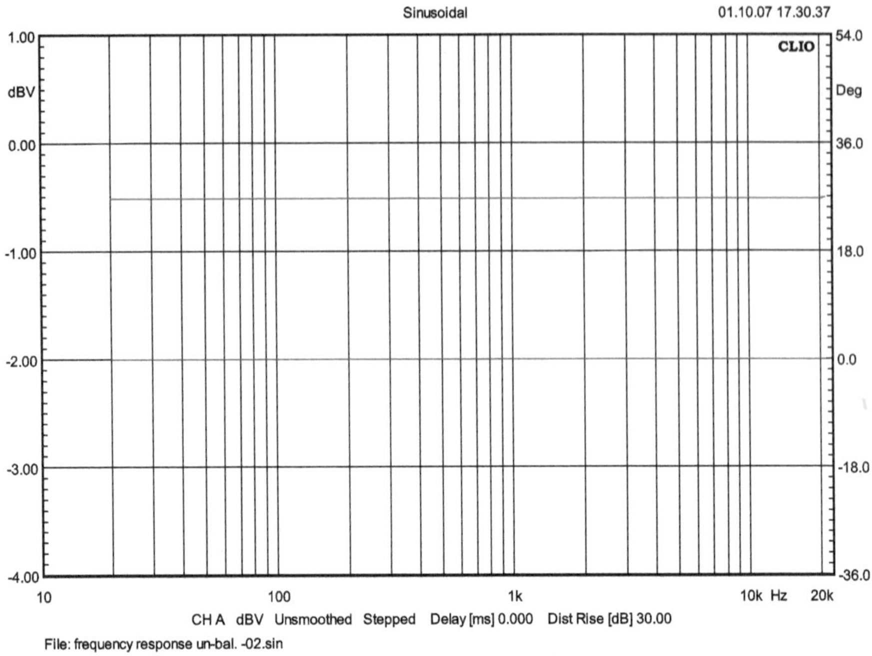


**Fig. 10.5** Frequency (*red – upper trace –* −0.5 dBV) and phase response (*blue – lower trace –* 0.0 Deg) of the CLIO PC-card via an un-balanced and direct coupled input (without measurement amp)

For exclusive frequency and phase response measurements of any kind of amps there is no need to use the balanced inputs nor to switch the measurement amp into the measurement chain. Via un-balanced and direct coupled inputs the respective frequency and phase responses are shown in Fig. 10.5, representing the following trace data:

frequency (upper trace) and phase (lower trace) response in $B_{20\,\mathrm{k}}$ look like:

- 20 Hz: $\pm 0.00$ dB/$\pm 0.00°$
- 1 kHz: $\pm 0.00$ dB/$\pm 0.00°$
- 20 kHz: $\pm 0.00$ dB/$\pm 0.00°$

# Chapter 11
# Measurement Amps

In 1989 M Wilfried Adam published a great article on the design of low-noise audio amps[1]. The complete circuitry of a noise measurement system also became part of that article. M Adam described all necessary measurement filters too. Instead of buying a very expensive all-in-one measurement instrument a few months later I decided to develop my own measurement equipment, mainly based on M Adam's ideas, but with much less noisy op-amps.

## Measurement Amps

In conjunction with Eq. (10.2) the usefulness of a measurement amp was described in the previous chapter. The main amplifier chain of Fig. 11.1 contains 4 separate gain stages. Stage gains are set to 0 dB, +20 dB, +40 dB, +60 dB ($P_{2,4,6}$), the unbalanced output stages in Fig. 11.2 (OP11 … 12) have a gain of 0 dB, the balanced one (OP13) a gain of +6.02 dB, thus, with the shown switching mode in mind, the maximal gain can be set to +106.02 dB. $P_{1,3,5,7,8,13,15}$ have to be adjusted to get a 0.0 V DC output level of the respective op-amps.

Because the +60 dB gain stage (OP1) is mostly in use I've chosen to switch it "on-off" with a separate switch ($S_1$) and not as part of the rotary switch ($S_3$) for the other gain stages. A third switch ($S_2$) allows to by-pass the filter section – excluding the S-filter. The LT1028[2] op-amps are capable to produce – with rather low $1/f$ frequencies – noise voltages of less than 1 nV/rtHz and noise currents of 1 pA/rtHz at 1 kHz, thus, becoming the ideal choice for low source resistances at their inputs – like those of the outputs of most pre-amps and power-amps.

Together with pcb2 that carries the output stages, S-filter, rectifier and logarithmic converter (Fig. 11.2) the measurement amp on pcb1 got fixed in a separate

[1] "Designing low-noise audio amplifier", EW&WW 06-1989

[2] Linear Technology data sheet in the 1990 Databook

[3] relays are Matsushita TQ2-12V


DOI: 10.1007/978-3-540-76884-5, ©Springer 2008

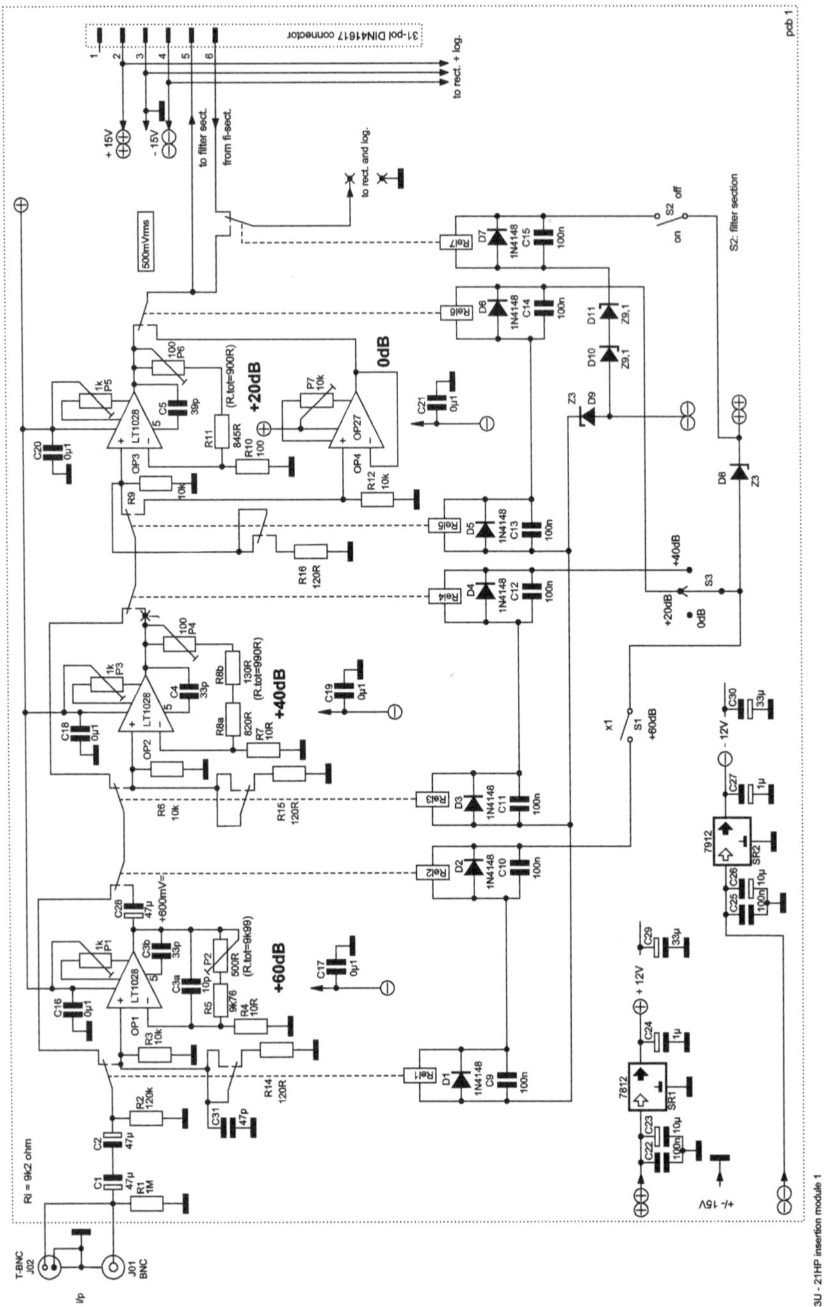


Fig. 11.1 Measurement amps[3]

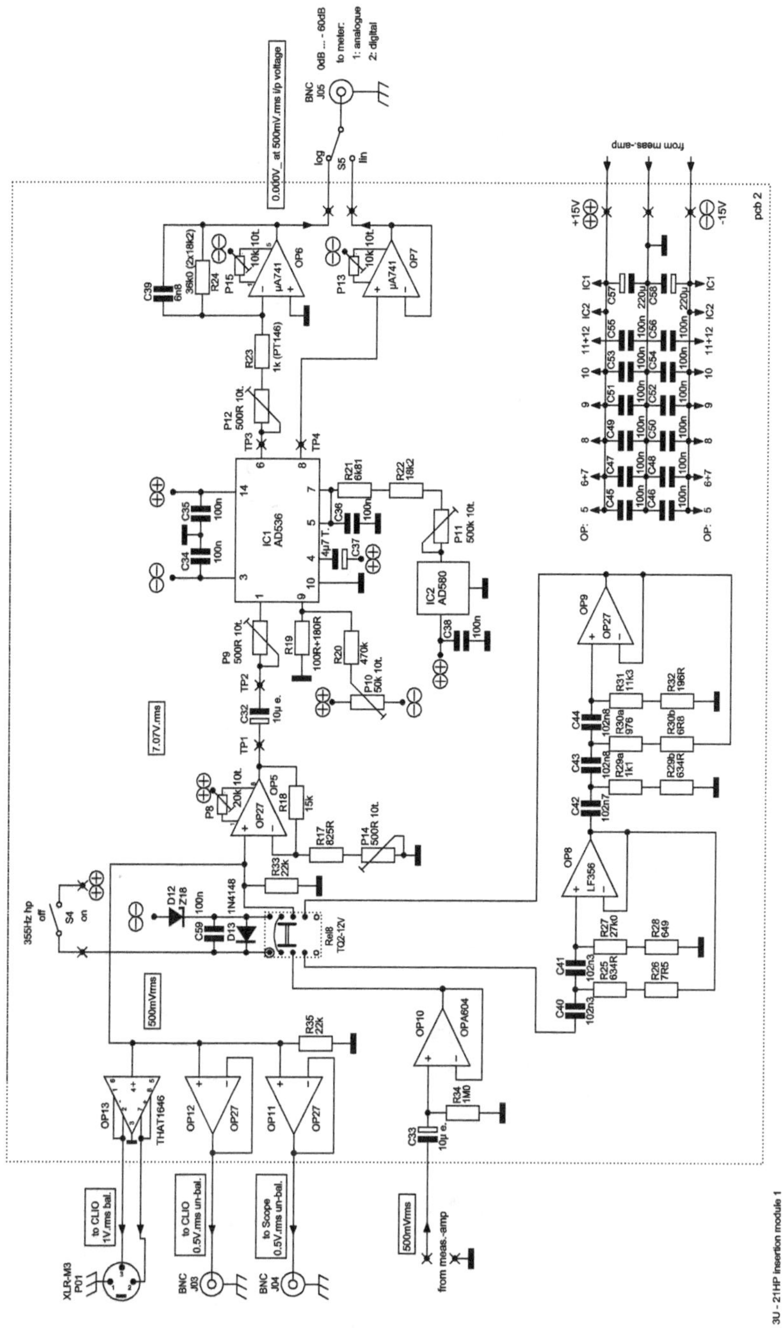


**Fig. 11.2** Rectifier, log-converter, S-filter and output stages

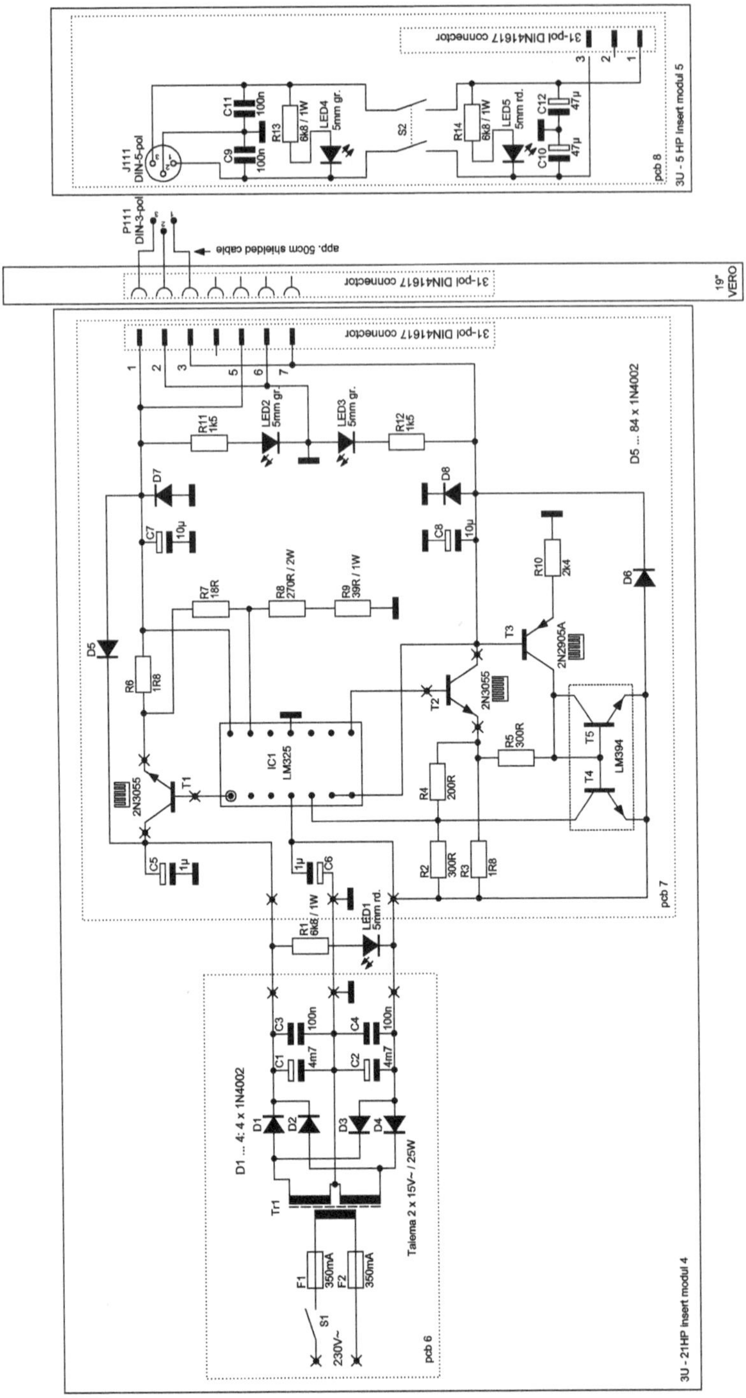


**Fig. 11.3** Power supply unit for measurement equipment

3U-21HP insertion module. To make the whole circuitry operational the following remarks describe essential points in a very short form:

- The circuit for the rectifier and log-converter unit is described in the data sheet of the AD536A[4] (dB connection) as well as the complete dB calibration process with $P_{10,11,12}$.
- $P_{14}$ must be trimmed to get 7.07 V$_{rms}$ at OP5's output – with a 500 mV$_{rms}$ and 1 kHz sinus at the input of OP5.
- With inputs shorted trimming of $P_{13,15}$ must ensure a 0.00 V DC output level of OP6, 7.
- For calibration purposes several test points ($T_{1...4}$) allow easy access into the circuitry.
- $S_4$ turns the S-filter on or off.
- $S_5$ makes it possible to switch between log [dB] or lin [V] output.
- The analogue meter is the ideal choice for calibration works (e.g. $P_1$ in Fig. 6.4). For easy reading its indicator should be as big as possible. I've found an old valve driven DC voltmeter that does the job very well – after replacement of the valve by an op-amp.
- The digital meter is a ±200 mV 3 1/2-digits meter block that should be set to a range of ±20 V via appropriate input voltage divider and decimal point shift.
- Shown in Fig. 11.3 he whole measurement system is fed by a short-circuit protected and well stabilized low-noise ±15 V/±400 mA DC power supply. This PSU is located far away from the amp section and is connected via a shielded cable to the MS.

The noise voltage relevant result of all these measures is shown in Fig. 11.4. With input shorted, $G_{MS}$ set to +106.02 dB and a balanced 1.5 m connection cable between measurement balanced output and TT balanced trafo input the measured $SN_{ne.MS}$ becomes:

$$\begin{aligned} SN_{ne.MS} &= \text{CLIO reading} + \text{trafo insertion loss} - G_{MS} \\ SN_{ne.MS} &= -29.97\,\text{dBV} + 0.1\,\text{dB} - 106.02\,\text{dB} \\ &= -135.89\,\text{dBV} \end{aligned} \tag{11.1}$$

In other words: not specifically taking into account $1/f$ effects the average input referred noise voltage density $e_{N.in.MS}$ for any frequency in $B_{20\,k}$ and for highest gain becomes:

$$\begin{aligned} e_{N.in.MS} &= \frac{10^{\frac{SN_{ne.MS}}{20}}}{\sqrt{B_{20\,k}}} \\ &= 1.14\,\text{nV/rtHz} \end{aligned} \tag{11.2}$$

The respective value for 1 kHz can be picked from the graph: −73.90 dBV. Taking into account the trafo insertion loss and the maximal gain of +106.02 dB

[4] "Special Linear Reference Manual" Analog Devices 1992

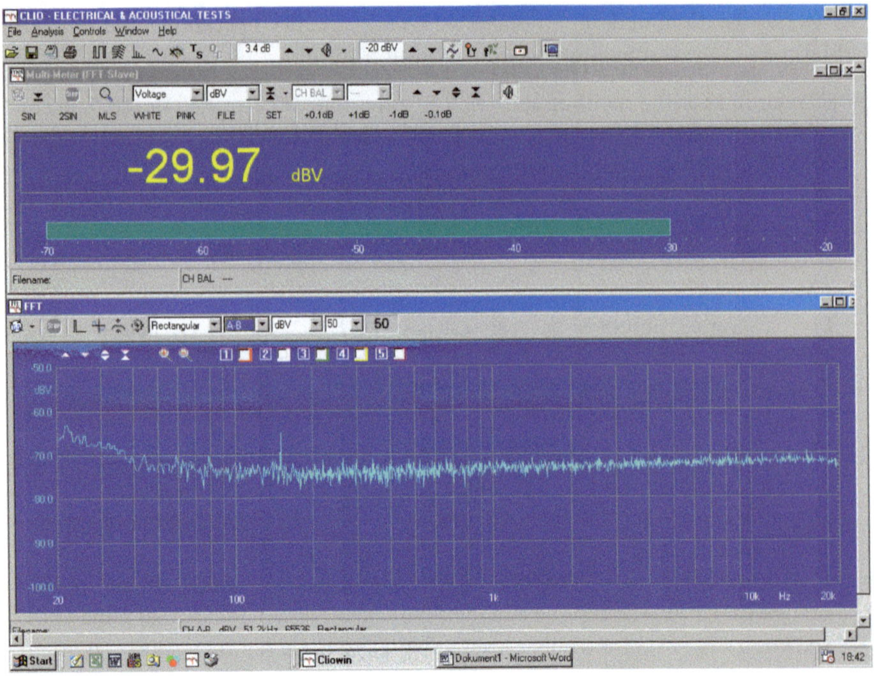


**Fig. 11.4** Spectral noise voltage density of measurement amp with gain set to +106.02 dB, input shorted, average 50, via balanced and potential-free connection (see text)

the noise voltage density for 1 kHz becomes calculated 1.02 nV/rtHz (from −179.8 dBV).

No mains induced hum can be detected in the measurement equipment. The exclusively PC induced hum spike at 150 Hz reads −171.0 dBV. This is 35.1 dB smaller than $SN_{\text{ne.MS}}$. That's why it has no influence on measurement results and it will totally be covered by the whole noise in $B_{20\,\text{k}}$.

The S-filter calculation will be described in Chap. 12 "Measurement Filters and Networks".

# Chapter 12
# Measurement Filters and Networks

## Measurement Filters and Networks

Measurement filters and electronic transfer ratio reproducing networks play an essential role for any audio measurement set-up – especially when talking about noise measurements and RIAA phono-amps. In this chapter I'll focus only a bit – but hopefully extensively enough – on that issue because most of the content was also published in EW[1,2].

The filter bank of Fig. 12.1 consists of a

1. 20 Hz ... 20 kHz band-pass (bp) with a sharp cut-off hp at 20 Hz (Chebyshev 0.1 dB, $n = 6$) and a smoother cut-off lp at 20 kHz (Butterworth, $n = 6$). The 20 Hz hp enables $B_{20\,\mathrm{k}}$ noise measurements without any rumble or tonearm resonance disturbance <20 Hz as well as without heavy $1/f$-effects. Because of the high order of the filters there is no need to correct the AC noise bandwidth[3] of it. In addition, with the 16-bit nature of the CLIO AD converter, the noise bandwidth of this measurement instrument is restricted to 21.3 kHz – obtained by a very sharp cut-off anti-aliasing filter.
2. 30 Hz low-pass (lp) to measure negative low-frequency noise effects like shot noise and rather low $1/f$ noise.
3. 355 Hz high-pass (hp) (= S-filter) to measure noise without any low-frequency effects like hum, $1/f$ noise, etc.
4. A-weighting filter to measure SNs according to the NAB/ANSI S1.4-1986 standard – mostly valid for consumer purposes only.
5. A-weighting filter to measure SNs according to the CCIR 468 standard – mostly valid for professional purposes only, e.g. studio equipment, microphones, etc.

---

[1] "Precision A-weighting filter", EW 12-2004

[2] "Precision Anti-RIAA", EW 05-2007

[3] "Intuitive IC Op-amps", Th. M. Frederiksen, National's Semiconductor Technology Series

[4] 20 Hz ... 20 kHz bp, 30 Hz lp and A-filter are improved versions of the W. Adam developments demonstrated in his EW&WW 06-1989 article on "Designing low-noise audio amplifiers"


DOI: 10.1007/978-3-540-76884-5, ©Springer 2008

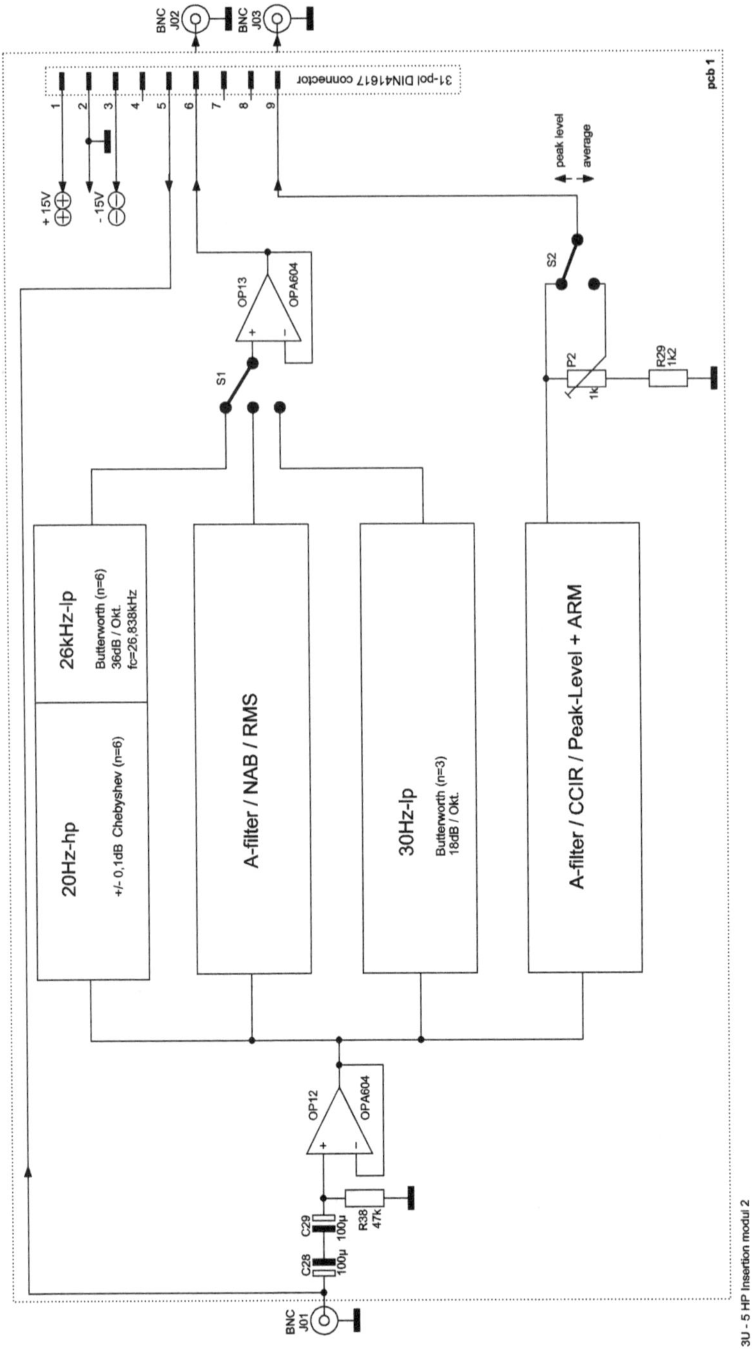


**Fig. 12.1** Filter bank with 20 kHz band-pass, 30 Hz low-pass, NAB and CCIR A-weighting

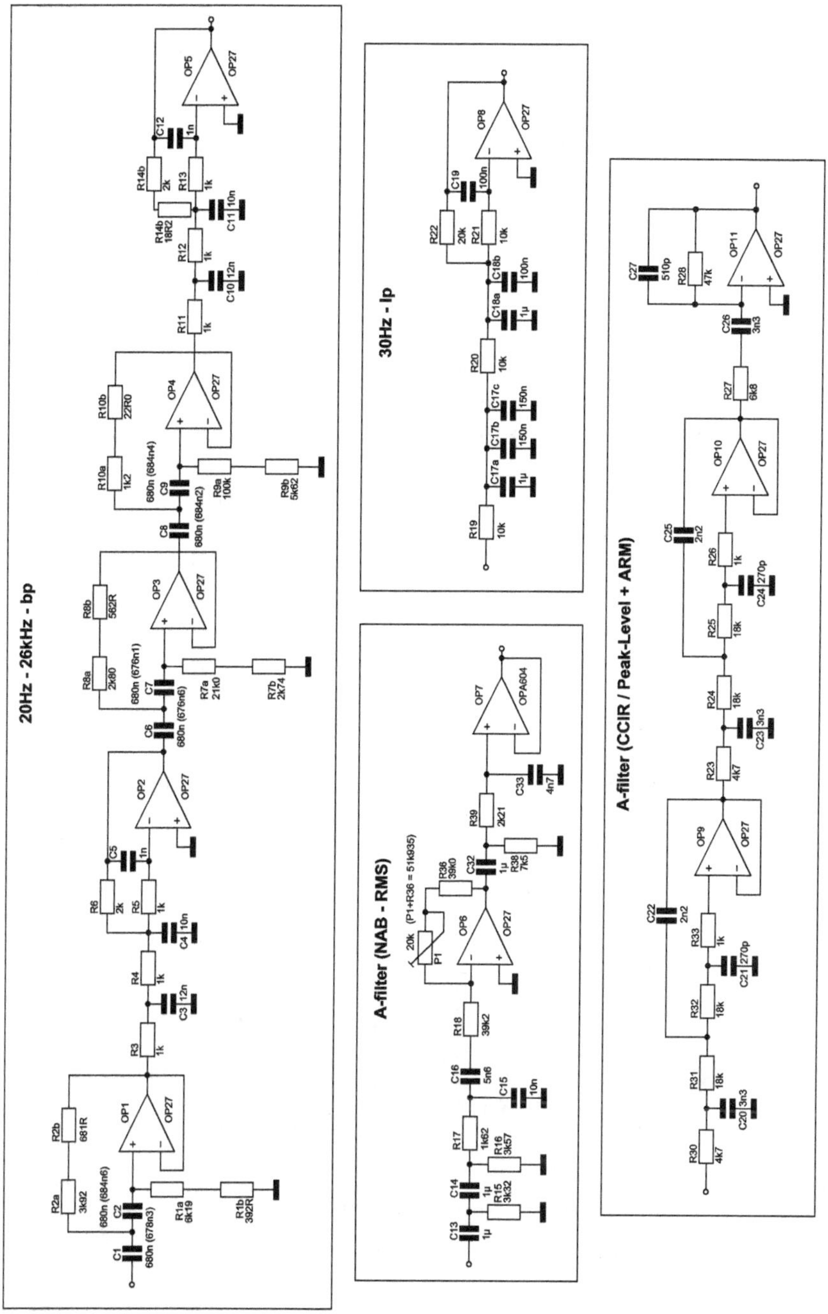


**Fig. 12.2** Circuitry details of the Fig. 12.1 filter bank[4]

$S_2$ of Fig. 11.1 switches the filter bank on and off. With the exception of the 30 Hz lp (gain of 0 dB at 10 Hz) the gain of the filters is always 0 dB at 1 kHz. The respective circuits are given in Fig. 12.2. The measured frequency responses are shown in Fig. 12.3 … 12.4. To avoid pushing on the 0 dBV level I've split the graph on three levels: −6 dBV for 30 Hz lp and 355 Hz hp, 0 dBV for $B_{20\,k}$, +6 dBV for "no filter". The reason to include the trace of the transfer of the NAB A-weighting filter for comparisons lies in the fact that it produces nearly the same RIAA equalized $SN_{ariaa}$ like the 355 Hz hp S-filter.

The circuit of the S-filter is already shown in Fig. 6.1 of Chap. 6. To calculate the values for the S-filter we need to know the filter coefficients for a 0.1 dB Chebyshev hp. They can be derived by application of the respective lp coefficients that are listed in Table 11-37 of Chapter 11 in the Electronic Filter Design Handbook[5].

For $C_{1_c \ldots 5_c} = 1\,\mathrm{F}$ (F = Farad) the coefficients for $R_{1_c \ldots 5_c}$ become:

$R_{1_c} = 1/\,0.1580\,\mathrm{F}$; $R_{2_c} = 1/6.810\,\mathrm{F}$; $R_{3_c} = 1/4.446\,\mathrm{F}$; $R_{4_c} = 1/2.520\,\mathrm{F}$; $R_{5_c} = 1/0.3804\,\mathrm{F}$.

Multiplication of $R_{1_c \ldots 5_c}$ with $\{1\,\mathrm{F} \times [2 \times \pi \times (f_{hp} = 355\,\mathrm{Hz}) \times (C = 100\,\mathrm{nF})]^{-1}\}$ will lead the Fig. 6.1 values for $R_{1 \ldots 5}$.

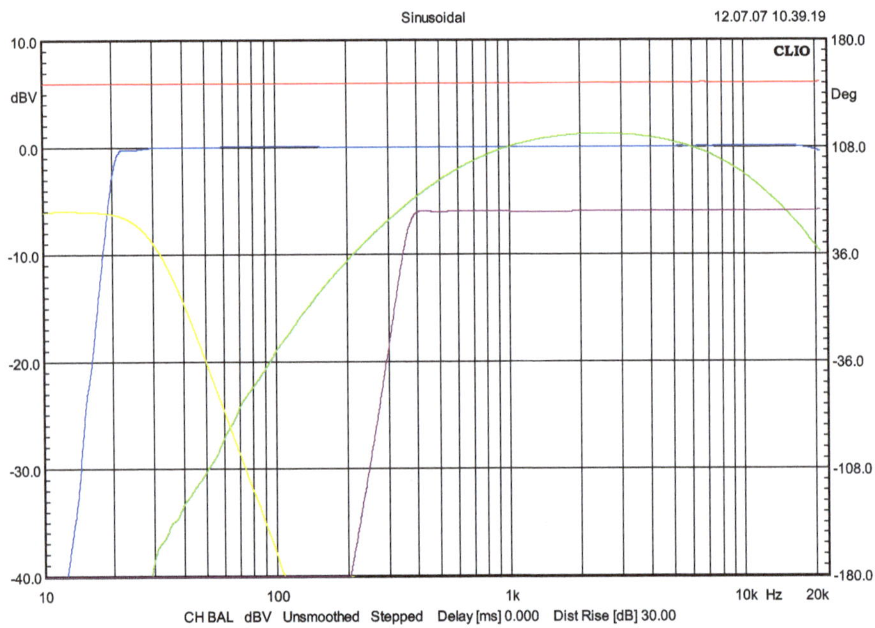


**Fig. 12.3** Frequency responses of all measurement filters:
*red trace* at +6 dBV: no filter
*blue trace* at 0 dBV: 20 Hz – 20 kHz bp
*violet trace* at −6 dBV: S-filter −355 Hz hp
*yellow trace* at −6 dBV: 30 Hz lp
*green trace* at 0 dBV/1 kHz: NAB/ANSI A-weighting filter

[5] "Electronic Filter Design Handbook", A. Williams & F. Taylor, McGraw-Hill, 4th edition, 2006

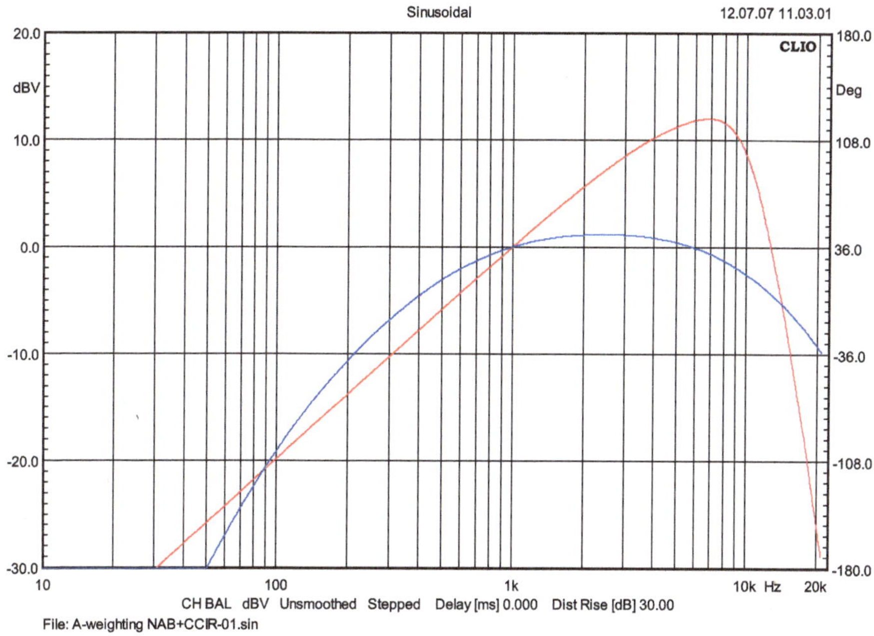


**Fig. 12.4** A-weighting filters:
*blue trace* at 0 dBV/1 kHz: NAB/ANSI
*red trace* at 0 dBV/1 kHz: CCIR

Figure 12.5 again shows the measured NAB A-weighting transfer, but in this case to compare it with the studio standard A-weighting CCIR filter transfer. In general, because of the lower gain at mid-frequencies (app. +1.7 dB at 2.5 kHz for NAB vs. +12 dB at 7 kHz for CCIR), the SNs measured with the NAB filter look always better than the ones measured with the CCIR filter. In average, by measuring the same amp, unlike the CCIR filter the NAB filter allows less noise energy to pass through it in a broad mid-frequency range from 1 kHz to app. 16 kHz, thus, pushing its *SN* result positively vs. the SN result of the CCIR filter. An additional role plays the rectifying circuit for the AC signals. In the CCIR case it's quasi peak, in the NAB case it's rms.

The deviation from the exact NAB/ANSI transfer is only ±0.1 dB and the tolerance of the CCIR filter is according to the standard.

The only transfer ratio reproducing networks that are needed for measuring purposes of RIAA phono-amps are:

1. RIAA decoder of Fig. 12.5a, that is a RIAA transfer network to perform e.g. MM cartridge measurements like the ones described in Chap. 4.
2. RIAA encoder of Fig. 12.5b, that is an Anti-RIAA transfer network to test the exactness of the transfer ratio of a RIAA phono-amp.

For phase measurements and to move to the right place the respective trace of the measurement graph both circuits were completed by an inverter (OP1 and OP4). The

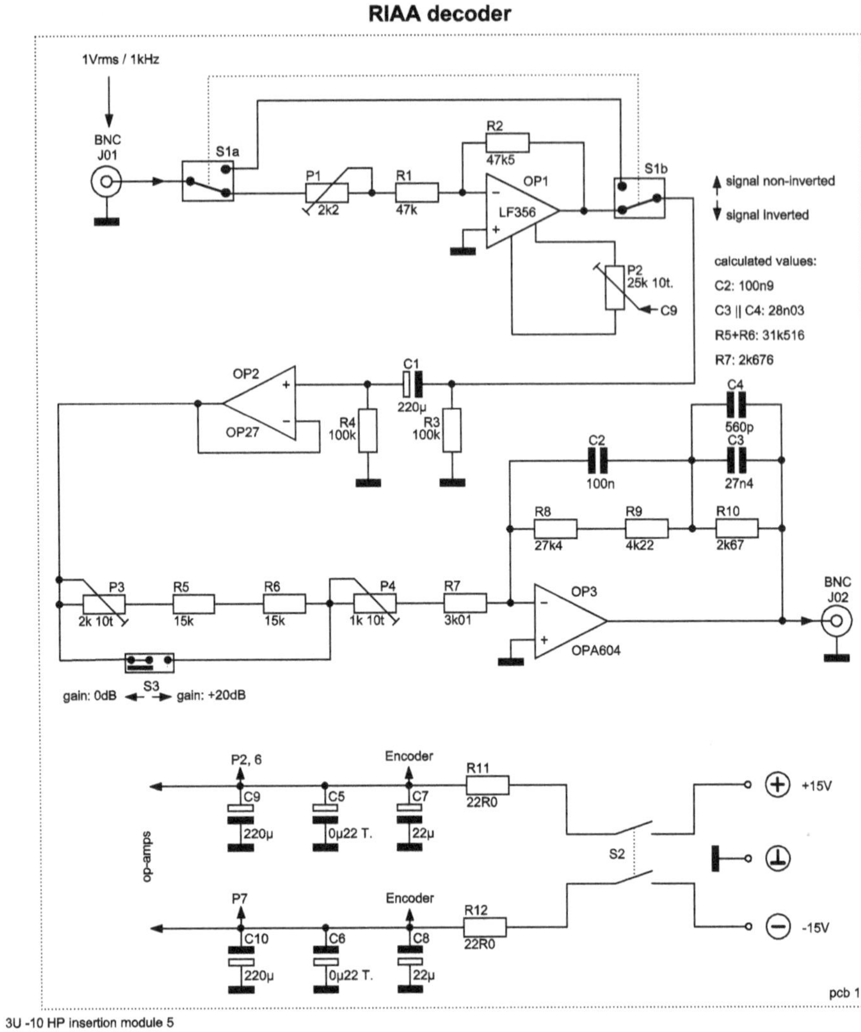


**Fig. 12.5a** Details of the RIAA decoder circuit (part of insertion module 5)

RIAA decoding function is performed by the circuitry around OP3. The mathematics to calculate the exact type $E_{ub}$ network values of $C_{2...4}$, $C_{13...14}$, $R_{8...10}$, $R_{19...20}$ can be found on Worksheet X of Chap. 9. For other measurement purposes with $S_3$ the gain of the OP3 stage can be switched between 0 dB and +20 dB. OP2 and OP5 ensure very low output impedances for the following frequency dependent circuitries. $R_{18}$ and $C_{12}$ ensure stable and ringing-free operation for the rather high capacitive load of OP5[6]. OP7 creates an output impedance low enough to drive a 50 R load with tiny MC input signals of app. 0.5 mV$_{rms}$/1 kHz. But the buffer's

[6] See OPA604 data sheet

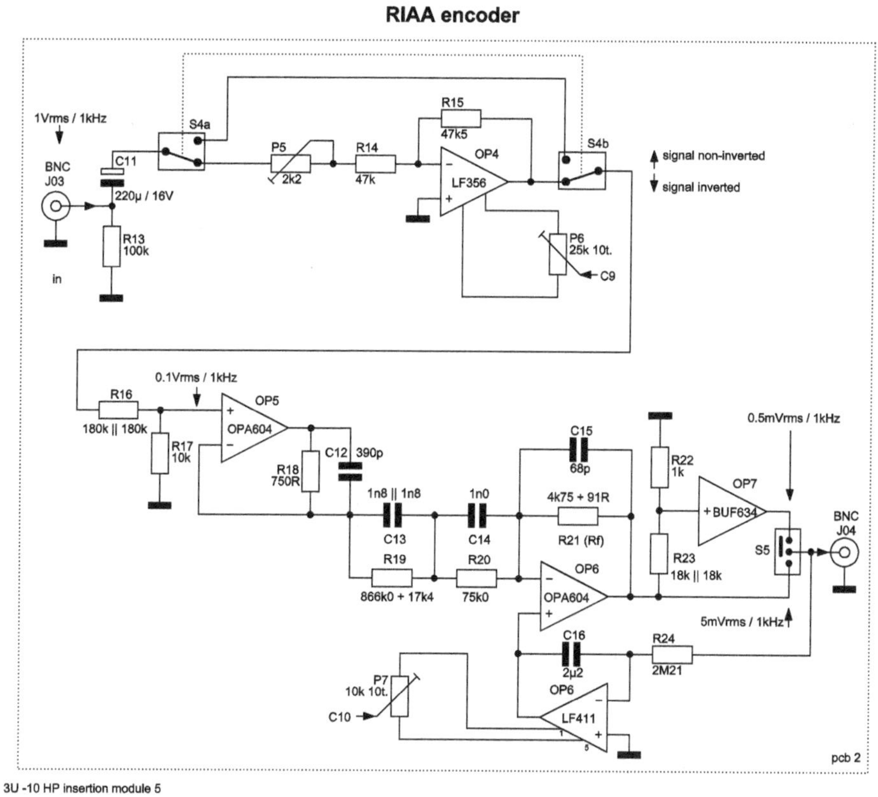


**Fig. 12.5b** Details of the RIAA encoder (Anti RIAA) circuit (part of insertion module 5)

open-loop output impedance is not 0 R! It lies at app. 80 R and looks constant in $B_{20\,\mathrm{k}}$[7]. The voltage divider effect between OP7 output impedance and phono-amp input impedance has to be taken into account when setting the encoder input voltage to get a specific phono-amp input voltage!

For the MM encoding situation the transfer plots are given in Fig. 12.6 and for the MC encoding situation in Fig. 12.7. In both cases, by connecting the output of the decoder to the input of the encoder, the phase responses should be flat. This is only possible with growing (up to extremely high) gain of the encoder at frequencies $\gg 20$ kHz. In reality, this cannot be handled because of the limits set by the supply voltages of the op-amps. Somewhere $>20$ kHz and depending on the output swing limits of the op-amps the encoder transfer trace (blue in both figures) will hit the supply voltage level, thus, producing a fourth time constant and lp corner frequency $f_{4\mathrm{th}}$. This $f_{4\mathrm{th}}$ very early starts to change the phase response of the encoder as well as the sum of the phase response of the decoder plus phase response of the encoder.

In any case, with 1% capacitors and 1% metal film resistors the decoder's and encoder's deviation from the exact RIAA or Anti-RIAA transfer lies within $\pm 0.1$ dB.

[7] See BUF603 data sheet

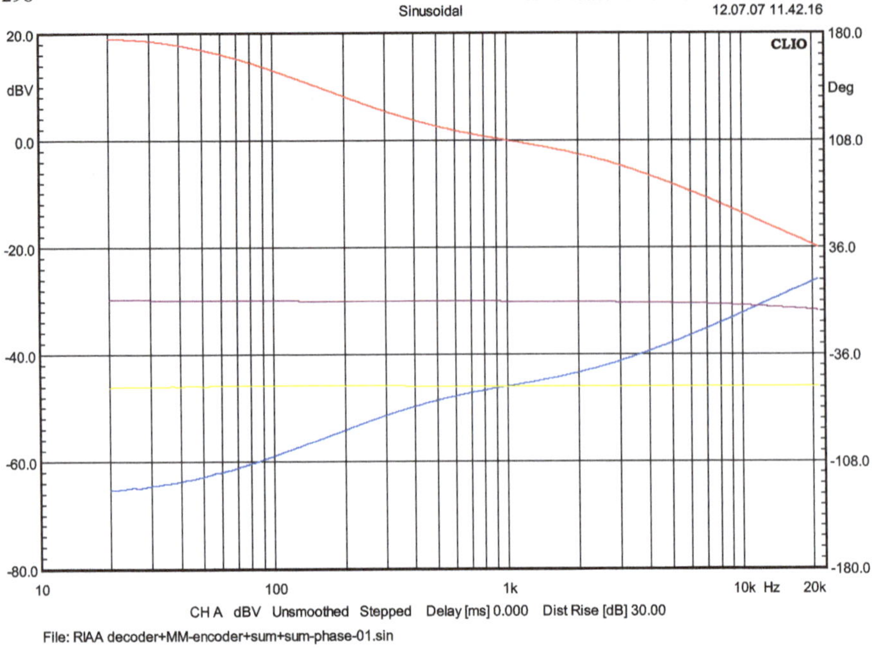


**Fig. 12.6** Frequency and phase responses of RIAA decoder and encoder:
*red trace* at 0 dBV/1 kHz: decoder output
*blue trace* at −46 dBV/1 kHz: encoder with MM output
*yellow trace* at −46 dBV/1 kHz: sum of decoder + encoder output
*violet trace* at 0°/1 kHz: phase response of sum of decoder + MM-encoder output

At the end of this chapter I have to deal with two rather disadvantageous but challenging issues.

1. At the beginning of Chap. 10 I've explained the reason to develop my own measurement set-up instead of buying a very expensive all-in-one solution à la Audio Precision. Of course, I was sure that this decision could also result in some troubles at corners on the development course I've never thought of before. One of the corners were the restrictions set by the CLIO measurement system with e.g. the fixed frequency wobble (sweep) speed, another was – and still is – the hum contamination level of the environment I'm (and we all are) working in that forces me (us) to react adequately with – sometimes enormous – amounts of shielding and other efforts.

The fixed frequency wobble speed does not harmonize with the chosen settling time of the DC voltage nulling circuitry around OP6 of Fig. 12.5b, thus, it produces trace disturbances of max. ±1.5 dB at low frequencies <100 Hz (see Fig. 12.7 and respective note).

2. To overcome most of the hum interference problems I strongly recommend to only ground the all-metal-case of a measurement- or phono-amp with one ground connector at the highest sensitive input of the amp. Although, the exis-

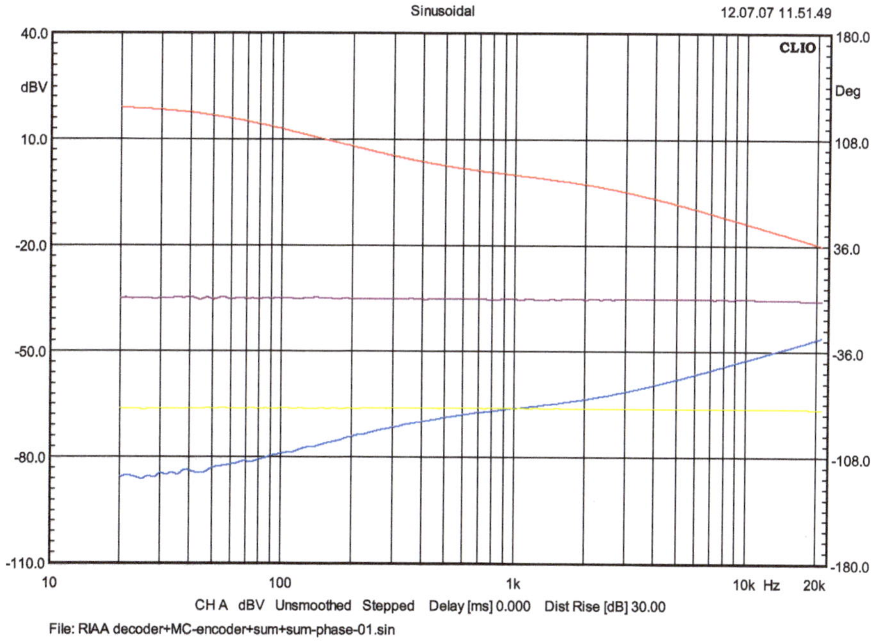


**Fig. 12.7** Frequency and phase responses of RIAA decoder and encoder:
*red trace* at 0 dBV/1 kHz: decoder output
*blue trace* at −66 dBV/1 kHz: encoder with MC output[10]
*yellow trace* at −66 dBV/1 kHz: sum of decoder + encoder output
*violet trace* at 0°/1 kHz: phase response of sum of decoder + MC-encoder output

tence of some devices on the high-end market with wooden or even plastic made cases: their non-shielding-capabilities trigger an absolute no-no to copy it!

Balanced cable connections – best-case: potential-free by insertion of transformers into the connection chain – should be used as much as possible[8]. It's a fact that one of the best sounding pop albums ever produced was mixed on a valve driven and transformer input and output connected 4-track 1″ tape recorder from Studer: "Sergeant Pepper's Lonely Hearts Club Band" from the fab 4 on the famous J37 (its main audio valves are: 10 × E188CC/7308). To ensure the best sound, in addition, valve driven Neumann U87 microphones with transformer outputs fed the equally valve driven mixing console[9] via rather long (>10 m) balanced cables. No hum can be heard on the quiet sections of this record.

8 "Balanced Lines in Audio Systems: Fact, Fiction, and Transformers", Bill Whitlock, JAES Vol. 43, No. 6, June 1995 – or Jensen Transformers, Inc. Ca. USA

9 "Here, there and everywhere", Geoff Emerick & Howard Massey, Gotham Books, Penguin Group, New York, USA

10 The disturbance at the lower end of the frequency response <100 Hz comes from the rather high wobble speed of the CLIO frequency generator that overtakes the DC voltage settling time of the MC output in conjunction with OP6 in Fig. 12.5b. The disturbance will disappear by application of a very much slower wobble or sweep by hand

# Part IV
# The RIAA Phono-Amp Engine

# Chapter 13
# Overview

## Purpose and Goals

The purpose of the RIAA Phono-Amp Engine is to have everything on one spot what is needed to demonstrate that measurement results (carried out in a home-office in a typical city housing area and not in a lab environment) are not far away from what could be figured out by theory based calculations. At the same time it helps to check and test a great variety of connectors and connection possibilities. Additionally, to also fulfil the needs of people who want to listen to music with rather low-impedance MC cartridges I've included the draft design of a special trafo driven MC phono-amp. With that in mind this engine fulfils the following goals:

1. serving all kinds of MM cartridges as the appropriate phono-amp to ensure maximal possible $SN_{ariaa}$ with an un-balanced phono-amp input and an input impedance of 47 k$||C_{opt}$
2. serving all kinds of MC cartridges as the appropriate phono-amp to ensure maximal possible $SN_{ariaa}$ with cartridge impedances ranging from 2 R to 50 R and balanced and un-balanced phono-amp inputs ranging from 25 R to 1 k – for solid-state and transformer solutions
3. the output configurations allow balanced and un-balanced operations as well as partly potential-free ones
4. switchable IEC 20 Hz hp
5. fully separate operation of the modules – including remote power supply units.

## Development Results

These goals can not be met by only one phono-amp. That's why I've developed several different devices:

a. one fully solid-state device for a MC cartridge impedance range of 2 R ... 50 R
b. one draft design circuit with transformer input for a MC cartridge impedance range of 2 R ... 20 R with switchable transformer turns ratios from e.g. tr = 1:2 to tr = 1:12

B. Vogel, *The Sound of Silence*
DOI: 10.1007/978-3-540-76884-5, ©Springer 2008

c. one device with transformer input (tr = 1:10) for a MC cartridge impedance range of 20 R ... 50 R
d. one fully solid-state device for MM cartridges with output voltages of rated 5 $mV_{rms}$/1 kHz/8 cm/s and minimal of 1.8 $mV_{rms}$/1 kHz/8 cm/s or high output MC cartridges for 47 k loads
e. one fully solid-state draft design circuit for MM cartridges with output voltages of rated 10 $mV_{rms}$/1 kHz/8 cm/s and minimal of 3.6 $mV_{rms}$/1 kHz/8 cm/s.

To sum up: I tried to cover nearly every possible phono-amp input situation set by different cartridges!

## Engine Functions

Figure 13.1 shows the general engine diagram. Three different stereo phono-amp modules and three different PSUs fulfil the above listed goals:

- module 1 consists of a balanced trafo input driven MC phono-amp (c.) that can be switched to a fully solid-state un-balanced input MM phono-amp (d.) fed by one IC based internal ±15 V DC power supply. Both phono-amps drive a potential-free balanced trafo coupled and a direct coupled un-balanced output per channel.
- module 2 consists of a fully solid-state MC phono-amp (a.) that drives a potential-free balanced trafo coupled and a direct coupled un-balanced output via a switchable volume potentiometer (linear with 23 steps) to allow fast adaptation of the phono-amp's input sensitivity to various input voltage ranges per channel. One rather complex low-noise μA723 based internal ±15 V DC power supply completes the whole circuitry.
- module 3 consists of (points 1. ... 5. per stereo channel)

    1. one phono-amp section b) and e) (similar to module 1 – without its output circuit)
    2. one switchable IEC 20 Hz hp
    3. three switchable inputs – one balanced, two un-balanced
    4. one un-balanced direct coupled output driver for rather short cable connections
    5. one balanced direct coupled output driver with gain of +6 dB to ensure a high common mode rejection ratio (for very long cable connections) and unity gain in conjunction with a −6 dB input stage gain of the following amp
    6. one IC based internal ±15 V DC power supply for the source selector and an equivalent one for the phono-amp section for both channels.

In addition, each module can be fed with supply voltage and current by a separate power supply unit PSU 1 ... 3. To check what happens with hum and noise I've configured the supply voltage in- and outputs of the modules in a way that operation of two modules with only one PSU could also be made possible. The circuit diagram of the PSUs is given in Fig. 6.6. They are housed in three separate 3UH-21HP 19″ insertion cases. Together, PSU-4 switches all PSUs on/off.

It is not a must to put all three 1HU 19″ modules into one plug-in case. The whole engine as well works very good with the modules closely put to turntables that are located somewhere in a house or room. Than, balanced outputs are strongly recommended and one of the two un-balanced inputs of module 3 should be reconfigured as a balanced one.

The respective input impedances and input and output sensitivities are given in Fig. 13.1 as well. The XLR input and output connectors come from Neutrik (DL series, nickel plated contacts). I also installed a pair with gold plated contacts at the MC input of module 1. From a hum and noise point of view I couldn't detect any advantage nor any difference. The same applies to the sound of that phono-amp. In all studios sound creation and manipulation is of such a great importance that I cannot see any advantage of gold and silver plated contacts, although, theoretically, it might help to give better contact in extremely low-voltage environments. In the times of Sergeant Peppers no one thought of gold plated connectors – but surprisingly enough, this record is of extremely good quality. BNC and Twin-BNC connectors come from various manufacturers like e.g. Vitelec (UK), L-Com (USA), etc. I never use cinch/RCA connectors. All cases are products from fischer elektronik's 19″ series.

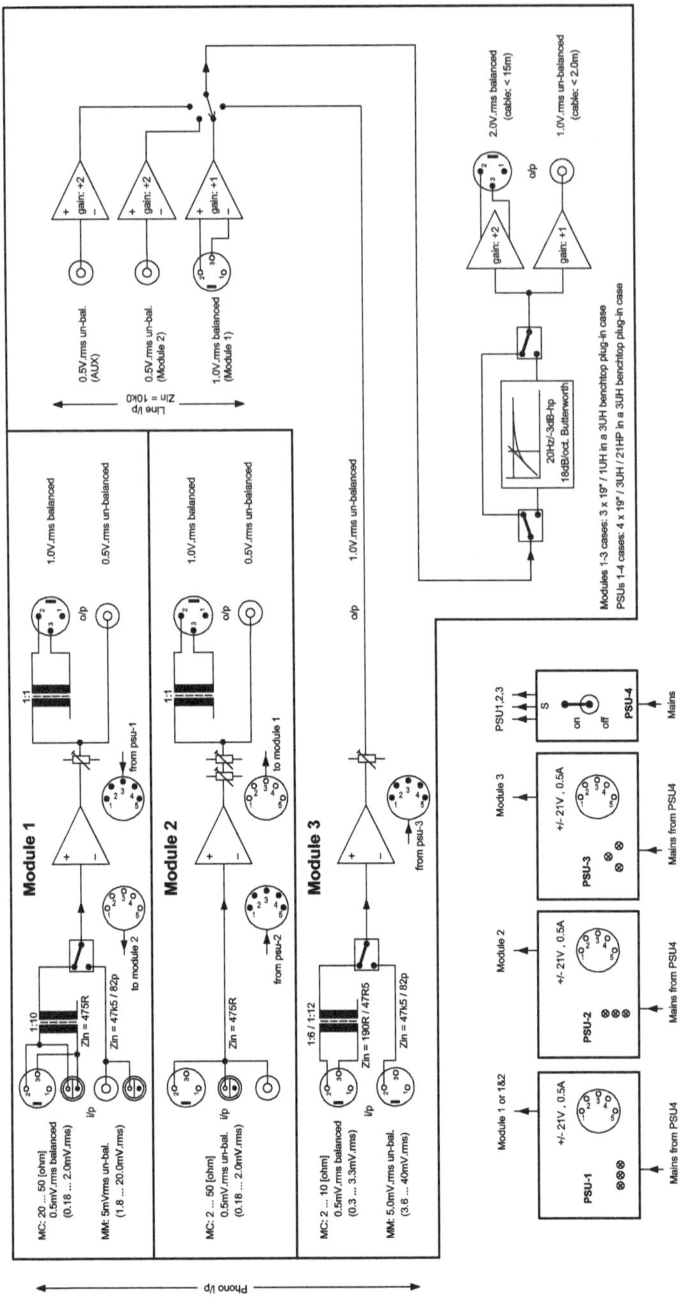


**Fig. 13.1** General diagram of the RIAA Phono-Amp Engine

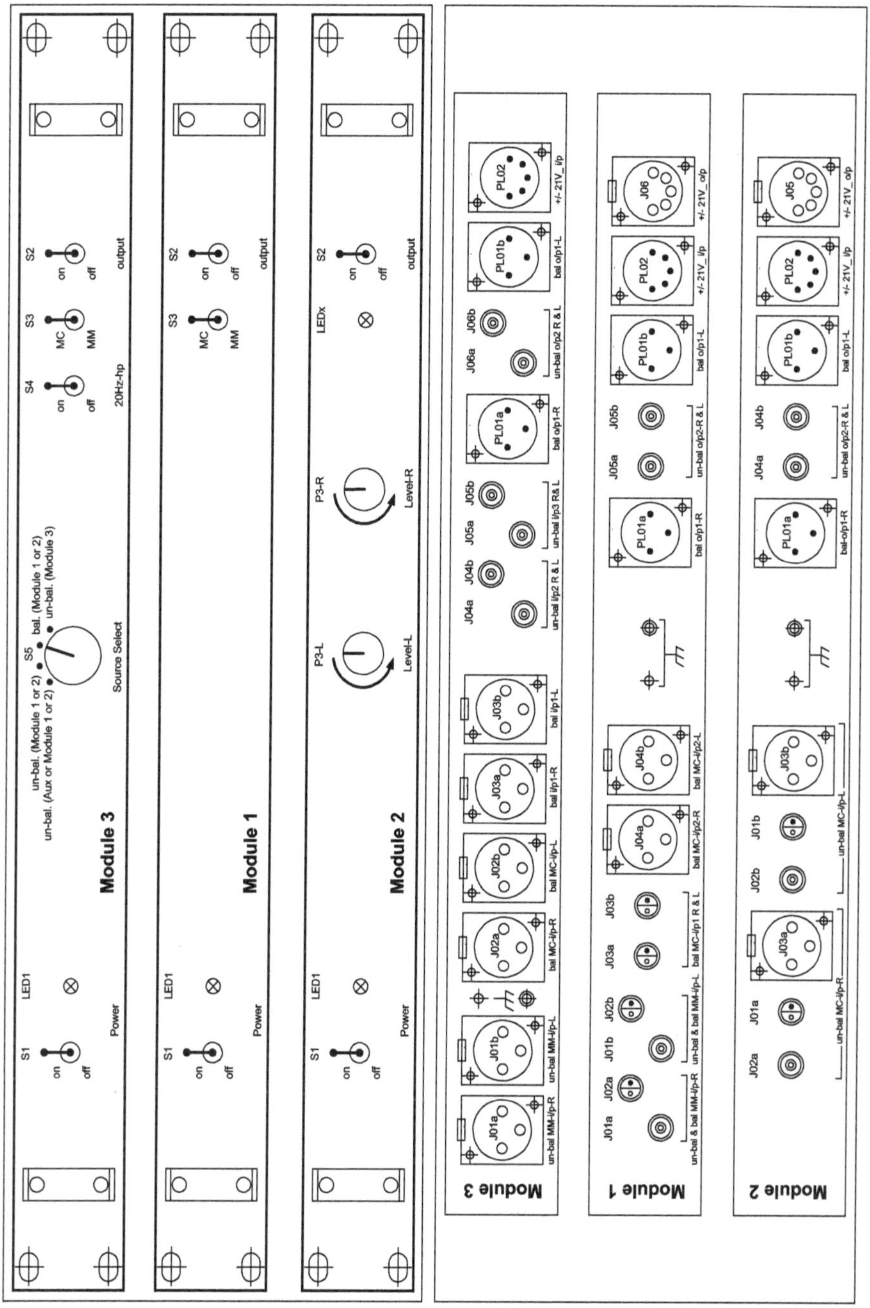


**Fig. 13.2** Front and rear view of the RIAA Phono-Amp Engine

# Chapter 14
# Module 1

## Intro and Function

The circuit diagram of the module 1 phono-amp section is shown in Figs. 6.3a … 6.3b and the respective – in-case – power supply unit circuit diagram is shown in Fig. 6.3c. I guess these diagrams are self-explanatory. Figure 14.1 shows the wiring of the input connectors and Fig. 14.2 shows the output stage.

Calibration of the OP4 DC offset setting $P_3$ needs some time because of the rather high time constant of $R_{27,28}$ and $C_{25,26}$. $P_6$ sets the stage gain at $+6\,\text{dB}$ – or any other gain to get an output voltage as indicated. It makes no sense to further

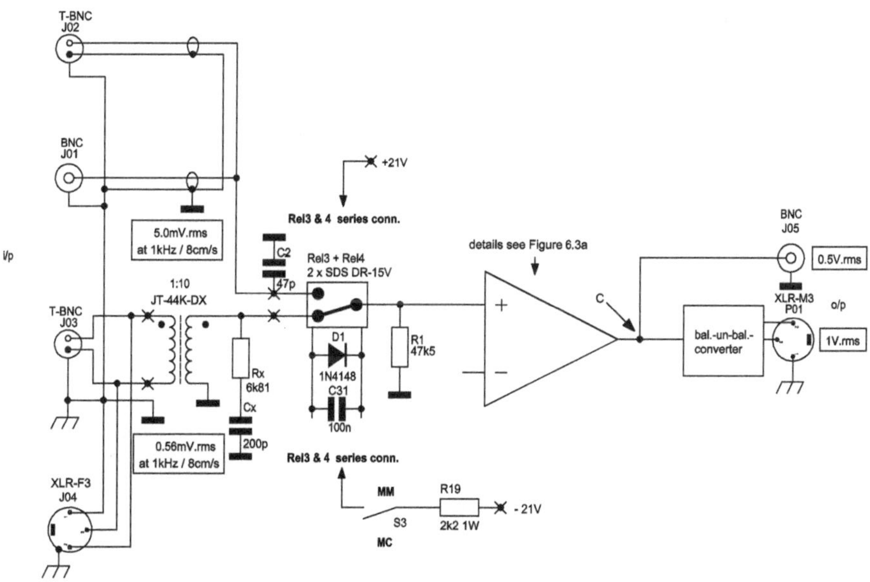


**Fig. 14.1** Module 1 wiring of input and output connectors

B. Vogel, *The Sound of Silence*
DOI: 10.1007/978-3-540-76884-5, ©Springer 2008

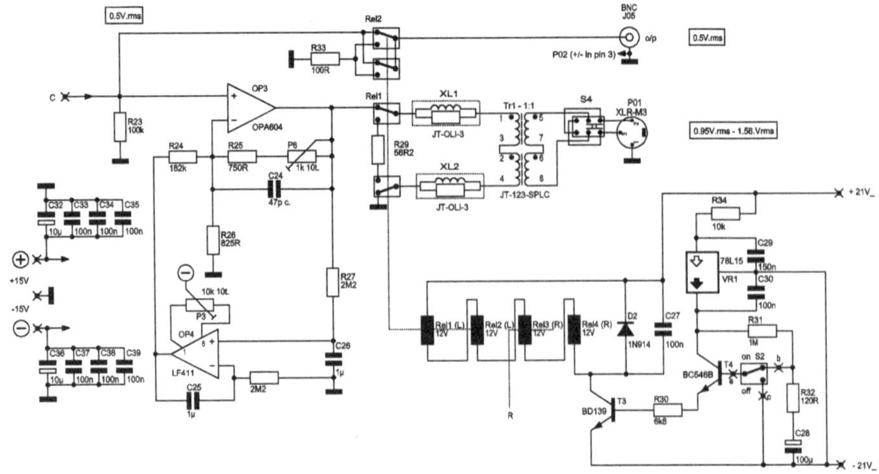


**Fig. 14.2** Module 1 output stage with trafo driven un-balanced – balanced conversion and delay circuit

increase the output voltage of Tr1 because of growing distortion[1]. $S_4$ allows a 180° phase change of the balanced output voltage. $S_2$ switches the two outputs on/off and $R_{31} + C_{28}$ ensure a delayed on-switch of the two outputs when turning the module on. Rel 1 … 2 are TQ2-12V relays from Matsushita.

[1] JT-123-SPLC data sheet, Jensen Transformers, USA

# Chapter 15
# Module 2

## Intro and Function

The circuit diagram of the module 2 phono-amp section is mainly based on the circuit of Fig. 6.4. It is shown in Fig. 15.1 and the respective – in-case – power supply unit circuit diagram is shown in Fig. 6.7. The output stage with the un-balanced – balanced converter looks like the one in Fig. 14.1 – with one exception: J05 becomes J04.

To allow to adequately react on a broad range of input sensitivities in contrast to Fig. 6.4 I've added an on-pcb switch $S_3$ to enable doubling of the gain of OP4 for very silent cartridges and I've included a 23 step rotary switch (ELMA) to allow damping of very loud MC cartridges.

## Power-Supply Issues

I guess Fig. 6.7 is also self-explanatory. It may look a bit strange that a real old-timer like the µA723 plays a significant role in today's low-noise supply voltage production but there is nothing on the market that is able to noise-wise outperform this IC. The Fairchild application[1] note I own was printed in 1968. The ripple rejection without filter effect of $R_{11}$ and $C_{29}$ is 74 dB. With filter effect included the output noise voltage becomes only $2.5\,\mu V_{rms}$ in a frequency band of 100 Hz ... 20 kHz[2] (=17.7 nV/rtHz).

There will be no danger of supply voltage noise disturbance as result of today's rather high supply voltage rejection ratios of modern op-amps. But this is not the case for the input transistors $T_{1...4}$. To keep them totally free of potential supply voltage noise they need a further voltage noise filtering by adding an adequate circuitry around $T_5$. It is claimed that this ensures a further improvement of 50 dB[3]. As a matter of fact and as shown in Fig. 6.8 hum and supply line noise don't have a chance to creep through the chosen configuration.

---

[1] Application Note on the µA723 Precision Voltage Regulator, Fairchild Semiconductor, 1968

[2] Linear Data Book, National Semiconductor Corp, 1982

[3] "AN 222", Linear Applications, National Semiconductor Corp., 1986

B. Vogel, *The Sound of Silence*
DOI: 10.1007/978-3-540-76884-5, ©Springer 2008

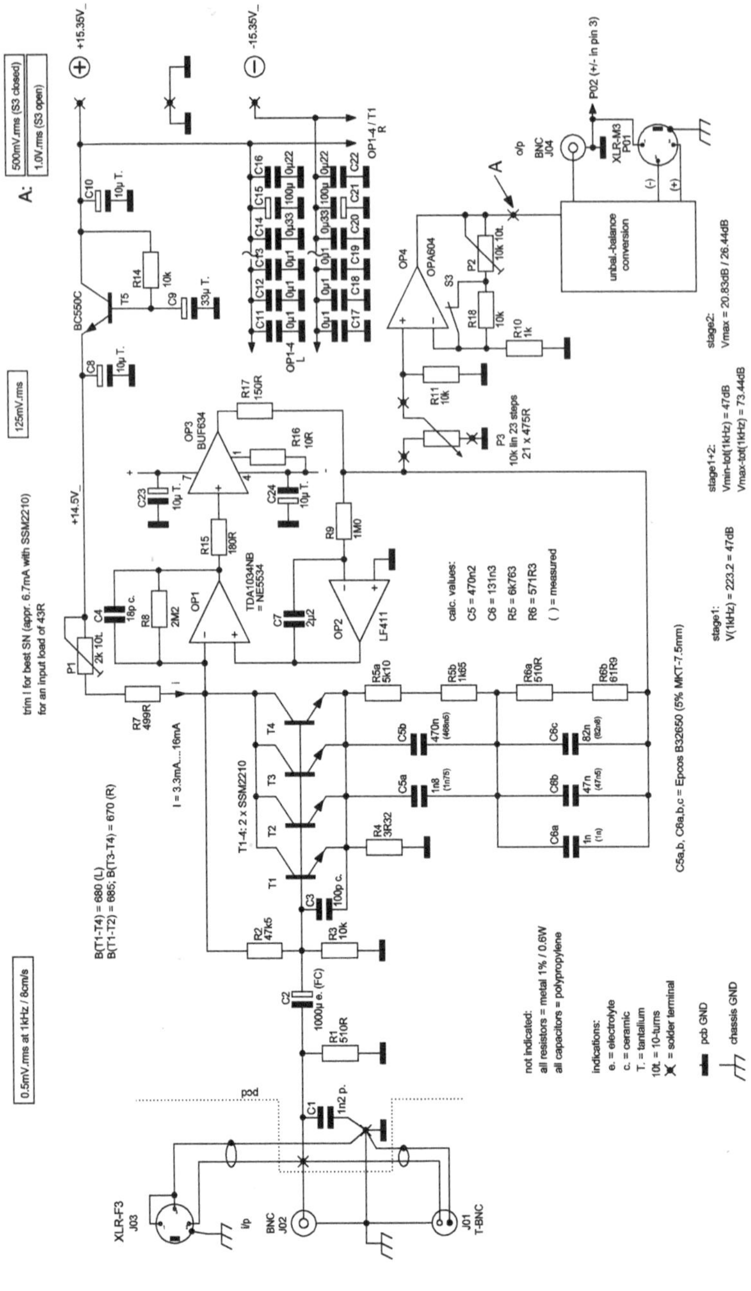


**Fig. 15.1** Module 2 phono-amp stage

# Chapter 16
# Module 3

## Intro and Function

The module 3 can be taken as a kind of pre-amp without volume control. It includes three line inputs and enough free space to fix a phono-amp section with two inputs for MC and MM cartridge purposes. Figure 16.1 shows all circuit details.

To enable a $1\,V_{rms}/1\,kHz$ output voltage two line inputs are configured as unbalanced +6 dB gain stages, one input is configured as a 0 dB/+40 dB (left channel via $S_{7a}$) or 0 dB/+60 dB (right channel via $S_{7b}$) gain stage. This configuration allows to use these two inputs as additional balanced input measurement amps without or with the gain specified. With input shorted and measured via un-balanced output (J06) the respective $SN_{ne}$ in $B_{20\,k}$ look as follows:

1. balanced i/p channel and gain = +0 dB: $SN_{ne.b} = -98.1\,dBV$
2. un-balanced i/p and gain = +6 dB: $SN_{ne.ub} = -115.2\,dBV$

The result of 1. is app. 1.7 dB better than calculated one with the data sheet[1] value for a gain of 1 at 1 kHz (−96.4 dB). I've selected the two SSM-2017 for lowest noise output from a bunch of twenty devices. With $1/f$ effect included into the calculation the difference would end up even worse. The result of 2. lies in the expected *SN* range and is 1 dB better than calculated (−114.2 dB without $1/f$ effect of the OPA627).

Because of the app. 17 dB lower $SN_{ne}$ the following question becomes valid: how much does the noise of the balanced input channel influence the SN results of a phono-amp that is connected to it?

Answer:

Taken from Table 6.2 we have e.g. the measured $SN_{riaa.trafo}$ of the trafo driven phono-amp with an i/p load of 43 R. With Eq. (3.340) we can calculate the

[1] "Audio/Video Reference Manual", Analog Devices 1992

B. Vogel, *The Sound of Silence*
DOI: 10.1007/978-3-540-76884-5, ©Springer 2008

difference of the sum of the two SNs minus the SN of the trafo driven phono-amp:

$$SN_{\text{riaa.res}} = 20\log\left(\sqrt{10^{\left(\frac{SN_{\text{riaa.trafo}}}{10}\right)} + 10^{\left(\frac{SN_{\text{ne.b}}}{10}\right)}}\right) - SN_{\text{riaa.trafo}}$$
$$= 0\,\text{dB} \tag{16.1}$$

In this case the noise contribution of the balanced input is 0 dB! In addition, the difference of the calculations of the A-weighted $SN_{\text{ariaa.trafo}}$ of the trafo driven phono-amp plus the A-weighted *SN* of the balanced input stage (−100.1 dB) as well do not show any difference to the A-weighted $SN_{\text{ariaa.trafo}}$ of the trafo driven phono-amp.

For the solid-state MC phono-amp and the MM phono-amp with e.g. a Shure V15V as i/p load (see Table 4.2) the resultant looks the same. Thus, with noisy input loads like those of phono-amps the rather lousy SSM-2017 *SN* for a gain of 0 dB can fully be ignored. But, set to a gain of +40 dB or +60 dB the SSM-2017 is an excellent low-noise balanced i/p amp for low-impedance output audio purposes with i/p noise voltage densities at 1 kHz of 1.95 nV/rtHz or 0.95 nV/rtHz and a noise current density of 2 pA/rtHz.

Operation of $S_6$ allows to overcome potential pin1 (ground) problems. It should be left open as long as all other potential hum problems could not be solved.

The input section is followed by a switchable 18 dB/oct. 20 Hz Butterworth hp that reproduces the 4th RIAA time constant defined by the IEC.

The output section consists of a simple gain 1 un-balanced o/p buffer for short cables ≤5 m and a gain 2 balanced o/p driver for very long cables ≤30 m. From a distortion point of view the THAT device might be slightly better than the devices from AD or TI, in any case, the OPA627 op-amps outperform any other device on the market.

## Phono-Amp Section

Figure 16.2 shows the draft design of phono-amp section of module 3. In contrast to the Jensen Transformers recommended AD797 as the RIAA gain stage for the chosen type of trafo, basically, it looks like the phono-amp circuit of module 1, but with several additional exceptions that are not mentioned in Chap. 13:

1. The JT-346-AXT transformer is a special device[2] for a rather low trafo o/p load of 6 k81, thus, in comparison with any other trafo for a 47 k5 o/p load it produces less SNs with the same i/p load.
2. The turns ratio of the input transformer can be switched from tr = 1:12 to tr = 1:2 via headers and jumpers (closely placed to the trafo input and output). Thus, enabling the connection of a broad range of low-impedance cartridges. Changing the turns ratio means automatically a change of input impedance of the trafo-phono-amp chain!

[2] JT-346-AXT data sheet, Jensen Transformers, USA

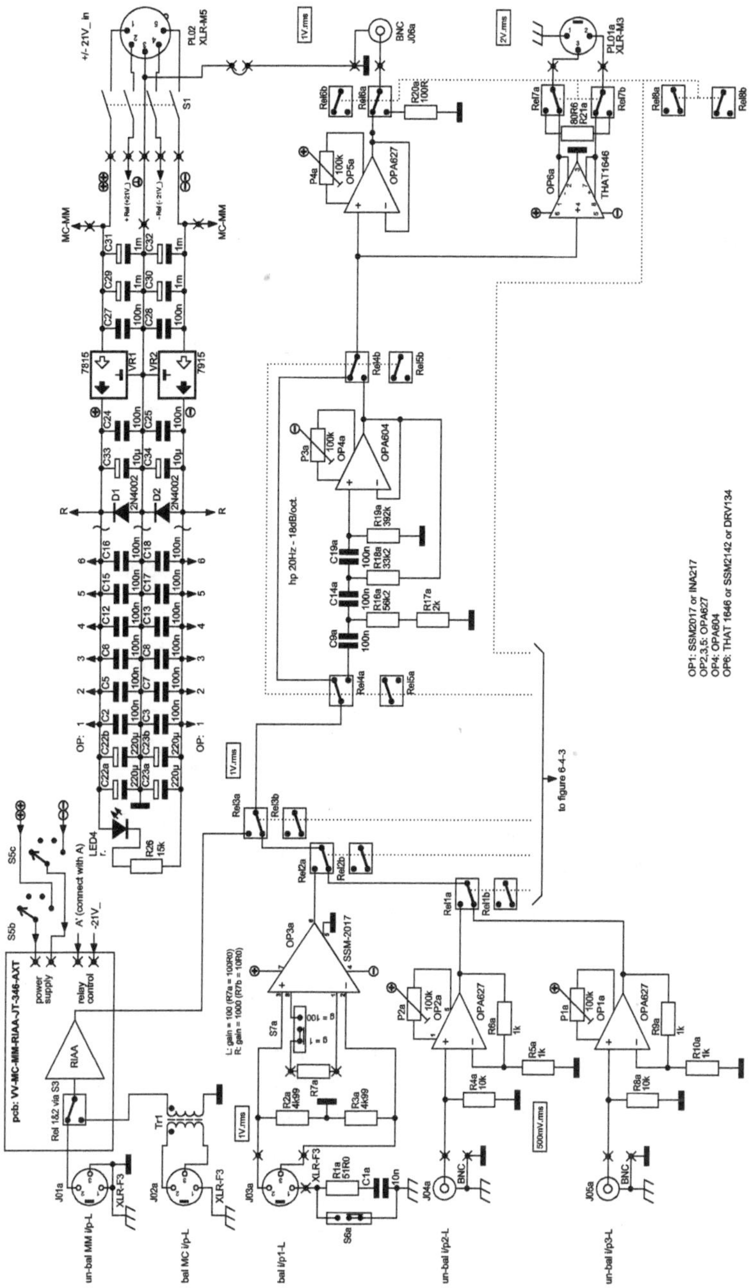

**Fig. 16.1** Module 3 (Source Selector) circuit diagram (without relay driver circuits)

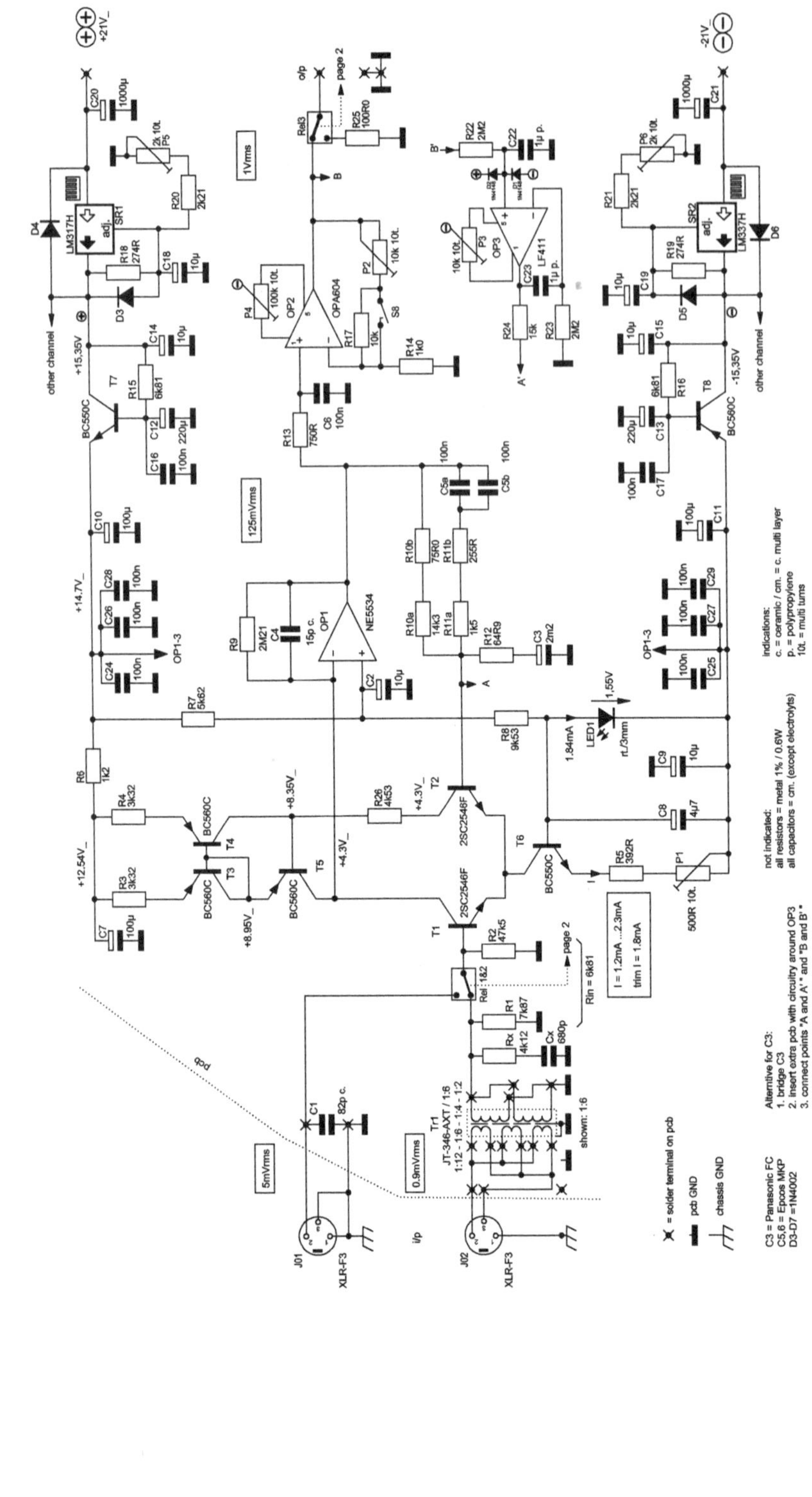


Fig. 16.2 Module 3 phono-amp section draft design

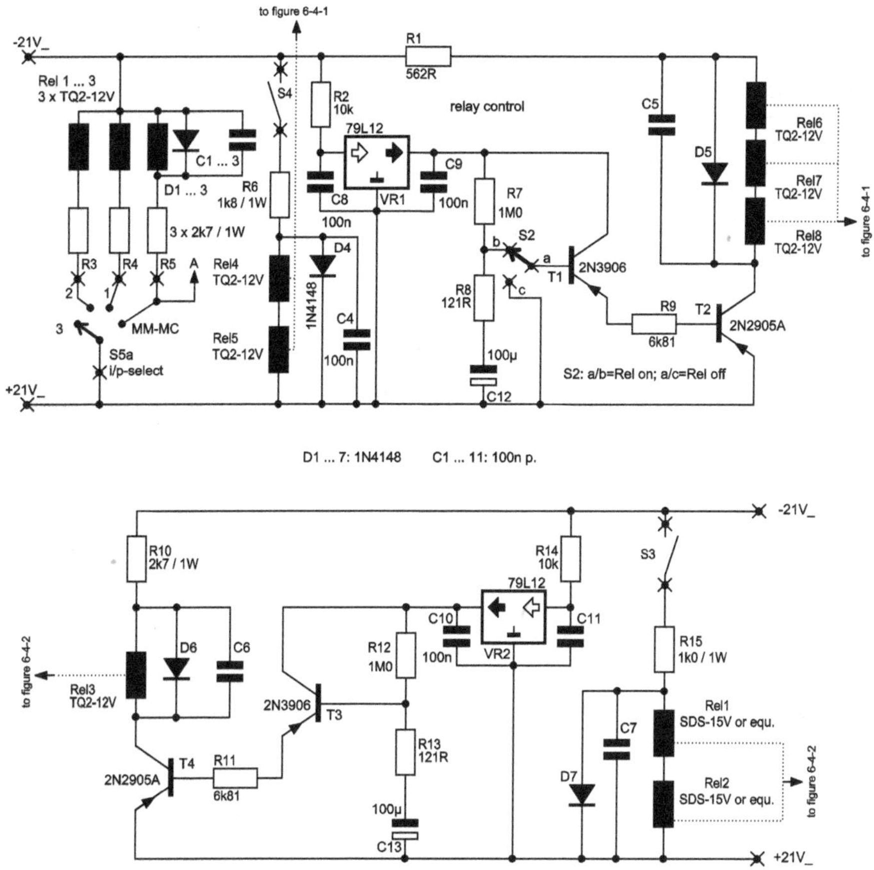


**Fig. 16.3a,b** (*top*) and (*bottom*) Relay drivers and soft start circuitries

3. The lower impedance of the feedback network of OP1 – resistors are halved, capacitors are doubled – produces less noise.
4. We have to find a compromise for the collector current $I_C$ of the long-tailed pair of the input BJTs. I've set it to 0.9 mA[3] for each transistor. This is nearly the optimal collector current with an input load of 4 R0 and a turns ratio of 1:12. Because of this rather high $I_C$ in comparison with the module 1 MM phono-amp the module 3 MM input produces less good *SN*s. That's why it's recommended for high-output MM cartridges (app. 8 … $10\,\mathrm{mV_{rms}}/1\,\mathrm{kHz}/8\,\mathrm{cm/s}$) only.
5. Sufficiently low-noise power-supplies around SR1 … 2 and $T_{7...8}$. A change of the $C_{18...19}$ values from 10 µ to a very much higher value would enable a soft start of the phono-amp without any o/p click!

[3] calculated *SN*s of this type of phono-amp are all app. 0.5 dB better than those with AD797: $SN_{ariaa}$: −87.329 dB (module 1 phono-amp with $I_C = 0.9$ mA) vs. −86.820 dB (AD797) with tr = 1:12 and i/p load = 4 R

### Relay Control

The relay control circuits are given in Fig. 16.3a for the source selector and in Fig. 16.3b for the phono-amp section. The following tasks have to be fulfilled:

- $S_1$ switches the whole module on-off
- $S_2$ switches the source-selector output on-off
- $S_3$ switches between MC and MM cartridge input
- $S_4$ switches the 20 Hz hp on-off
- $S_5$ selects the input sources and switches the phono-amp section on-off
- $T_{1...4}$, $R_7 + C_{12}$, $R_{12} + C_{13}$ enable a delayed turn-on of the phono-amp and source selector output.
- Point A must be connected with point A′ of the phono-amp section (see Fig. 16.1).

# Chapter 17
# Engine Performance

## Graphs

To demonstrate the low-noise quality of the stand-alone MC phono-amps their FFT spectral noise density graphs were already given in Figs. 6.8 ... 6.9. The following graphs show the engine's broadband quality with frequency ($F$) and phase ($P$) responses ($F$ = red/top trace + $P$ = blue/bottom trace) of several input–output combinations as well as some additional spectral noise density charts for the whole amp chain: from input connector via phono-amp and output stage to source-selector balanced output with +6 dB gain.

The following remarks sum-up a bit the results:

1. Compared with the equally rather flat $B_{20\,k}F_s + P_s$ of the un-balanced outputs the $F_s + P_s$ of the trafo driven balanced outputs of the phono-amps of module 1 and 2 do not look much different[1].
2. A change from the module 3 source selector un-balanced input/output configuration to the balanced one does only slightly influence $F_s + P_s$.
3. Already demonstrated with Eq. (16.1) spectral noise density of Fig. 17.7 (phono-amp plus source selector) compared with the one of Fig. 6.9 (phono-amp alone) shows no significant difference – with the exception that Fig. 17.7 was measured without measurement amp but with gain set to +6 dB and Fig. 6.9 was measured with a +60 dB gain set of the measurement amp (y-ordinate in dBmV!).
4. To demonstrate the influence of the tiny 150 Hz[2] hum on the character of different types of spectral noise density representation the spectral noise density charts of Figs. 17.7 ... 17.9 show three different traces:
   - blue (top): 3rd octave
   - violet (middle): 6th octave
   - red (bottom): full 20 kHz FFT

[1] See Chap. 12, footnote 6, for the trace disturbances reason <100 Hz in Figs. 17.1 ... 17.4

[2] See Chap. 10, Fig. 10.2 for more on that


DOI: 10.1007/978-3-540-76884-5, ©Springer 2008

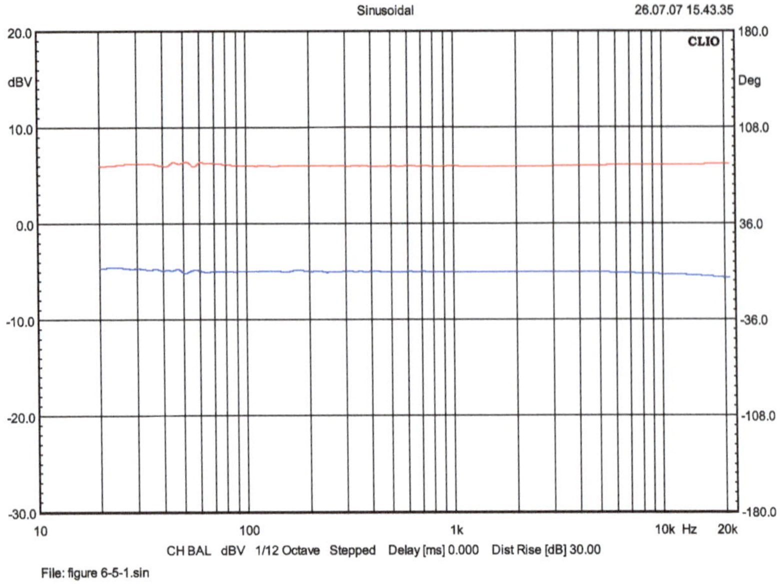


**Fig. 17.1** $F + P$ of the Fig. 15.1 module 2 BJT input MC phono-amp via un-balanced output and module 3 un-balanced input/un-balanced output
3 dB corner frequencies: Phase at 20 Hz: $+2.9° \rightarrow f_{c.hp} = 1$ Hz
Phase at 20 kHz: $-4.2° \rightarrow f_{c.lp} = 272$ kHz

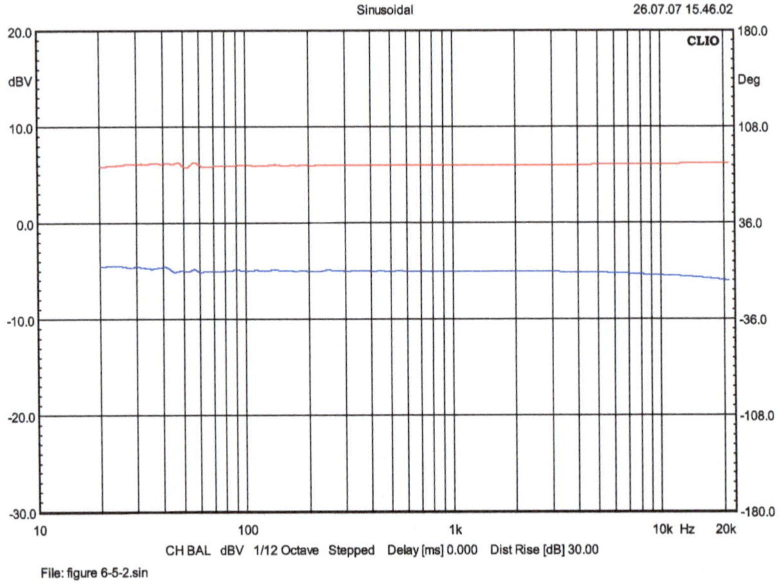


**Fig. 17.2** $F + P$ of the Fig. 15.1 module 2 BJT input MC phono-amp via balanced trafo output and module 3 balanced input and balanced output
3 dB corner frequencies: Phase at 20 Hz: $+3.5° \rightarrow f_{c.hp} = 1.2$ Hz
Phase at 20 kHz: $-6.5° \rightarrow f_{c.lp} = 176$ kHz

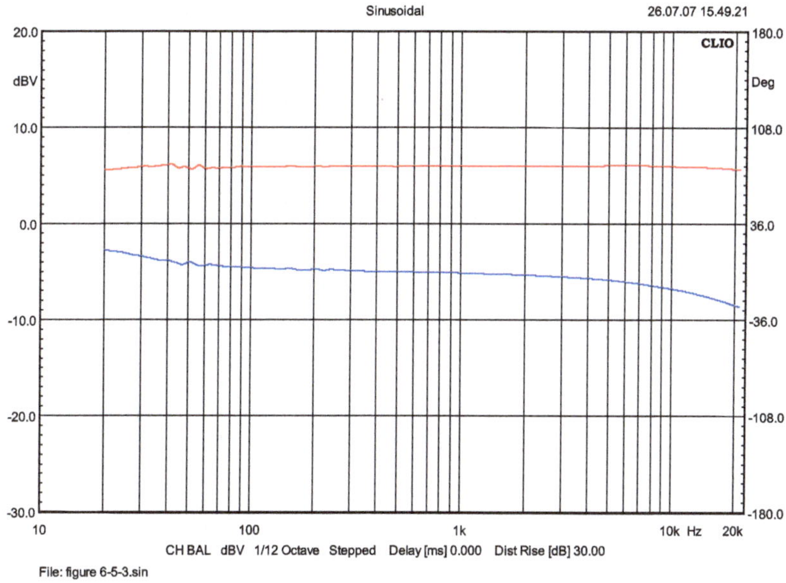


**Fig. 17.3** $F+P$ of the Fig. 14.1/6.3a module 1 balanced trafo input MC phono-amp via un-balanced output and module 3 un-balanced input/un-balanced output
3 dB corner frequencies: Phase at 20 Hz: $+16.0^\circ \rightarrow f_{c.hp} = 5.7$ Hz
Phase at 20 kHz: $-24.1^\circ \rightarrow f_{c.lp} = 44.7$ kHz

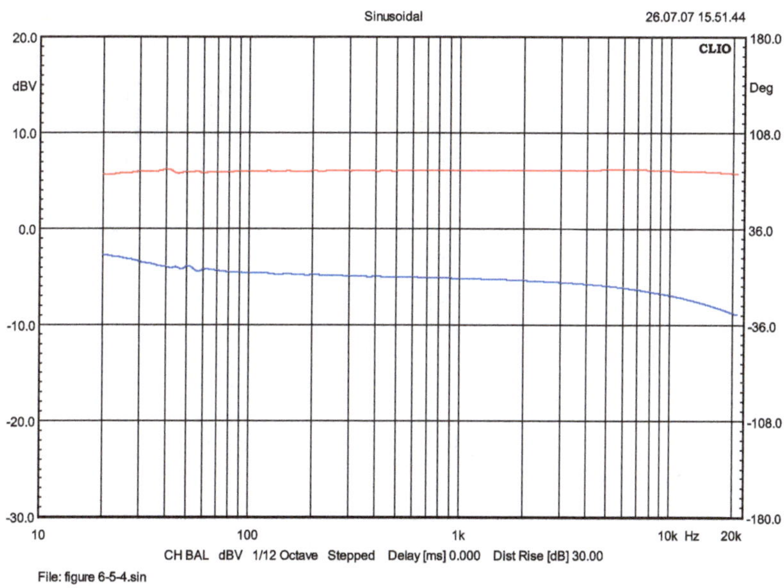


**Fig. 17.4** $F+P$ of the Fig. 14.1/6.3a module 1 balanced trafo input MC phono-amp via balanced trafo output and module 3 balanced input/balanced output
3 dB corner frequencies: Phase at 20 Hz: $+16.2^\circ \rightarrow f_{c.hp} = 5.8$ Hz
Phase at 20 kHz: $-26.7^\circ \rightarrow f_{c.lp} = 39.8$ kHz

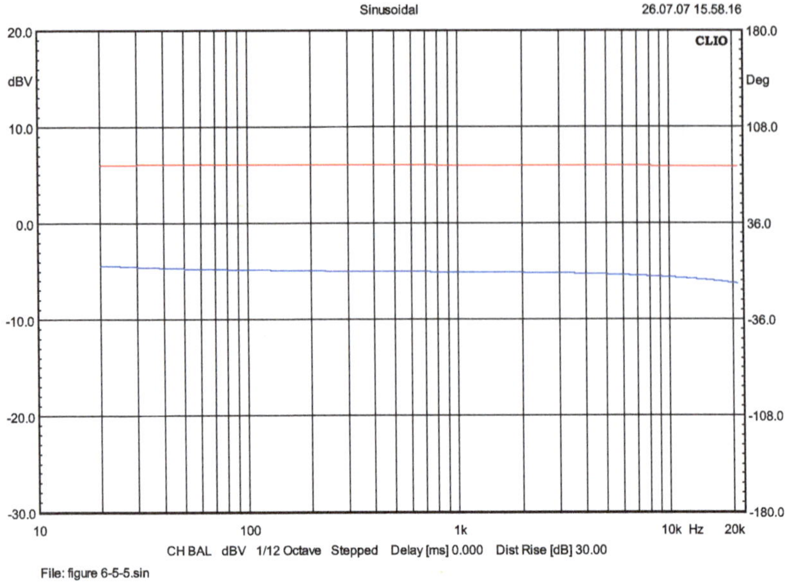


**Fig. 17.5** $F+P$ of the Fig. 14.1/6.3a module 1 MM input phono-amp via un-balanced output and module 3 un-balanced input/un-balanced output
3 dB corner frequencies: Phase at 20 Hz: +4.4° $\rightarrow f_{\text{c.hp}} = 1.5$ Hz
Phase at 20 kHz: −8.8° $\rightarrow f_{\text{c.lp}} = 129$ kHz

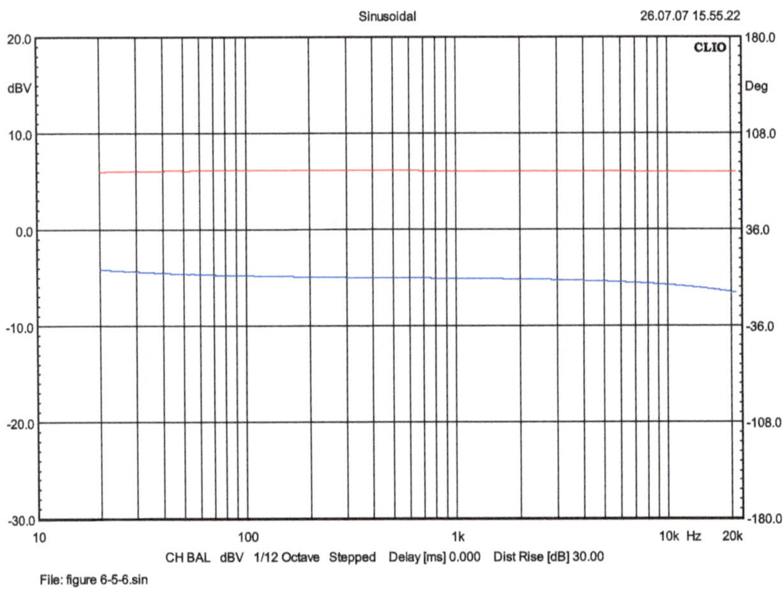


**Fig. 17.6** $F+P$ of the Fig. 14.1/6.3a module 1 MM input phono-amp via balanced trafo output and module 3 balanced input/balanced output
3 dB corner frequencies: Phase at 20 Hz: +6.1° $\rightarrow f_{\text{c.hp}} = 2.1$ Hz
Phase at 20 kHz: −10.7° $\rightarrow f_{\text{c.lp}} = 106$ kHz

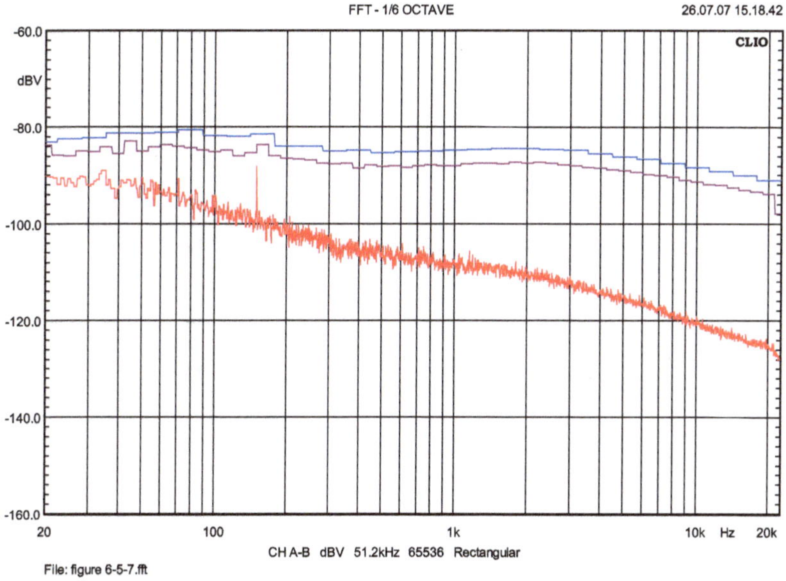


**Fig. 17.7** Spectral noise voltage density of module 1 balanced trafo MC phono-amp input with 43 R load measured via module 3 un-balanced input/balanced output and gain of +6 dB

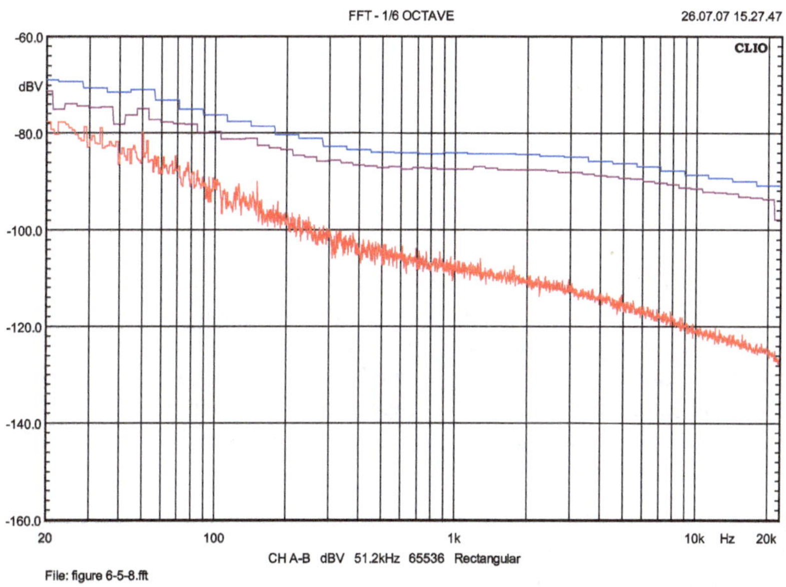


**Fig. 17.8** Three different spectral noise voltage density traces of module 1 BJT MM phono-amp input with 1 k load measured via module 3 un-balanced input/balanced output and gain of +6 dB

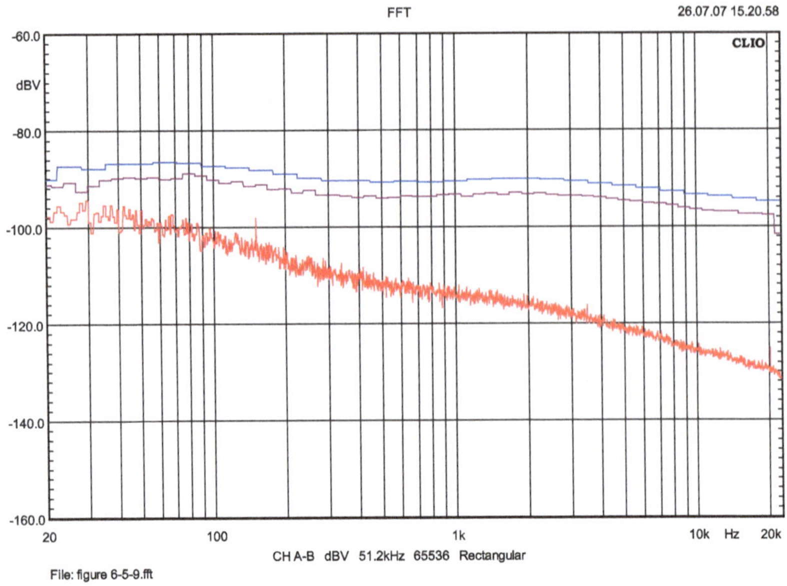


**Fig. 17.9** Three different spectral noise voltage density traces of module 2 BJT MC phono-amp input with 43 R load measured via module 3 un-balanced input/balanced output and gain of +6 dB

The clearly visible computer induced 150 Hz spikes are fully covered by the whole $B_{20\,\mathrm{k}}$ noise floor of the respective phono-amp. That's why they practically do not worsen the *SN*s.

## Sound

Finally, I feel that I have to add (a purely subjective selection, of course) some informations on three very low-noise vinyl records that are capable to uncover hidden weaknesses of the cartridge – phono-amp – pre- & power-amp – loudspeaker chain. For comparison reasons each of the three vinyl records is also available in a CD version. They all sound extremely well. Differences between vinyl and CD are detectable, but – I guess – marginal, because they heavily depend on the listener's actual mood when listening.

1. Traditional lacquer technology with full speed mastering:
   "Saitensprung", Friedemann, 180 g
   biber records 2007, www.in-akustik.com
   Cutting company: SST Brüggemann GmbH, Frankfurt, Germany
2. Advanced lacquer technology with half speed mastering:
   "Road to Escondido", JJ Cale & Eric Clapton, 2 × 180 g
   Reprise Records 2006, www.repriserecords.com
   Cutting company: Stan Ricker Mastering, Ridgecrest Ca. USA

3. Direct-to-Disc DMM full speed technology:
   "Friends of Carlotta", B. & C. Eiben – Jess – Kroner – Moeller – Zepf, 180 g
   Stockfisch Records 1999, www.stockfisch-records.de
   Cutting company: Pauler Acoustics, Northeim, Germany

# Part V
# Book-Ending Sections

# List of Figures

# List of Tables

# Constants, Abbreviations, Symbols

| | |
|---|---|
| A | A-weighting (special noise measurement filter) |
| a | valve anode or plate |
| AC | alternating current |
| ampx | amplifier *x* |
| b | balanced (suffix) |
| B | BJT base |
| $B$ | bandwidth (in general) |
| $B_1$ | bandwidth of 1 Hz |
| $B_{20\,k}$ | bandwidth of 20 Hz ... 20 kHz (19,980 Hz) |
| BJT | bipolar junction transistor |
| bp | band-pass filter |
| bv | big volume |
| C | BJT collector |
| $C$ | capacitance or capacitor |
| c | valve cathode |
| ca | contribution allowed |
| CCIR | Commité Consultatif International des Radicommunications |
| $c_1$ | proportional factor for RIAA network type (E) calculations |
| D | FET drain |
| dB | decibel |
| DC | direct current |
| diff. | difference, different |
| D/S | Douglas Self (author) |
| d.u.t. | device under test |
| E | BJT emitter |
| e | AC voltage |
| e | $20 \times \log(xyz)$ (suffix) |
| EW | Electronics World (magazine) |
| EW&WW | Electronics World & Wireless World (older version of EW) |
| ex | excess noise voltage (suffix) |
| ex | excluding rumble (suffix) |

| | |
|---|---|
| $f$ | frequency |
| $F$ | frequency response |
| FET | Field Effect Transistor |
| G | FET gate |
| $G_x$ | gain of stage x |
| g | valve grid |
| $h_{FE}$ | BJT current gain |
| HU | Hight Unit (of 19″ case) |
| HP | Width Unit (of 19″ case) |
| hp | high-pass filter |
| Hz | Hertz |
| I | DC current |
| i | AC current |
| id | ideal |
| JFET | junction field effect transistor |
| K | Kelvin [K] |
| $k$ | $1.38065\,\mathrm{V\,A\,s\,K^{-1}}$ (Boltzmann's constant) |
| $L$ | inductance |
| LP | Long play vinyl record |
| lp | low-pass filter |
| lv | low volume |
| MC | moving coil (cartridge) |
| M/C | Motchenbacher/Connelly (authors) |
| MCD | Mathcad |
| mcd | Mathcad |
| MM | moving magnet (cartridge) |
| MS | measurement system |
| MSR | Maxi single vinyl record |
| N | noise |
| n | noise |
| $n$ | secondary trafo turns divided by primary turns, thus, tr becomes $1:n$ |
| NAB | National Association of Radio and Television Broadcasters (ex NARTB = a US organisation) |
| $NF$ | noise factor |
| $NF_e$ | noise figure ($20\log(NF)$) |
| nom | nominal |
| $P$ | potentiometer |
| P | phase response |
| p | pentode (suffix) |
| ppa | pre-pre-amp |
| par | parallel |
| PSU | power supply unit |
| $q$ | $1.6022\times10^{-19}\,\mathrm{A\,s}$ (electron charge) |
| $R$ | resistance or resistor (equivalent unit symbol for ohm [Ω]) |
| $r_{bb'}$ | BJT base spreading resistance |

| | |
|---|---|
| RIAA | Radio Industry Association of America, a standard setting US organisation |
| re | real |
| ref. | reference, referenced |
| $r_N$ | valve (tube) equivalent noise resistance |
| rt | root |
| $r_1$ | proportional factor for RIAA network type (E) calculations |
| S | Source of a FET |
| S | S-filter (special noise measurement hp-filter) |
| $S_x$ | switch *x* |
| s | second |
| sec | second |
| seq | sequence or sequential |
| ser | serial |
| *SN* | signal-to-noise-ratio |
| $SN_a$ | *SN* after A-weighting |
| $SN_{ariaa}$ | *SN* after RIAA equalization and A-weighting |
| $SN_{ne}$ | *SN* non-equalized and non-weighted |
| $SN_{riaa}$ | *SN* after RIAA equalization |
| $SN_{sriaa}$ | *SN* after RIAA equalization and S-weighting |
| sol. | solution |
| SR | Single vinyl record |
| succ-app(s) | successive approximation(s) |
| *T* | room temperature 300 K |
| t | triode (suffix) |
| tr | transformer turns ratio (e.g.: 3:11) |
| trafo | transformer |
| T/S | Tietze/Schenk (authors) |
| TT | test terminal |
| $T_x$ | time constant *x* |
| ub | un-balanced (suffix) |
| $V_{DC}$ | DC voltage |
| $V_{cc}$ | DC supply voltage positive |
| $V_{ee}$ | DC supply voltage negative |
| VR | vinyl record |
| $V_x$ | amplifying stage or device *x* |
| W | white noise region (suffix) |
| WW | Wireless World (oldest version of EW magazine) |
| Z | impedance formed of different components (*R* and/or *C* and/or *L*) |
| $\|\|$ | parallel |

# Index

**W**



**X**

# Epilogue

In the early 60ies of last century I went to a school in Hamburg – located not far away from the Reeperbahn. I never forget my first 1962 visits to the Top Ten and Star Club discotheques. In those days they where the centres of Rock Music in Germany. The Beatles at the Star Club became my favourite band and with that push into the modern music world I started buying records and did what every rock music fan did in those days: saving money by copying hit-parades on valve driven tape recorders from stations like AFN (American Forces Network) and BFN (British Forces Network – today BFBS).

The rock music virus never got lost. To finance my electronic studies at Darmstadt's Technische Universität (University of Technology) I worked 4 years as a DJ at a discotheque in Mannheim. Very big and powerful KT88 driven power amps and Shure M44-G MM cartridges on SME tonearms fixed on Thorens turntables produced a top sound. A strong competition among Mannheim's two top discotheques forced us to an ongoing search for improvements of the sound equipment as well as of record selection. This became the beginning of my interest in noise. Not only electronic made noise: any kind of thermal noise. It was the universe's background noise that triggered my intention to study the mathematical background of Einstein's theories.

Unfortunately or fortunately – who knows, finally all my studies led to an industry career path that I couldn't foresee in those days: managing companies that produce solutions on the electronics, telecommunications, IT and media sectors. Far away from my interests in noise. But, during the last 15 years I worked as an Interim Manager, managing more than a dozen of different companies and organisations in Germany, Switzerland and Austria.

At last, between the different interim management assignments I found enough time to dive deeper into the amplifier noise question. In the companies I had to manage I always met employees with a great interest in music production and reproduction. Very fruitful discussions on these issues could be carried out. The idea to write a book on noise in phono-amps was born as a result of these discussions. It was clear to most of my discussion partners that the vinyl record and the valve driven amp would never die. But someone should collect – on one spot – all the (engineer's nightmare producing) phono-amp noise mechanics know-how that threatens to get lost by total digitalization. From valve to most modern IC – but not too complicate and from a math point of view easy to follow. That's what I tried to accomplish with this book for practical men and women – after 5 years of developing and building up nearly all presented electronic circuits.

Today, I live in a region of southern Germany that could be called the centre of top quality automobile and parts production in Europe. It's the home base of Mercedes Benz, Porsche and Bosch. More and more, these companies were confronted with the type of customer that owns one or several vintage cars of one of these famous car brands. The know-how to repair or even totally rebuild these cars after crashes etc. must be kept over an extremely long period of time. That's why it was not a surprise that, not long ago, quite new education courses were started to "produce" graduates with perfect old-timer know-how.

I guess, with growing sales revenues of vintage electronics and the die out of the ones who knew how to handle it, the time is not far away that we will also need fresh graduates who understand not only valve and other old-time audio and video technologies. An equally big problem can be found in the first generations of computer storage equipment. It's really time to do the right thing! Coming out end of 2008 my next writing project will be a book on triode driven valve pre-amps.

FSC
www.fsc.org
MIX
Papier | Fördert
gute Waldnutzung
FSC® C083411

Zeitfracht Medien GmbH
Ferdinand-Jühlke-Straße 7
99095 Erfurt, Deutschland
produktsicherheit@kolibri360.de